Fundamental Mathematics through Applications

Fundamental Mathematics through Applications

Fourth Edition

Geoffrey Akst   Sadie Bragg
Borough of Manhattan Community College,
City University of New York

Boston • San Francisco • New York • London • Toronto
Sydney • Tokyo • Singapore • Madrid • Mexico City
Munich • Paris • Cape Town • Hong Kong • Montreal

Editorial Director	Christine Hoag
Editor in Chief	Maureen O'Connor
Production Manager	Ron Hampton
Executive Project Manager	Kari Heen
Associate Editor	Joanne Doxey
Developmental Editor	Laura Wheel
Senior Designer	Barbara T. Atkinson
Text and Cover Design	Leslie Haimes
Composition	Pre-Press PMG
Media Producer	Ashley O'Shaughnessy
Software Development	TestGen: Mary Durnwald; MathXL: Jozef Kubit
Marketing Manager	Michelle Renda
Marketing Coordinator	Nathaniel Koven
Senior Prepress Supervisor	Caroline Fell
Manufacturing Manager	Evelyn Beaton
Senior Media Buyer	Ginny Michaud
Cover Photo	© Peter Stroumtos/Alamy

About the Cover: Well-designed sunglasses are not just cool—they also make it easy to see the world more clearly. Akst and Bragg's distinctive side-by-side example/practice exercise format encourages students to practice as they learn, while the authors' use of motivating applications connects mathematics to real life. With clear and simple explanations that make mathematics understandable, the Akst/Bragg series brings the world of developmental mathematics into focus for students.

Photo Credits: p. 1: PhotoDisc (PP); p. 11, 19, 27, 37, 42, 74, 250, 279, 284, 312: Digital Vision; p. 11: Jon Feingersh/zefa/CORBIS; p. 12: AP Photo/Al Behrman; p.12, 111, 509: PhotoLink; p. 13: Neal Preston/CORBIS; p. 29: epa/CORBIS; p. 37, 218: NASA; p. 55, 175, 238, 257: Getty Editorial; p. 61: Photographer's Choice RF/Getty Royalty Free; p. 75: AP Photo; p. 81: NASA Headquarters; p. 82, 156, 239: National Geographic/Getty Royalty Free; p. 85: Radius Images/Punchstock Royalty Free; p. 111: Jeff Haynes/ Pool/Reuters/CORBIS; p. 156: Jaysen F. Snow/Airliners.net; p. 156: DAJ/Getty Royalty Free; p. 161: Blend Images/Getty Royalty Free; p. 162, 170, 174, 192, 255, 278, 300: PhotDisc; p. 173: Photo Researchers; p. 174: Indianapolis Speedway; p. 209: Peter Guttman/CORBIS; p. 219: fStop Getty Royalty Free; p. 245: Hisham F. Ibrahim; p. 250: Stockbyte/Getty Royalty Free; p. 255: NASA/Johnson Space Center; p. 257: Purestock/Getty Royalty Free; p. 259: PhotoDisc Red; p. 265: Bob Krist/CORBIS; p. 278: Blend Images/Getty Royalty Free; p. 287: AP Wideworld Photos; p. 291, 295: CORBIS Royalty Free; p. 299, 307, 311, 315: Shutterstock; p. 324: G. Baden/zefa/CORBIS; p. 325: Scot Frei/CORBIS; p. 330: Bettmann/CORBIS; p. 331: DAJ/Getty Royalty Free

Library of Congress Cataloging-in-Publication Data
Akst, Geoffrey.
Fundamental mathematics through applications / Geoffrey Akst, Sadie Bragg.—4th ed.
p. cm.
Includes index.
ISBN-13: 978-0-321-49690-4 ISBN-10: 0-321-49690-6 (student ed. : alk. paper)
1. Mathematics—Textbooks. I. Bragg, Sadie II. Title.
QA39.3.A49 2004
510—dc22 2007060133

8 17

Contents

Preface

From the Authors

Our goal in writing *Fundamental Mathematics through Applications,* Fourth Edition, was to help motivate students and to establish a strong foundation for their success in a developmental mathematics program. Our text provides the appropriate coverage and review of whole numbers, fractions, decimals, ratio and proportion, and percents. Compared to other texts on the market, we have introduced algebra earlier in our text to stress the importance of algebraic concepts and skills.

For all topics covered in this text, we have carefully selected applications that we believe are relevant, interesting, and motivating. This thoroughly integrated emphasis on applications reflects our view that college students need to master basic mathematics not so much for its own sake but rather to be able to apply this understanding to their everyday lives and to the demands of subsequent college courses.

Our goal throughout the text has been to address many of the issues raised by the American Mathematical Association of Two-Year Colleges and the National Council of Teachers of Mathematics by writing a flexible, approachable, and readable text that reflects:

- an emphasis on applications that model real-world situations
- explanations that foster conceptual understanding
- exercises with connections to other disciplines
- an early introduction to algebraic concepts and skills
- appropriate use of technology
- an emphasis on estimation
- the integration of geometric visualization with concepts and applications
- exercises in student writing and groupwork that encourage interactive and collaborative learning
- the use of real data in charts, tables, and graphs

This text is part of a series that includes the following books:

- *Fundamental Mathematics through Applications,* Fourth Edition
- *Basic Mathematics through Applications,* Fourth Edition
- *Introductory Algebra through Applications,* Second Edition
- *Intermediate Algebra through Applications,* Second Edition
- *Introductory and Intermediate Algebra through Applications,* Second Edition

The following key content and key features stem from our strong belief that mathematics is logical, useful, and fun.

Geoffrey Akst and Sadie Bragg

Key Content

Applications One of the main reasons to study mathematics is its application to a wide range of disciplines, to a variety of occupations, and to everyday situations. Each chapter begins with a real-world application to show the usefulness of the topic under discussion and to motivate student interest. These opening applications vary widely from fractions and cooking to percents and surveys (see pages 85 and 287). Applications, appropriate to the section content, are highlighted in section exercise sets with an Applications heading (see pages 54, 110, and 278).

Concepts Explanations in each section foster intuition by promoting student understanding of underlying concepts. To stress these concepts, we include discovery-type exercises on reasoning and pattern recognition that encourage students to be logical in their problem-solving techniques, promote self-confidence, and allow students with varying learning styles to be successful (see pages 32, 289, and 330).

Skills Practice is necessary to reinforce and maintain skills. In addition to comprehensive chapter problem sets, chapter review exercises include mixed applications that require students to choose among and use skills covered in that chapter (see pages 215, 254, and 330).

Writing Writing both enhances and demonstrates students' understanding of concepts and skills. In addition to the user-friendly worktext format, open-ended questions throughout the text give students the opportunity to explain their answers in full sentences. Students can build on these questions by keeping individual journals (see pages 47, 103, and 319).

Estimation Students need to develop estimation skills to distinguish between reasonable and unreasonable solutions, as well as to check their solutions. The chapters on whole numbers (Chapter 1), fractions (Chapter 2), decimals (Chapter 3), and percents (Chapter 6) cover these skills (see pages 1, 85, 161, and 287).

Use of Geometry Students need to develop their abilities to visualize and compare objects. Throughout this text, students have opportunities to use geometric concepts and drawings to solve problems (see pages 36, 228, and 245).

Use of Technology Each student should be familiar with a range of computational techniques—mental, paper-and-pencil, and calculator arithmetic—depending on the problem and the student's level of mathematical preparation. This text includes optional calculator inserts (see pages 63, 192, and 309), which provide explanations for calculator techniques. These inserts also feature paired side-by-side examples and practice exercises, as well as a variety of optional calculator exercises in the section exercise sets that use the power of scientific calculators to perform arithmetic operations (see pages 195, 208, and 280). Both calculator inserts and calculator exercises are indicated by a calculator icon.

Key Features

Pretests and Posttests To promote individualized learning—particularly in a self-paced or lab environment—pretests and posttests help students gauge their level of understanding of chapter topics both at the beginning and at the end of each chapter. The pretests and posttests also allow students to target topics for which they may need to do extra work to achieve mastery. All answers to pretests and posttests are given in the answer section of the student edition. Students can watch an instructor working through the full solutions to the Posttest exercises on the Pass the Test CD.

Section Objectives At the beginning of each section, clearly stated learning objectives help students and instructors identify and organize individual competencies covered in the upcoming content.

Side-by-Side Example/Practice Format A distinctive side-by-side format pairs each numbered example with a corresponding practice exercise, encouraging students to get actively involved in the mathematical content from the start. Examples are immediately followed by solutions so that students can have a ready guide to follow as they work (see pages 17, 91, and 179).

Tips Throughout the text, students find helpful suggestions for understanding concepts, skills, or rules, and advice on avoiding common mistakes (see pages 93, 166, and 275).

Cultural Notes To show how mathematics has evolved over the centuries—in many cultures and throughout the world—each chapter features a compelling Cultural Note that investigates and illustrates the origins of mathematical concepts. Cultural notes give students further evidence that mathematics grew out of a universal need to find efficient solutions to everyday problems. Diverse topics include the evolution of digit notation, the popularization of decimals, and the role that taxation played in the development of the percent concept (see pages 176, 265, and 301).

Mindstretchers For every appropriate section in the text, related investigation, critical thinking, mathematical reasoning, pattern recognition, and writing exercises—along with corresponding group work and historical connections—are incorporated into one broad-ranged problem set called mindstretchers. Mindstretchers target different levels and types of student understanding and can be used for enrichment, homework, or extra credit (see pages 112, 196, and 313).

For Extra Help These boxes, found at the top of the first page of every section's exercise set, direct students to helpful resources that will aid in their study of the material.

Key Concepts and Skills At the end of each chapter, a comprehensive chart organized by section relates the key concepts and skills to a corresponding description and example, giving students a unique tool to help them review and translate the main points of the chapter.

Chapter Review Exercises Following the Key Concepts and Skills at the end of each chapter, a variety of relevant exercises organized by section helps students test their comprehension of the chapter content. As mentioned earlier, included in these exercises are mixed applications, which give students an opportunity to practice their reasoning skills by requiring them to choose and apply an appropriate problem-solving method from several previously presented (see pages 218, 253, and 283).

Cumulative Review Exercises At the end of Chapter 2, and for every chapter thereafter, Cumulative Review Exercises help students maintain and build on the skills learned in previous chapters.

What's New in the Fourth Edition?

NEW Math Study Skills Foldout This full-color foldout provides students with tips on organization, test preparation, time management, and more.

NEW Scientific Notation Appendix A brief appendix on scientific notation introduces the basic mathematics student to this alternative way of writing numbers.

NEW Tab Your Way to Success Guide The Tab Your Way to Success guide provides students with color-coded Post-it® tabs to mark important pages of the text that they may need to revisit for review work, test preparation, instructor help, and so forth.

NEW Mathematically Speaking Exercises Located at the beginning of nearly every section's exercise set, these new exercises have been added to help students understand and use standard mathematical vocabulary.

NEW Mixed Practice Exercises Mixed Practice exercises, located in nearly every section's exercise set, reinforce the student's knowledge of topics and problem solving skills covered in the section.

Updated Exercise Sets This revision includes over 600 new exercises, including the new Mathematically Speaking and Mixed Practice exercises.

NEW Pass the Test CD Included with every new copy of the book, the Pass the Test CD includes video footage of an instructor working through the complete solutions for all Posttest exercises for each chapter, vocabulary flashcards, interactive Spanish glossary, and additional tips on improving time management.

What Supplements Are Available?

For a complete list of the supplements and study aids that accompany *Fundamental Mathematics through Applications*, Fourth Edition, see pp. xi through xiii.

Acknowledgments

We are grateful to everyone who has helped to shape this textbook by responding to questionnaires, participating in telephone surveys and focus groups, reviewing the manuscript, and using the text in their classes. We wish to thank all of them, especially the following diary and manuscript reviewers who provided feedback for this revision: Yon Kim, *Passaic County Community College;* James Morgan, *Holyoke Community College;* Margaret Patin, *Vernon College;* Lee H. LaRue, *Paris Junior College;* Susan Santolucito, *Delgado Community College;* Marcia Swope, *Santa Fe Community College;* Carol Marinas, *Barry College;* Kate Horton, *Portland Community College;* James Cochran, *Kirkwood Community College;* Sylvia Brown, *Mountain Empire Community College;* and LeAnn L. Lotz-Todd, *Metropolitan Community College–Longview.* In addition, we would like to extend our gratitude to our accuracy checkers: Robert Holt, *Queensborough Community College, City University of New York;* Sharon Testone, *Onondaga Community College;* Sharon O'Donnell, *Chicago State University;* Janis Cimperman, *St. Cloud State University;* and Perian Herring, *Okaloosa-Walton College.*

Writing a textbook requires the contributions of many individuals. Special thanks go to Greg Tobin, our publisher at Addison-Wesley, for encouraging and supporting us throughout the entire process. We are very grateful to Laura Wheel, our developmental editor, who assisted us in more ways than one could imagine and whose unwavering support made our work more manageable. We thank Kari Heen for her patience and tact, Michelle Renda and Maureen O'Connor for keeping us abreast of market trends, Joanna Doxey for attending to the endless details connected with the project, Ron Hampton and Laura Houston for their support throughout the production process, Leslie Haimes for the text and cover design, and the entire Addison-Wesley developmental mathematics team for helping to make this text one of which we are very proud.

Geoffrey Akst
gakst@nyc.rr.com

Sadie Bragg
sbragg@bmcc.cuny.edu

For Mag Dora Chavis and Harriet Young, and to the memory of Maxine Jefferson and Anne Akst

Student Supplements

Student's Solutions Manual

By Beverly Fusfield

- Provides detailed solutions to the odd-numbered exercises in each exercise set and solutions to all chapter pretests and posttests, practice exercises, review exercises, and cumulative review exercises

ISBN-10: 0-321-50051-2 ISBN-13: 978-0-321-50051-9

New Video Lectures on CD or DVD

- Complete set of digitized videos on CD-ROM or DVD for students to use at home or on campus
- Includes a full lecture for each section of the text
- Optional captioning in English is available

CD: ISBN-10: 0-321-49871-2 and ISBN-13: 978-0-321-49871-7

DVD: ISBN-10: 0-321-53580-4 and ISBN-13: 978-0-321-53580-1

Instructor Supplements

Annotated Instructor's Edition

- Provides answers to all text exercises in color next to the corresponding problems
- Includes teaching tips

ISBN-10: 0-321-50058-X ISBN-13: 978-0-321-50058-8

Instructor's Solutions Manual

- Provides complete solutions to even-numbered section exercises
- Contains answers to all Mindstretcher problems

ISBN-10: 0-321-50052-0 ISBN-13: 978-0-321-50052-6

Instructor and Adjunct Support Manual

- Includes resources designed to help both new and adjunct faculty with course preparation and classroom management, including sample syllabi, tips for using supplements and technology, and useful external resources
- Offers helpful teaching tips correlated to the sections of the text

ISBN-10: 0-321-50054-7 ISBN-13: 978-0-321-50054-0

Student Supplements *(continued)*

New Pass the Test CD

Automatically included with the book, this CD-ROM contains

- Video footage of an instructor working through the complete solutions for all Posttest problems
- Vocabulary flashcards
- An interactive Spanish glossary
- A short video offering tips on time management

ISBN-10: 0-321-53965-6 ISBN-13: 978-0-321-53965-6

New Worksheets for Classroom or Lab Practice

By Mark Stevenson, *Oakland Community College*

- Provides one worksheet for each section of the text, organized by section objective
- Each worksheet lists the associated objectives from the text, provides fill-in-the-blank vocabulary practice, and exercises for each objective.

ISBN-10: 0-321-53631-2 ISBN-13: 978-0-321-53631-0

MathXL® Tutorials on CD

- Provides algorithmically-generated practice exercises that correlate at the objective level to the exercises in the textbook.
- Includes an example and a guided solution for every exercise; selected exercises also include a video clip
- Provides helpful feedback for incorrect answers and generates printed summaries of students' progress

ISBN-10: 0-321-50055-5 ISBN-13: 978-0-321-50055-7

Math Tutor Center

- Staffed by qualified mathematics instructors
- Provides tutoring on examples and odd-numbered exercises from the textbook through a registration number with a new textbook or if purchased separately
- Accessible via toll-free telephone, toll-free fax, e-mail, or the Internet

www.aw-bc/tutorcenter

Instructor Supplements *(continued)*

Printed Test Bank

By Kay Haralson, *Austin Peay State University* and Nancy Matthews, *Montgomery Central Middle School*

- Contains three free-response and one multiple-choice test forms per chapter, and two final exams

ISBN 10: 0-321-50050-4 ISBN-13: 978-0-321-50050-2

PowerPoint Lecture Slides (available online)

- Present key concepts and definitions from the text

New Active Learning Lecture Slides (available online)

- Provide several multiple-choice questions for each section of the book, allowing instructors to quickly assess mastery of material in class
- Available in PowerPoint for use with classroom response systems

TestGen® (available online)

- *New* Now includes a premade test for each chapter that has been correlated problem-by-problem to the chapter tests in the book
- Enables instructors to build, edit, print, and administer tests using a computerized bank of questions developed to cover all text objectives
- Algorithmically based, TestGen allows instructors to create multiple but equivalent versions of the same question or test with the click of a button.
- Instructors can also modify test bank questions or add new questions.
- Tests can be printed or administered online.
- Software and test bank are available for download from the Pearson Education online catalog.
- Available on a dual-platform Windows/Macintosh CD-ROM.

Student Supplements (*continued*)

InterAct Math Tutorial Website
www.interactmath.com

- Get practice and tutorial help online
- Provides algorithmically generated practice exercises that correlate directly to the textbook exercises
- Retry an exercise as many times as desired with new values each time for unlimited practice and mastery
- Every exercise is accompanied by an interactive guided solution that gives the student helpful feedback when an incorrect answer is entered
- View the steps of a worked-out sample problem similar to the one that has been worked on

Instructor Supplements (*continued*)

Adjunct Support Center

The Math Adjunct Support Center is staffed by qualified mathematics instructors with over 50 years combined experience at both the community college and university level. Assistance is provided for faculty in the following areas:

- Suggested syllabus consultation
- Tips on using materials packaged with the text
- Book-specific content assistance
- Teaching suggestions including advice on classroom strategies

For more information visit
www.aw-bc.com/tutorcenter/math-adjunct.html

Available for Students and Instructors

MathXL® MathXL® is a powerful online homework, tutorial, and assessment system that accompanies Pearson Education's textbooks in mathematics or statistics. With MathXL, instructors can create, edit, and assign online homework and tests using algorithmically-generated exercises correlated at the objective level to the textbook. Instructors can also create and assign their own online exercises and import TestGen tests for added flexibility. All student work is tracked in MathXL's online gradebook. Students can take chapter tests in MathXL and receive personalized study plans based on their test results. The study plan diagnoses weaknesses and links students directly to tutorial exercises for the objectives they need to study and retest. Students can also access supplemental video clips and animations directly from selected exercises.

MyMathLab® MyMathLab is a series of text-specific, easily customizable online courses for Pearson Education's textbooks in mathematics and statistics. Powered by CourseCompass™ (our online teaching and learning environment) and MathXL® (our online homework, tutorial, and assessment system), MyMathLab gives you the tools you need to deliver all or a portion of your course online, whether your students are in a lab setting or working from home. MyMathLab provides a rich and flexible set of course materials, featuring free-response exercises that are algorithmically generated for unlimited practice and mastery. Students can also use online tools, such as video lectures, animations, and a multimedia textbook, to independently improve their understanding and performance. Instructors can use MyMathLab's homework and test managers to select and assign online exercises correlated directly to the textbook, and they can also create and assign their own online exercises and import TestGen tests for added flexibility. MyMathLab's online gradebook—designed specifically for mathematics and statistics—automatically tracks students' homework and test results and gives the instructor control over how to calculate final grades. Instructors can also add offline (paper-and-pencil) grades to the gradebook. MyMathLab is available to qualified adopters. For more information, visit our Web site at www.mymathlab.com or contact your sales representative.

Feature Walk-Through

CHAPTER 2

Fractions

Fractions and Cooking

Using recipes in cooking has a very long and complicated history that goes back thousands of years. For instance, archaeologists have found Mesopotamian recipes written on clay tablets dating from 1700 B.C. Until the Industrial Revolution, at which time standard measurements and precise cooking directions were introduced, recipes gave just a list of ingredients and a general description for cooking a dish.

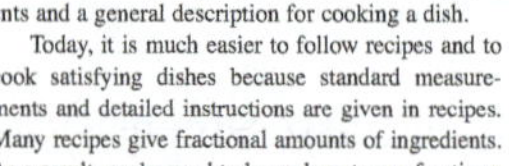

Today, it is much easier to follow recipes and to cook satisfying dishes because standard measurements and detailed instructions are given in recipes. Many recipes give fractional amounts of ingredients. As a result, cooks need to know how to use fractions.

For example, consider the ingredients given in a recipe for Italian-style meatloaf. The recipe, which serves 8 people, calls for $1\frac{1}{2}$ pounds of ground meat, of which $\frac{1}{3}$ is pork and $\frac{2}{3}$ is beef. To follow this recipe, a cook uses multiplying fractions to determine that for $1\frac{1}{2}$ pounds of ground meat, the amount of ground pork needed is $\frac{1}{3} \times 1\frac{1}{2}$, or $\frac{1}{2}$ pound, and the amount of ground beef needed is $\frac{2}{3} \times 1\frac{1}{2}$, or 1 pound. Multiplying fractions is also used if the cook wants to increase or decrease the number of servings that the recipe yields. (*Source:* http://www.foodtimeline.org)

85

Chapter Openers The focus of this textbook is applications, and you will find them everywhere. Each chapter opener introduces students to the material that lies ahead through an interesting real-life application that grabs students' attention and helps them understand the relevance of topics they are learning.

86 Chapter 2 | Fractions

Chapter 2 PRETEST

To see if you have already mastered the topics in this chapter, take this test.

1. Find all the factors of 20.
2. Express 72 as the product of prime factors.
3. What fraction does the shaded part of the diagram represent?

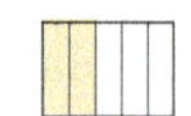

4. Write $20\frac{1}{3}$ as an improper fraction.
5. Express $\frac{31}{30}$ as a mixed number.
6. Write $\frac{9}{12}$ in simplest form.
7. What is the least common multiple of 10 and 4?
8. Which is greater, $\frac{1}{8}$ or $\frac{1}{9}$?

Add.

9. $\frac{1}{2} + \frac{7}{10}$
10. $7\frac{1}{3} + 5\frac{1}{2}$

Subtract.

11. $8\frac{1}{4} - 6$
12. $12\frac{1}{2} - 7\frac{7}{8}$

Multiply.

13. $2\frac{1}{3} \times 1\frac{1}{2}$
14. $\frac{5}{8} \times 96$
15. Divide: $3\frac{1}{3} \div 5$
16. Calculate: $2 + 1\frac{1}{3} + \frac{4}{5}$

Solve. Write your answer in simplest form.

17. In 2006, the Pittsburgh Steelers won their fifth Super Bowl championship game. If 40 Super Bowl championship games have been played, what fraction of these games has this team won? (*Source:* National Football League)
18. In a biology class, three-fourths of the students received a passing grade. If there are 24 students in the class, how many students received failing grades?
19. Find the perimeter of the traffic island shown.

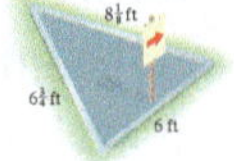

20. According to the nutrition information given, one serving of Honey Nut Cheerios® contains 22 grams of carbohydrates. If one serving is $\frac{3}{4}$ cup, what amount of carbohydrates is contained in $2\frac{1}{4}$ cups of Honey Nut Cheerios? (*Source:* General Mills)

• Check your answers on page A-3.

Pretests Pretests, found at the beginning of each chapter, help students gauge their understanding of the chapter ahead. Answers can be found in the back of the book.

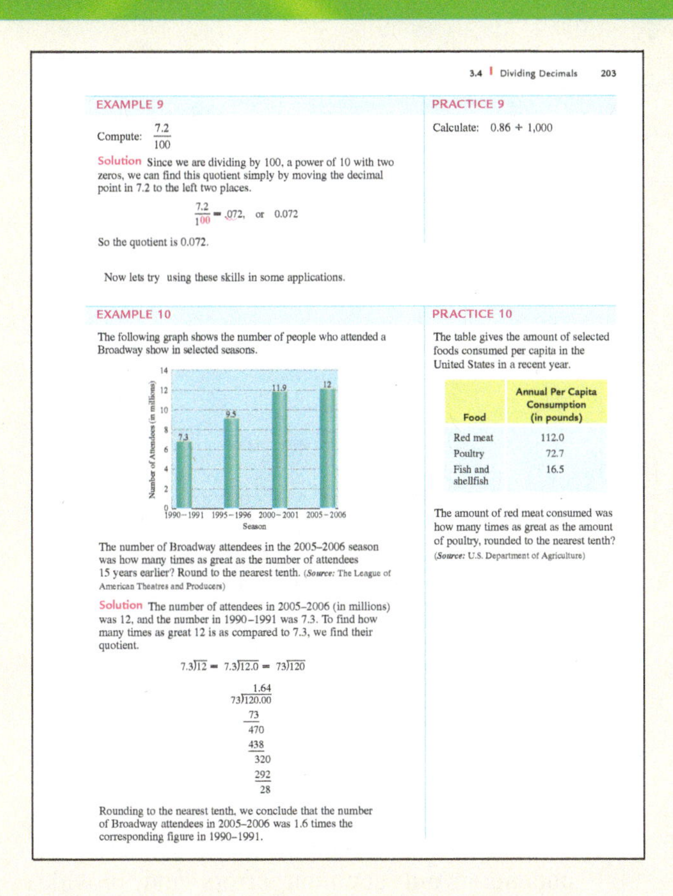

EXAMPLE 9

Compute: $\dfrac{7.2}{100}$

Solution Since we are dividing by 100, a power of 10 with two zeros, we can find this quotient simply by moving the decimal point in 7.2 to the left two places.

$$\frac{7.2}{100} = .072, \quad \text{or} \quad 0.072$$

So the quotient is 0.072.

PRACTICE 9

Calculate: $0.86 + 1{,}000$

Now lets try using these skills in some applications.

EXAMPLE 10

The following graph shows the number of people who attended a Broadway show in selected seasons.

The number of Broadway attendees in the 2005–2006 season was how many times as great as the number of attendees 15 years earlier? Round to the nearest tenth. (*Source:* The League of American Theatres and Producers)

Solution The number of attendees in 2005–2006 (in millions) was 12, and the number in 1990–1991 was 7.3. To find how many times as great 12 is as compared to 7.3, we find their quotient.

$$7.3\overline{)12} = 7.3\overline{)12.0} = 73\overline{)120}$$

$$\begin{array}{r} 1.64 \\ 73\overline{)120.00} \\ \underline{73} \\ 470 \\ \underline{438} \\ 320 \\ \underline{292} \\ 28 \end{array}$$

Rounding to the nearest tenth, we conclude that the number of Broadway attendees in 2005–2006 was 1.6 times the corresponding figure in 1990–1991.

PRACTICE 10

The table gives the amount of selected foods consumed per capita in the United States in a recent year.

Food	Annual Per Capita Consumption (in pounds)
Red meat	112.0
Poultry	72.7
Fish and shellfish	16.5

The amount of red meat consumed was how many times as great as the amount of poultry, rounded to the nearest tenth? (*Source:* U.S. Department of Agriculture)

Side-by-Side Format This format pairs examples and their step-by-step solutions side by side with corresponding practice exercises, encouraging active learning from the start. Students use this format for solving skill exercises, application problems, and technology exercises throughout the text.

Using the Tabs

Customize Your Textbook and Make It Work for You!

These removable and reusable tabs offer you five ways to be successful in your math course by letting you bookmark pages with helpful reminders.

- **Important** — Use these tabs to flag anything your instructor indicates is important.
- **Review This** — Mark important definitions, procedures, or key terms to review later.
- **Ask for Help** — Not sure of something? Need more instruction? Place these tabs in your textbook to address any questions with your instructor during your next class meeting or with your tutor during your next tutoring session.
- **On the Test** — If your instructor alerts you that something will be covered on a test, use these tabs to bookmark it.
- Write your own notes or create more of the preceding tabs to help you succeed in your math course.

EAN

***New* Tab Your Way to Success** The Tab Your Way to Success guide provides students with color-coded Post-it® tabs to mark important pages of the text that they may need to revisit for review work, test preparation, instructor help, and so forth.

***New* Math Study Skills Foldout** This insert, found at the very front of the book, provides students with tips on organization, test preparation, time management, and more.

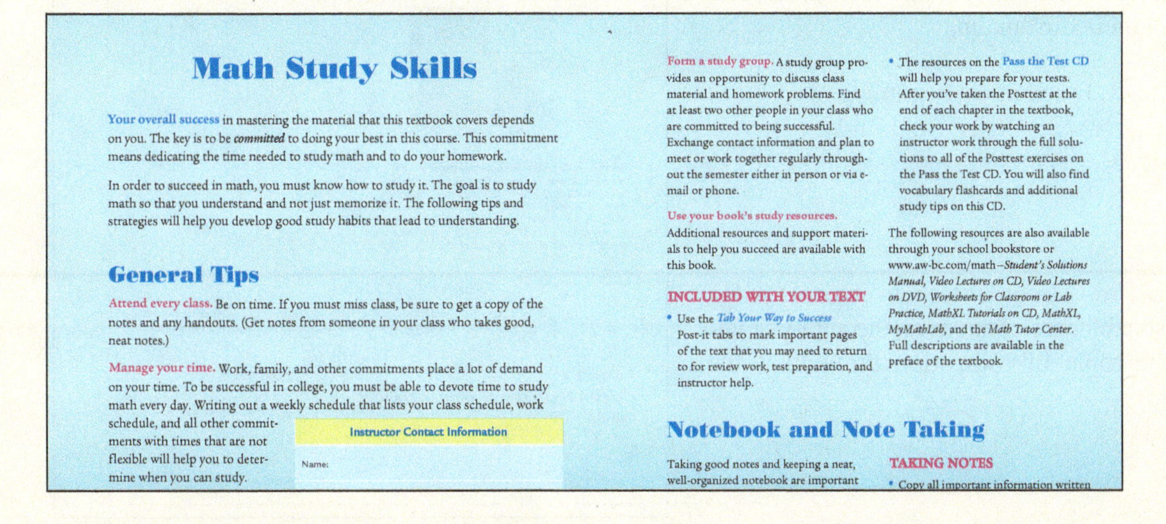

Math Study Skills

Your overall success in mastering the material that this textbook covers depends on you. The key is to be ***committed*** to doing your best in this course. This commitment means dedicating the time needed to study math and to do your homework.

In order to succeed in math, you must know how to study it. The goal is to study math so that you understand and not just memorize it. The following tips and strategies will help you develop good study habits that lead to understanding.

General Tips

Attend every class. Be on time. If you must miss class, be sure to get a copy of the notes and any handouts. (Get notes from someone in your class who takes good, neat notes.)

Manage your time. Work, family, and other commitments place a lot of demand on your time. To be successful in college, you must be able to devote time to study math every day. Writing out a weekly schedule that lists your class schedule, work schedule, and all other commitments with times that are not flexible will help you to determine when you can study.

Instructor Contact Information
Name:

Form a study group. A study group provides an opportunity to discuss class material and homework problems. Find at least two other people in your class who are committed to being successful. Exchange contact information and plan to meet or work together regularly throughout the semester either in person or via e-mail or phone.

Use your book's study resources. Additional resources and support materials to help you succeed are available with this book.

INCLUDED WITH YOUR TEXT

- Use the *Tab Your Way to Success* Post-it tabs to mark important pages of the text that you may need to return to for review work, test preparation, and instructor help.
- The resources on the Pass the Test CD will help you prepare for your tests. After you've taken the Posttest at the end of each chapter in the textbook, check your work by watching an instructor work through the full solutions to all of the Posttest exercises on the Pass the Test CD. You will also find vocabulary flashcards and additional study tips on this CD.

The following resources are also available through your school bookstore or www.aw-bc.com/math–*Student's Solutions Manual, Video Lectures on CD, Video Lectures on DVD, Worksheets for Classroom or Lab Practice, MathXL Tutorials on CD, MathXL, MyMathLab*, and the *Math Tutor Center*. Full descriptions are available in the preface of the textbook.

Notebook and Note Taking

Taking good notes and keeping a neat, well-organized notebook are important

TAKING NOTES

- Copy all important information written

Teaching Tips These tips, found only in the Annotated Instructor's Edition, help instructors with explanations, reminders of previously covered material, and tips on encouraging students to write in a journal.

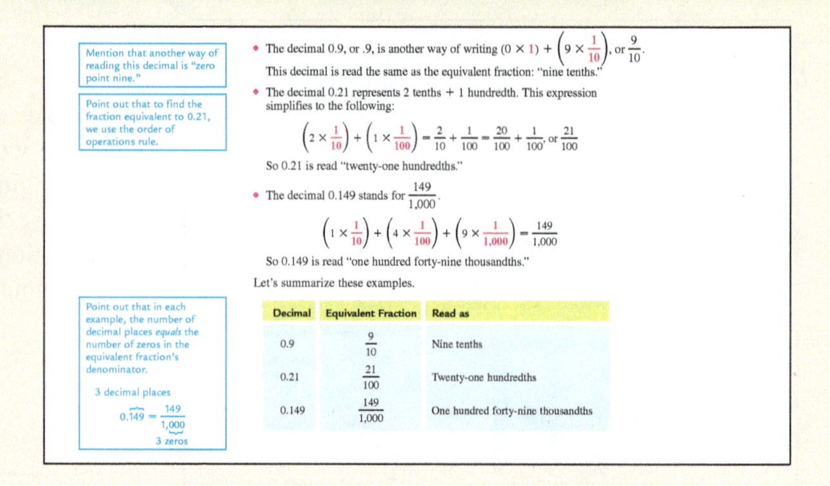

Mention that another way of reading this decimal is "zero point nine."

- The decimal 0.9, or .9, is another way of writing $(0 \times 1) + \left(9 \times \frac{1}{10}\right)$, or $\frac{9}{10}$. This decimal is read the same as the equivalent fraction: "nine tenths."

Point out that to find the fraction equivalent to 0.21, we use the order of operations rule.

- The decimal 0.21 represents 2 tenths + 1 hundredth. This expression simplifies to the following:

$$\left(2 \times \frac{1}{10}\right) + \left(1 \times \frac{1}{100}\right) = \frac{2}{10} + \frac{1}{100} = \frac{20}{100} + \frac{1}{100}, \text{ or } \frac{21}{100}$$

So 0.21 is read "twenty-one hundredths."

- The decimal 0.149 stands for $\frac{149}{1,000}$.

$$\left(1 \times \frac{1}{10}\right) + \left(4 \times \frac{1}{100}\right) + \left(9 \times \frac{1}{1,000}\right) = \frac{149}{1,000}$$

So 0.149 is read "one hundred forty-nine thousandths."

Let's summarize these examples.

Point out that in each example, the number of decimal places *equals* the number of zeros in the equivalent fraction's denominator.

3 decimal places $0.149 = \frac{149}{1,000}$ 3 zeros

Decimal	Equivalent Fraction	Read as
0.9	$\frac{9}{10}$	Nine tenths
0.21	$\frac{21}{100}$	Twenty-one hundredths
0.149	$\frac{149}{1,000}$	One hundred forty-nine thousandths

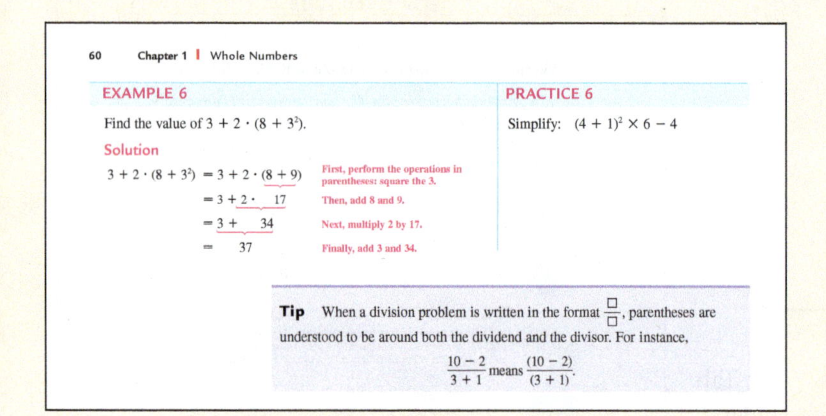

60 Chapter 1 | Whole Numbers

EXAMPLE 6

Find the value of $3 + 2 \cdot (8 + 3^2)$.

Solution

$3 + 2 \cdot (8 + 3^2) = 3 + 2 \cdot (8 + 9)$ First, perform the operations in parentheses: square the 3.

$= 3 + 2 \cdot 17$ Then, add 8 and 9.

$= 3 + 34$ Next, multiply 2 by 17.

$= 37$ Finally, add 3 and 34.

PRACTICE 6

Simplify: $(4 + 1)^2 \times 6 - 4$

Tip When a division problem is written in the format $\frac{\square}{\square}$, parentheses are understood to be around both the dividend and the divisor. For instance,

$$\frac{10 - 2}{3 + 1} \text{ means } \frac{(10 - 2)}{(3 + 1)}.$$

Student Tips These insightful tips help students avoid common errors and provide other helpful suggestions to foster understanding of the material at hand.

Geometry The authors integrate geometry throughout the text so students can see its relevance to their surroundings.

Estimation In order to help students judge the reasonableness of answers, the authors integrate the topic of estimation throughout the text.

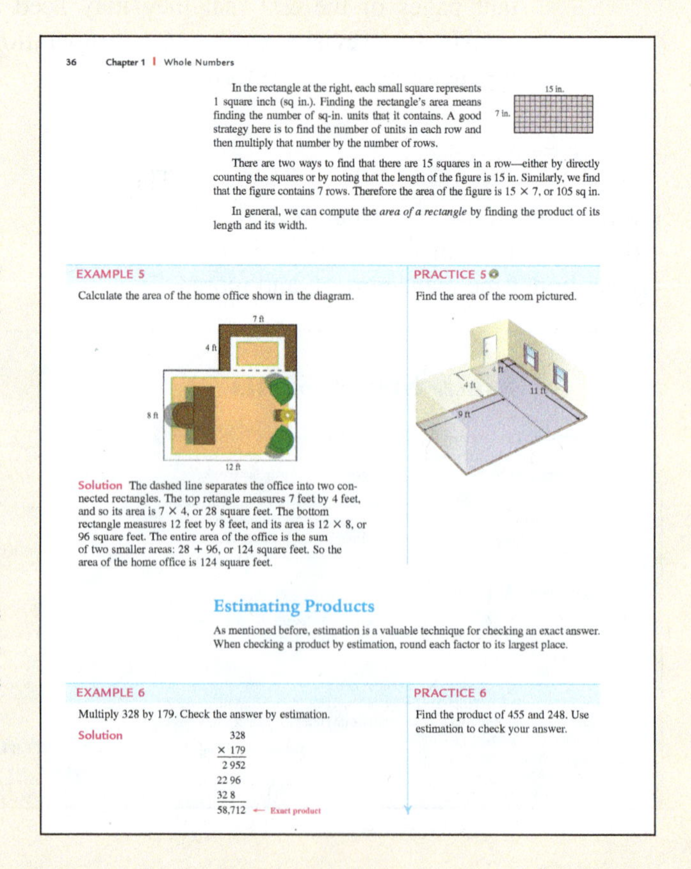

36 Chapter 1 | Whole Numbers

In the rectangle at the right, each small square represents 1 square inch (sq in.). Finding the rectangle's area means finding the number of sq-in. units that it contains. A good strategy here is to find the number of units in each row and then multiply that number by the number of rows.

There are two ways to find that there are 15 squares in a row—either by directly counting the squares or by noting that the length of the figure is 15 in. Similarly, we find that the figure contains 7 rows. Therefore the area of the figure is 15×7, or 105 sq in.

In general, we can compute the *area of a rectangle* by finding the product of its length and its width.

EXAMPLE 5

Calculate the area of the home office shown in the diagram.

Solution The dashed line separates the office into two connected rectangles. The top retangle measures 7 feet by 4 feet, and so its area is 7×4, or 28 square feet. The bottom rectangle measures 12 feet by 8 feet, and its area is 12×8, or 96 square feet. The entire area of the office is the sum of two smaller areas: $28 + 96$, or 124 square feet. So the area of the home office is 124 square feet.

PRACTICE 5

Find the area of the room pictured.

Estimating Products

As mentioned before, estimation is a valuable technique for checking an exact answer. When checking a product by estimation, round each factor to its largest place.

EXAMPLE 6

Multiply 328 by 179. Check the answer by estimation.

Solution

```
    328
  × 179
  2 952
  22 96
  32 8
 58,712  ← Exact product
```

PRACTICE 6

Find the product of 455 and 248. Use estimation to check your answer.

Cultural Notes Necessity is the mother of invention, and mathematics was created out of a need to solve problems in everyday life. Cultural Notes investigate the origins of mathematical concepts, discussing and illustrating the evolution of mathematics throughout the world.

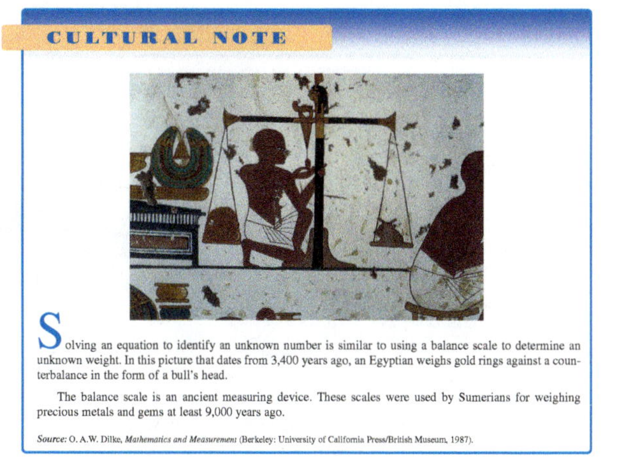

CULTURAL NOTE

Solving an equation to identify an unknown number is similar to using a balance scale to determine an unknown weight. In this picture that dates from 3,400 years ago, an Egyptian weighs gold rings against a counterbalance in the form of a bull's head.

The balance scale is an ancient measuring device. These scales were used by Sumerians for weighing precious metals and gems at least 9,000 years ago.

Source: O. A.W. Dilke, *Mathematics and Measurement* (Berkeley: University of California Press/British Museum, 1987).

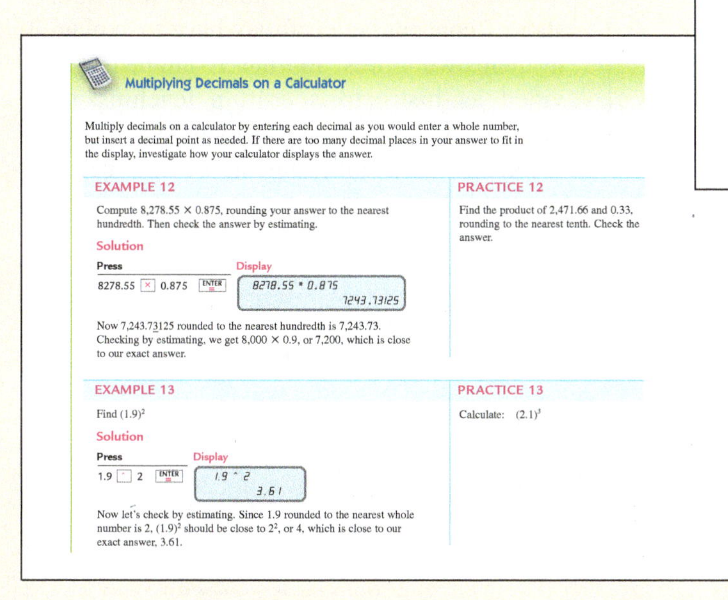

Multiplying Decimals on a Calculator

Multiply decimals on a calculator by entering each decimal as you would enter a whole number, but insert a decimal point as needed. If there are too many decimal places in your answer to fit in the display, investigate how your calculator displays the answer.

EXAMPLE 12

Compute 8,278.55 × 0.875, rounding your answer to the nearest hundredth. Then check the answer by estimating.

Solution

Press	Display
8278.55 [×] 0.875 [ENTER]	8278.55 * 0.875 7243.73125

Now 7,243.73125 rounded to the nearest hundredth is 7,243.73. Checking by estimating, we get 8,000 × 0.9, or 7,200, which is close to our exact answer.

PRACTICE 12

Find the product of 2,471.66 and 0.33, rounding to the nearest tenth. Check the answer.

EXAMPLE 13

Find $(1.9)^2$

Solution

Press	Display
1.9 [^] 2 [ENTER]	1.9 ^ 2 3.61

Now let's check by estimating. Since 1.9 rounded to the nearest whole number is 2, $(1.9)^2$ should be close to 2^2, or 4, which is close to our exact answer, 3.61.

PRACTICE 13

Calculate: $(2.1)^3$

Technology Inserts In order to familiarize students with a range of computational methods—mental, paper and pencil, and calculator arithmetic—the authors include optional technology inserts that instruct students on how to use the scientific calculator to perform arithmetic operations. The side-by-side format is also used here to provide consistency across the text.

Exercise Variety The Akst/Bragg texts provide instructors with a variety of exercise types.

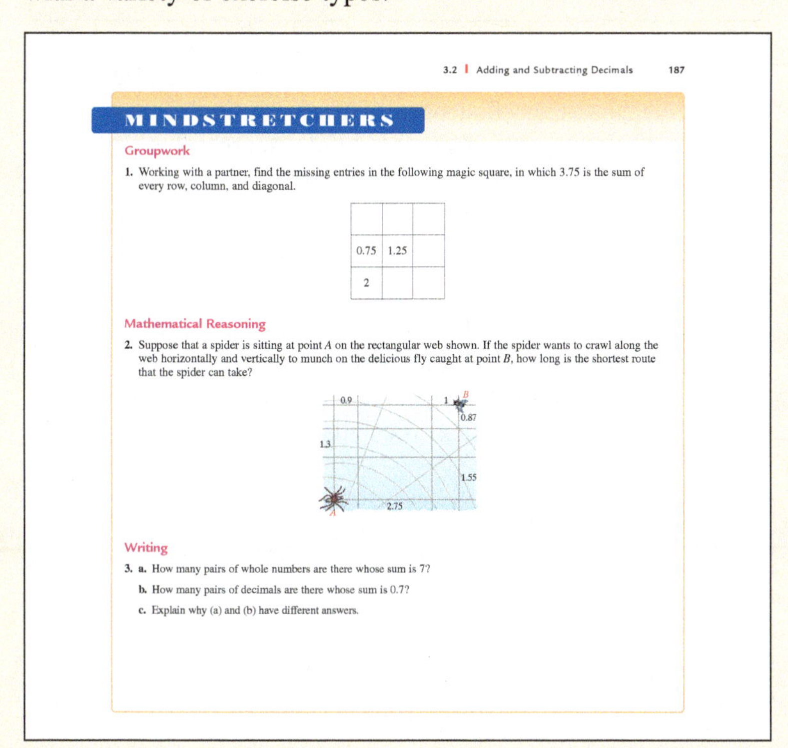

3.2 | Adding and Subtracting Decimals 187

MINDSTRETCHERS

Groupwork

1. Working with a partner, find the missing entries in the following magic square, in which 3.75 is the sum of every row, column, and diagonal.

0.75	1.25	
2		

Mathematical Reasoning

2. Suppose that a spider is sitting at point *A* on the rectangular web shown. If the spider wants to crawl along the web horizontally and vertically to munch on the delicious fly caught at point *B*, how long is the shortest route that the spider can take?

Writing

3. a. How many pairs of whole numbers are there whose sum is 7?
b. How many pairs of decimals are there whose sum is 0.7?
c. Explain why (a) and (b) have different answers.

Mindstretchers Found at the end of most sections, Mindstretchers are engaging activities that incorporate investigation, critical thinking, reasoning, pattern recognition, and writing exercises along with corresponding group work and historical connections in one comprehensive problem set. These problem sets target different levels and types of student understanding.

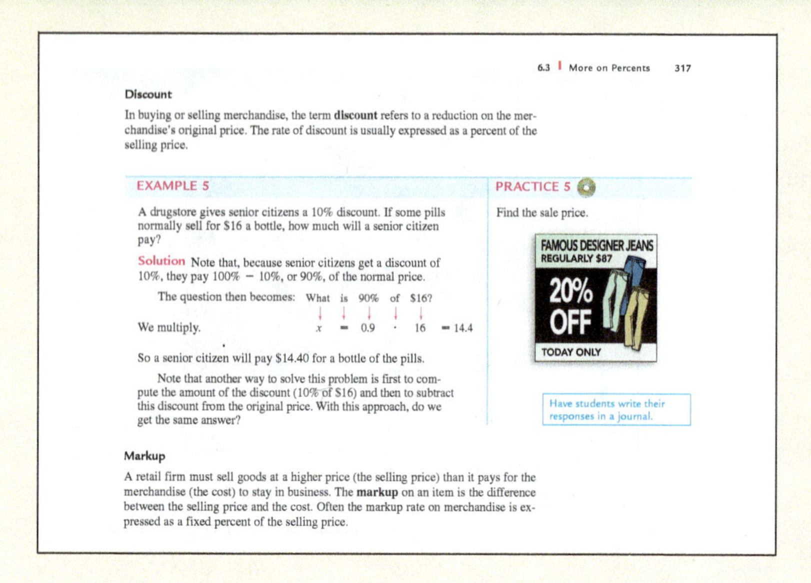

6.3 | More on Percents 317

Discount

In buying or selling merchandise, the term **discount** refers to a reduction on the merchandise's original price. The rate of discount is usually expressed as a percent of the selling price.

EXAMPLE 5

A drugstore gives senior citizens a 10% discount. If some pills normally sell for \$16 a bottle, how much will a senior citizen pay?

Solution Note that, because senior citizens get a discount of 10%, they pay 100% − 10%, or 90%, of the normal price.

The question then becomes: What is 90% of \$16?

We multiply. $x = 0.9 \cdot 16 = 14.4$

So a senior citizen will pay \$14.40 for a bottle of the pills.

Note that another way to solve this problem is first to compute the amount of the discount (10% of \$16) and then to subtract this discount from the original price. With this approach, do we get the same answer?

PRACTICE 5

Find the sale price.

Have students write their responses in a journal.

Markup

A retail firm must sell goods at a higher price (the selling price) than it pays for the merchandise (the cost) to stay in business. The **markup** on an item is the difference between the selling price and the cost. Often the markup rate on merchandise is expressed as a fixed percent of the selling price.

Video Lectures on CD The icon indicates examples, practice exercises, and exercises that are covered in the Video Lectures on CD.

Writing Exercises Students will understand a concept better if they have to explain it in their own words. Journal assignments (provided as instructor's edition teaching tips) allow students to work on their mathematical communication skills, thus improving their understanding of concepts.

Calculator Exercises These optional exercises, denoted with a icon, can be found in the exercise sets, giving students the opportunity to use a calculator to solve a variety of real-life applications.

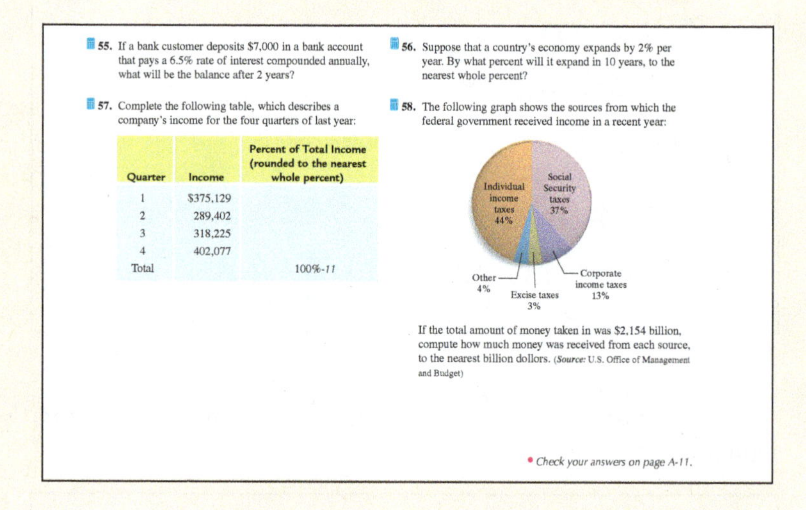

55. If a bank customer deposits \$7,000 in a bank account that pays a 6.5% rate of interest compounded annually, what will be the balance after 2 years?

56. Suppose that a country's economy expands by 2% per year. By what percent will it expand in 10 years, to the nearest whole percent?

57. Complete the following table, which describes a company's income for the four quarters of last year:

Quarter	Income	Percent of Total Income (rounded to the nearest whole percent)
1	\$375,129	
2	289,402	
3	318,225	
4	402,077	
Total		100%

58. The following graph shows the sources from which the federal government received income in a recent year:

If the total amount of money taken in was \$2,154 billion, compute how much money was received from each source, to the nearest billion dollars. (*Source:* U.S. Office of Management and Budget)

• *Check your answers on page A-11.*

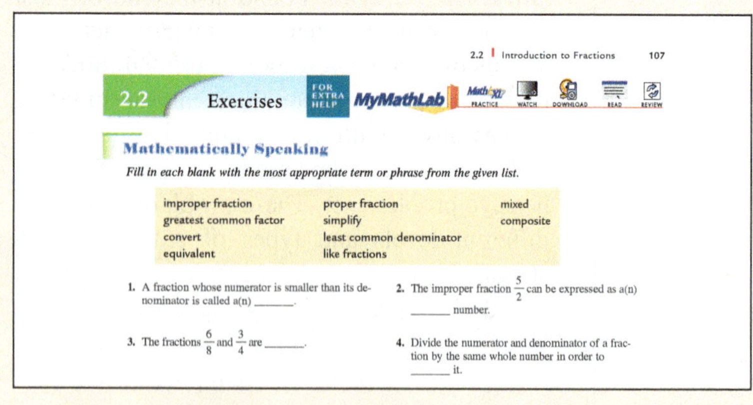

2.2 | Introduction to Fractions 107

2.2 Exercises FOR EXTRA HELP MyMathLab Math XL PRACTICE WATCH DOWNLOAD READ REVIEW

Mathematically Speaking

Fill in each blank with the most appropriate term or phrase from the given list.

improper fraction	proper fraction	mixed
greatest common factor	simplify	composite
convert	least common denominator	
equivalent	like fractions	

1. A fraction whose numerator is smaller than its denominator is called a(n) ______.

2. The improper fraction $\frac{5}{2}$ can be expressed as a(n) ______ number.

3. The fractions $\frac{6}{8}$ and $\frac{3}{4}$ are ______.

4. Divide the numerator and denominator of a fraction by the same whole number in order to ______ it.

For Extra Help These boxes, found at the top of the first page of every section's exercise set, direct students to helpful resources that will aid in their study of the material.

***New* Mathematically Speaking Exercises** Located at the begining of nearly every exercise section of the text, Mathematically Speaking exercises help students understand and master mathematical vocabulary.

***New* Mixed Practice Exercises** Mixed Practice exercises reinforce skills within the section and encourage review.

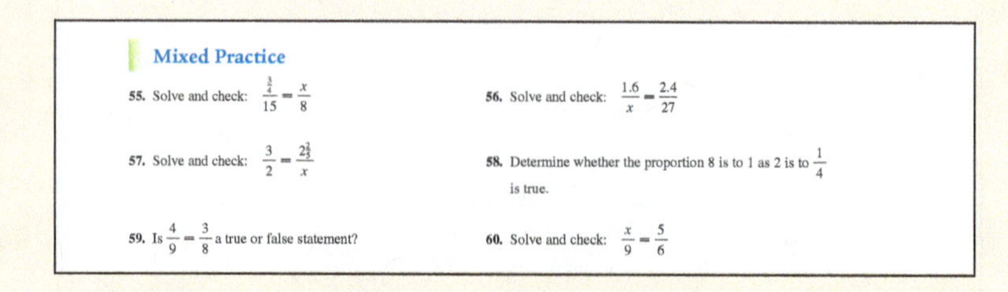

Mixed Practice

55. Solve and check: $\frac{\frac{1}{2}}{15} = \frac{x}{8}$

56. Solve and check: $\frac{1.6}{x} = \frac{2.4}{27}$

57. Solve and check: $\frac{3}{2} = \frac{2\frac{1}{4}}{x}$

58. Determine whether the proportion 8 is to 1 as 2 is to $\frac{1}{4}$ is true.

59. Is $\frac{4}{9} = \frac{3}{8}$ a true or false statement?

60. Solve and check: $\frac{x}{9} = \frac{5}{6}$

End-of-Chapter Material At the end of each chapter, students will find a wealth of review- and retention-oriented material to reinforce the concepts presented in current and previous chapters.

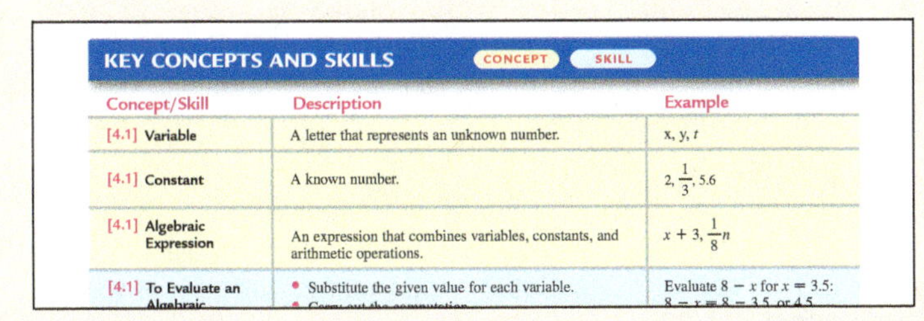

KEY CONCEPTS AND SKILLS CONCEPT SKILL

Concept/Skill	Description	Example
[4.1] Variable	A letter that represents an unknown number.	x, y, t
[4.1] Constant	A known number.	$2, \frac{1}{3}, 5.6$
[4.1] Algebraic Expression	An expression that combines variables, constants, and arithmetic operations.	$x + 3, \frac{1}{8}n$
[4.1] To Evaluate an Algebraic…	• Substitute the given value for each variable.	Evaluate $8 - x$ for $x = 3.5$:

Key Concepts and Skills These give students quick reminders of the chapter's most important elements and provide a one-stop quick review of the chapter material. Each concept/skill is keyed to the section in which it was introduced, and students are given a brief description and example for each.

Chapter 4 **Review Exercises**

To help you review this chapter, solve these problems.

[4.1] *Translate each algebraic expression to words.*

1. $x + 1$ 2. $y + 4$ 3. $w - 1$ 4. $s - 3$

5. $\frac{c}{7}$ 6. $\frac{a}{10}$ 7. $2x$ 8. $6y$

9. $y \div 0.1$ 10. $n \div 1.6$ 11. $\frac{1}{3}x$ 12. $\frac{1}{10}w$

Chapter Review Exercises These exercises are keyed to the corresponding sections for easy student reference. Numerous mixed application problems complete each of these exercise sets, reinforcing the applicability of what students are learning.

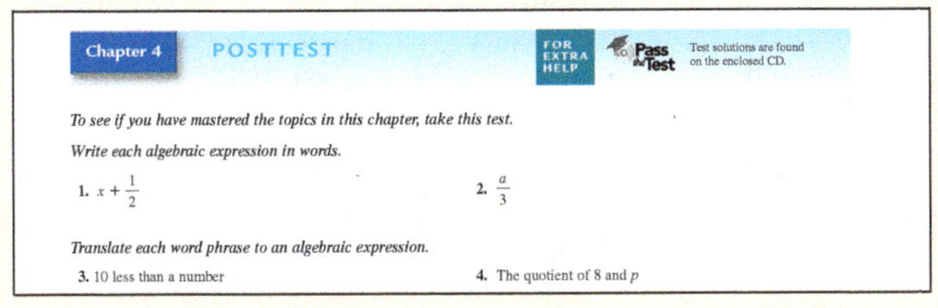

Chapter 4 POSTTEST FOR EXTRA HELP Pass the Test Test solutions are found on the enclosed CD.

To see if you have mastered the topics in this chapter, take this test.

Write each algebraic expression in words.

1. $x + \frac{1}{2}$ 2. $\frac{a}{3}$

Translate each word phrase to an algebraic expression.

3. 10 less than a number 4. The quotient of 8 and p

Chapter Posttest Just as every chapter begins with a Pretest to test student understanding *before* attempting the material, every chapter ends with a Posttest to measure student understanding *after* completing the chapter material. Answers to these tests are provided in the back of the book. Students can watch an instructor working through the full solutions to the Posttest exercises on the Pass the Test CD.

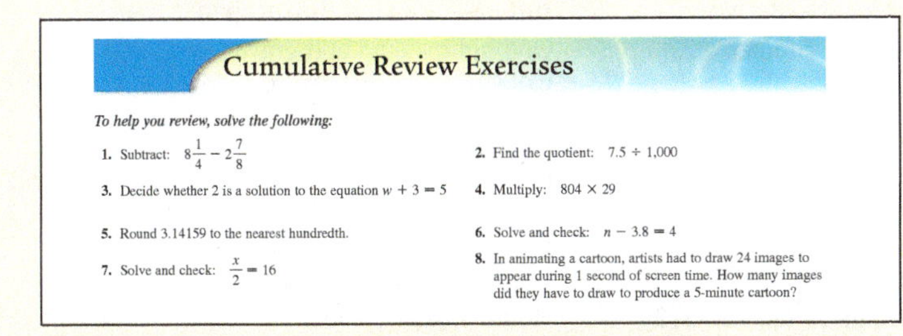

Cumulative Review Exercises

To help you review, solve the following:

1. Subtract: $8\frac{1}{4} - 2\frac{7}{8}$
2. Find the quotient: $7.5 \div 1{,}000$
3. Decide whether 2 is a solution to the equation $w + 3 = 5$
4. Multiply: 804×29
5. Round 3.14159 to the nearest hundredth.
6. Solve and check: $n - 3.8 = 4$
7. Solve and check: $\frac{x}{2} = 16$
8. In animating a cartoon, artists had to draw 24 images to appear during 1 second of screen time. How many images did they have to draw to produce a 5-minute cartoon?

Cumulative Review Exercises Beginning at the end of Chapter 2, students have the opportunity to maintain their skills by completing the Cumulative Review Exercises. These exercises are invaluable, especially when students need to recall a previously learned concept or skill before beginning the next chapter, or when studying for midterm and final examinations.

***New* Pass the Test CD** Included with every new copy of the book, the Pass the Test CD includes the following resources:

- video footage of an instructor working through the complete solutions for all Posttest exercises for each chapter
- vocabulary flashcards
- a video that offers an interactive Spanish glossary
- additional tips on improving time management

Index of Applications

Health/Life Sciences

Labor

Miscellaneous

Physics

Sports

Statistics/Demographics

Technology

Transportation

CHAPTER 1

Whole Numbers

Whole Numbers and the Census

Every ten years, the U.S. Bureau of the Census attempts to count and gather information about each man, woman, and child in the nation. The government then uses this information both to reapportion the 435 seats in the House of Representatives and to reallocate billions of dollars in federal funds.

The census also paints a picture of the nation, showing how it has changed since the last count. For example, in the decade that elapsed between the 1990 census and the 2000 census, the number of males rose from 94 million to 101 million, and the number of females grew from 101 million to 108 million. The total of family households swelled from 93 million to 103 million, and the number of students enrolled in college increased from 14 million to 15 million. (*Source:* U.S. Bureau of the Census)

Chapter 1 PRETEST

To see if you have already mastered the topics in this chapter, take this test.

1. Insert commas as needed in the number 2 0 5 0 0 7. Then write the number in words.

2. Write the number one million, two hundred thirty-five thousand in standard form.

3. What place does the digit 8 occupy in 805,674?

4. Round 8,143 to the nearest hundred.

5. Add: $38 + 903 + 7{,}285$

6. Subtract 286 from 5,000.

7. Subtract: $734 - 549$

8. Find the product of 809 and 36.

9. Find the quotient: $27\overline{)7{,}020}$

10. Divide: $13{,}558 \div 44$

11. Write $2 \cdot 2 \cdot 2$, using exponents.

12. Evaluate: 6^2

Simplify.

13. $26 - 7 \cdot 3$

14. $3 + 2^3 \cdot (8 - 3)$

Solve and check.

15. The mathematician Benjamin Banneker was born in 1731 and died in 1806. About how old was he when he died? (*Source: The New Encyclopedia Britannica*)

16. At a certain college, students pay \$75 for each college credit. If a student takes 9 credits, how much will it cost?

17. Tiger Woods had scores of 74, 66, 65, and 71 for his four rounds of golf at a Masters Tournament. What was his average score for a round of golf? (*Source: The Augusta Chronicle*, 2005)

18. The HP Photosmart 8050 photo printer can print a 4-inch by 6-inch photo in 27 seconds, and the 8250 model printer can print the same size photo in 14 seconds. How much longer would it take the HP 8050 printer to print 12 4-inch by 6-inch photos? (*Source:* Hewlett-Packard)

19. New London County Insurance Company (NLC) offers an installment plan for paying auto insurance premiums. For a \$540 policy, NLC requires a down payment of \$81. The balance is paid in 9 equal installments of \$54, which includes a billing fee. How much would be saved by paying for this policy without using the installment plan? (*Source:* New London County Insurance Company, 2006)

20. Which of the rooms pictured has the largest area? (feet = ft)

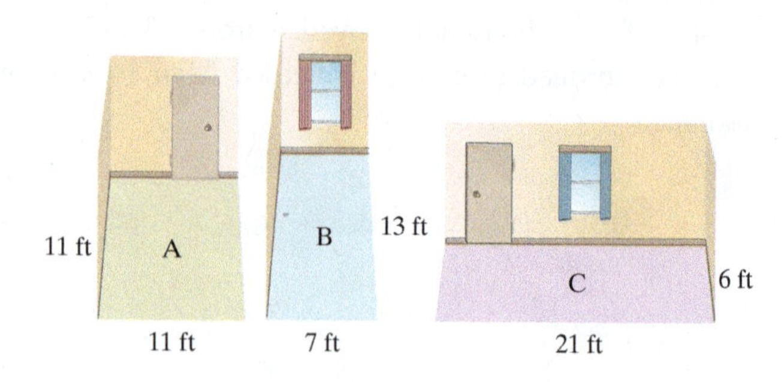

• *Check your answers on page A-1.*

1.1 Introduction to Whole Numbers

OBJECTIVES

- To read and write whole numbers
- To write whole numbers in expanded form
- To round whole numbers

What the Whole Numbers Are and Why They Are Important

We use whole numbers for counting, whether it is the number of *e*'s on this page, the number of stars in the sky, or the number of runs, hits, and errors in a baseball game.

The whole numbers are 0, 1, 2, 3, 4, 5, 6, 7, 8, 9, 10, 11, 12, 13, ... An important property of whole numbers is that there is always a next whole number. This property means that they go on without end, as the three dots above indicate.

Every whole number is either *even* or *odd*. The even whole numbers are 0, 2, 4, 6, 8, 10, 12, ... The odd whole numbers are 1, 3, 5, 7, 9, 11, 13, ...

We can represent the whole numbers on a number line. Similar to a ruler, the number line starts with 0 and extends without end to the right, as the arrow indicates.

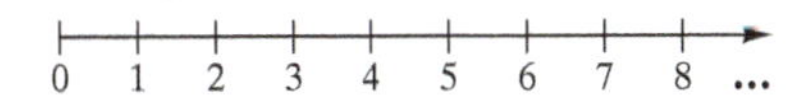

Reading and Writing Whole Numbers

Generally speaking, we *read* whole numbers in words, but we use the **digits** 0, 1, 2, 3, 4, 5, 6, 7, 8, and 9 to *write* them. For instance, we read the whole number *fifty-one* but write it *51*, which we call **standard form**.

Each of the digits in a whole number in standard form has a **place value**. Our **place value** system is very important because it underlies both the way we write and the way we compute with numbers.

The following chart shows the place values in whole numbers up to 12 digits long. For instance in the number 1,234,056 the digit 2 occupies the hundred thousands place. Study the place values in the chart now.

BILLIONS			MILLIONS			THOUSANDS			ONES			← Period
Hundreds	Tens	Ones	Hundreds	Tens	Ones	Hundreds	Tens	Ones	Hundreds	Tens	Ones	← Place value
					1	2	3	4	0	5	6	
		8	1	6	8	9	3	1	0	4	7	

Tip We read whole numbers from left to right, but it is easier in the place value chart to learn the names of the places *from right to left.*

When we write a large whole number in standard form, we insert *commas* to separate its digits into groups of three, called **periods**. For instance the number 8,168,931,047 has four periods: *ones, thousands, millions*, and *billions*.

EXAMPLE 1

In each number, identify the place that the digit 7 occupies.

a. 207

b. 7,654,000

c. 5,700,000,001

Solution

a. The ones place

b. The millions place

c. The hundred millions place

PRACTICE 1

What place does the digit 8 occupy in each number?

a. 278,056

b. 803,746

c. 3,080,700,059

The following rule provides a shortcut for *reading a whole number*.

To Read a Whole Number

Working from left to right,

- read the number in each period and then
- name the period in place of the comma.

For instance, 1,234,056 is read "one million, two hundred thirty-four thousand, fifty-six." Note that the ones period is not read.

EXAMPLE 2

How do you read the number 422,000,085?

Solution Beginning at the left in the millions period, we read this number as "four hundred twenty-two million, eighty-five." Note that because there are all zeros in the thousands period, we do not read "thousands."

PRACTICE 2

Write 8,000,376,052 in words.

EXAMPLE 3

The display on a calculator shows the answer 3578002105. Insert commas in this answer and then read it.

Solution The number with commas is 3,578,002,105. It is read "three billion, five hundred seventy-eight million, two thousand, one hundred five."

PRACTICE 3

A company is worth $7372050. After inserting commas, read this amount.

Until now, we have discussed how to *read* whole numbers in standard form. Now let's turn to the question of how they are *written* in standard form. We simply reverse the process just described. For instance, the number eight billion, one hundred sixty-eight million, nine hundred thirty-one thousand, forty-seven in standard form is 8,168,931,047. Here, we use the 0 as a **placeholder** in the hundreds place because there are no hundreds.

To Write a Whole Number

Working from left to right,

- write the number named in each period and
- replace each period name with a comma.

When writing large whole numbers in standard form, we must remember that the number of commas is always one less than the number of periods. For instance, the number one million, two hundred thirty-four thousand, fifty-six—1,234,056—has three periods and two commas. Similarly, the number 8,168,931,047 has four periods and three commas.

EXAMPLE 4

Write the number eight billion, seven in standard form.

Solution This number involves billions, so there are four periods—billions, millions, thousands, and ones—and three commas. Writing the number named in each period and replacing each period name with a comma, we get 8,000,000,007. Note that we write three 0's when no number is named in a period.

PRACTICE 4

Use digits and commas to write the amount ninety-five million, three dollars.

EXAMPLE 5

The treasurer of a company writes a check in the amount of four hundred thousand seven hundred dollars. Using digits, how would she write this amount on the check?

Solution This quantity is written with one comma, because its largest period is thousands. So the treasurer writes $400,700, as shown on the check below.

UNITED INDUSTRIES
Atlanta, Georgia 2066
DATE January 26, 2008
PAY TO THE ORDER OF American Vendors, Inc. $ 400,700—
Four Hundred Thousand Seven Hundred and 00/100 DOLLARS
SB Southern Bank
MEMO Maxine Jefferson
⑆72110756⑆ 0220006587⑈ 2066

PRACTICE 5

A rich alumna donates three hundred seventy-five thousand dollars to her college's scholarship fund.

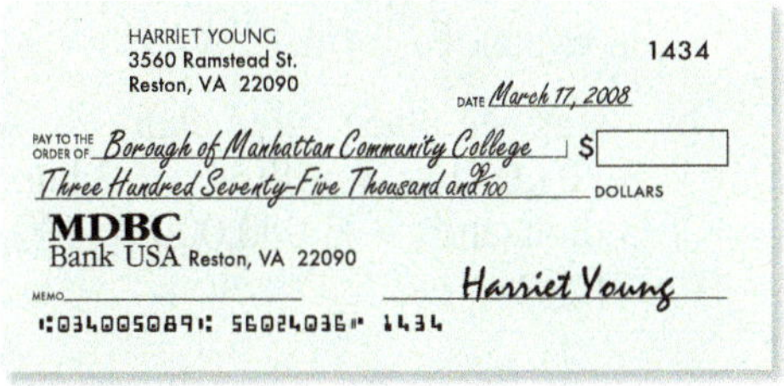

HARRIET YOUNG
3560 Ramstead St.
Reston, VA 22090 1434
DATE March 17, 2008
PAY TO THE ORDER OF Borough of Manhattan Community College $
Three Hundred Seventy-Five Thousand and 00/100 DOLLARS
MDBC
Bank USA Reston, VA 22090
MEMO Harriet Young
⑆034005089⑆ 5602403⑈ 1434

Using digits, how would she write this amount on the check?

____

When writing checks, we write the amount in both digits and words. Why do we do this?

Writing Whole Numbers in Expanded Form

We have just described how to write whole numbers in standard form. Now let's turn to how we write these numbers in **expanded form.**

Let's consider the whole number 4,325 and examine the place value of its digits.

$$4{,}325 = 4 \text{ thousands} + 3 \text{ hundreds} + 2 \text{ tens} + 5 \text{ ones}$$

This last expression is called the expanded form of the number, and it can be written as follows

$$4{,}000 + 300 + 20 + 5$$

The expanded form of a number spells out its value in terms of place value, helping us understand what the number really means. For instance, think of the numbers 92 and 29. By representing them in *expanded* form, can you explain why they differ in value even though their *standard* form consists of the same digits?

EXAMPLE 6

Write in expanded form:

a. 906

b. 3,203,000

Solution

a. The 6 is in the ones place, the 0 is in the tens place, and the 9 is in the hundreds place.

ONES		
Hundreds	Tens	Ones
9	0	6

So 906 is 9 hundreds + 0 tens + 6 ones = 900 + 0 + 6 in expanded form.

b. Using the place value chart, we see that 3,203,000 = 3 millions + 2 hundred thousands + 3 thousands = 3,000,000 + 200,000 + 3,000.

PRACTICE 6

Express in expanded form.

a. 27,013

b. 1,270,093

Rounding Whole Numbers

Most people equate mathematics with precision, but some problems require sacrificing precision for simplicity. In this case, we use the technique called **rounding** to approximate the exact answer with a number that ends in a given number of zeros. Rounded numbers have special advantages: They seem clearer to us than other numbers, and they make computation easier—especially when we are trying to compute in our heads.

Of these two headlines, which do you prefer? Why?

Study the following chart to see the connection between place value and rounding

Rounding to the nearest	Means that the rounded number ends in at least
10	One 0
100	Two 0's
1,000	Three 0's
10,000	Four 0's
100,000	Five 0's
1,000,000	Six 0's

Note in the chart that the place value tells us how many 0's the rounded number must have at the end. Having more 0's than indicated is possible. Can you think of an example?

When rounding, we use an underlined digit to indicate the place to which we are rounding.

Now let's consider the following rule for rounding whole numbers.

To Round a Whole Number

- Underline the place to which you are rounding.
- Look at the digit to the right of the underlined digit, called the **critical digit.** If this digit is 5 or more, add 1 to the underlined digit; if it is less than 5, leave the underlined digit unchanged.
- Replace all the digits to the right of the underlined digit with zeros.

EXAMPLE 7

Round 79,630 to

a. the nearest thousand

b. the nearest hundred.

Solution

a. $79{,}630 = 7\underline{9}{,}630$ ← Underline the digit in the thousands place.

$= 7\underline{9}{,}630$ ← The critical digit 6 is greater than 5; add 1 to the underlined digit.

$\approx 80{,}000$ ← Change the digits to the right of the underlined digit to 0's.

↑ This symbol means "is approximately equal to."

Note that adding 1 to the underlined digit gave us 10 and forced us to write 0 and carry 1 to the next column, changing the 7 to 8.

b. First, we underline the 6 because that digit occupies the hundreds place: $79{,}\underline{6}30$. The critical digit is 3: $79{,}\underline{6}30$. Since 3 is less than 5, we leave the underlined digit unchanged. Then, we replace all digits to the right with 0's, getting 79,600. We write $79{,}630 \approx 79{,}600$, meaning that 79,630 when rounded to the nearest hundred is 79,600.

PRACTICE 7

Round 51,760 to

a. the nearest thousand

b. the nearest ten thousand.

For Example 7, consider this number line.

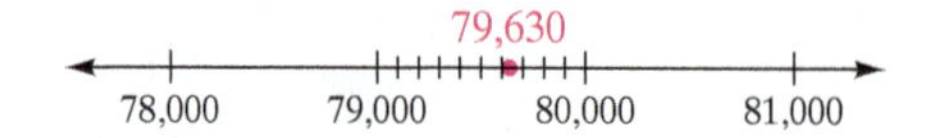

The number line shows that 79,630 lies between 79,000 and 80,000 and that it is closer to 80,000, as the rule indicates.

EXAMPLE 8

In an anatomy and physiology class, a student learned that the adult human skeleton contains 206 bones. How many bones is this to the nearest hundred bones?

Solution We first write 206. The critical digit 0 is less than 5, so we do *not* add 1 to the underlined digit. However, we do change both the digits to the right of the 2 to 0's. So 206 ≈ 200, and there are approximately 200 bones in the human body.

PRACTICE 8

Based on current population data, the U.S. Bureau of the Census projects that the U.S. resident population will be 419,845,000 in the year 2050. What is the projected population to the nearest million?

EXAMPLE 9

The following table lists five of the highest-grossing movies of all time, and the amount of money they took in.

Film	Year	World Total (in U.S. dollars)
Titanic	1997	1,845,034,188
Star Wars: Episode I, The Phantom Menace	1999	925,600,000
Harry Potter and the Sorcerer's Stone	2001	985,817,659
The Lord of the Rings: The Two Towers	2002	926,287,400
The Lord of the Rings: The Return of the King	2003	1,118,888,979

(**Source:** *The Top Ten of Everything 2006*)

a. Write in words the amount of money taken in by the film with the largest world total.

b. Round to the nearest ten million dollars the world total for *Harry Potter and the Sorcerer's Stone.*

Solution

a. *Titanic* has the largest world total. This total is read "one billion, eight hundred forty-five million, thirty-four thousand, one hundred eighty-eight dollars."

b. The world total for *Harry Potter and the Sorcerer's Stone* is $985,817,659. To round, we underline the digit in the ten millions place: 985,817,659. Since the critical digit is 5, we add 1 to the underlined digit, and change the digits to the right to 0's. So the rounded total is $990,000,000.

PRACTICE 9

This chart gives the number of male and female faculty members in U.S. colleges during a recent year.

Faculty Members	Number
Men	382,808
Women	248,788

(**Source:** *The Chronicle of Higher Education,* August 25, 2006)

a. Write in words the number of female college faculty members.

b. What is the number of male college faculty members rounded to the nearest hundred thousand?

1.1 Exercises

PRACTICE

WATCH

DOWNLOAD

READ

REVIEW

Mathematically Speaking

Fill in each blank with the most appropriate term or phrase from the given list.

calculated	rounded	periods	odd
even	digits	whole numbers	standard form
placeholder	place value	expanded form	

1. The _______ are 0, 1, 2, 3, 4, 5, … .

2. The numbers 0, 2, 4, 6, 8, 10, … are _______.

3. The numbers 1, 3, 5, 7, 9, … are _______.

4. The whole numbers are written with the _______ 0, 1, 2, 3, 4, 5, 6, 7, 8, and 9.

5. The number thirty-seven, when written as 37, is said to be in _______.

6. In the number 528, the _______ of the 5 is hundreds.

7. In the number 206, the 0 is used as a _______ in the tens place.

8. Commas separate the digits in a large whole number into groups of three called _______.

9. When the number 973 is written as 9 hundreds + 7 tens + 3 ones, it is said to be in _______.

10. The number 545 _______ to the nearest hundred is 500.

Underline the digit that occupies the given place.

11. 4,867 Thousands place

12. 975 Hundreds place

13. 316 Tens place

14. 41,722 Ten thousands place

15. 28,461,013 Millions place

16. 762,800 Hundred thousands place

Identify the place occupied by the underlined digit.

17. $\underline{6}$91,400

18. 7$\underline{2}$,109

19. 7,$\underline{3}$80

20. 3$\underline{5}$1

21. $\underline{8}$,450,000,000

22. $\underline{3}$5,832,775

Insert commas as needed, and then write the number in words.

23. 4 8 7 5 0 0

24. 5 2 8 0 5 0

25. 2 3 5 0 0 0 0

26. 1 3 5 0 1 3 2

27. 9 7 5 1 3 5 0 0 0

28. 2 1 0 0 0 1 3 2

29. 2 000 000 352

30. 4 100 000 007

31. 1 000 000 000

32. 379 052 000

Write each number in standard form.

33. Ten thousand, one hundred twenty

34. Three billion, seven hundred million

35. One hundred fifty thousand, eight hundred fifty-six

36. Twenty million, five thousand

37. Six million, fifty-five

38. Two million, one hundred twenty-two

39. Fifty million, six hundred thousand, one hundred ninety-five

40. Nine hundred thousand, eight hundred eleven

41. Four hundred thousand, seventy-two

42. Nine hundred billion

Write each number in expanded form.

43. 3

44. 6,300

45. 858

46. 9,000,000

47. 2,500,004

48. 7,251,380

Round to the indicated place.

49. 671 to the nearest ten

50. 838 to the nearest hundred

51. 7,103 to the nearest hundred

52. 46,099 to the nearest thousand

53. 28,241 to the nearest ten thousand

54. 7,802,555 to the nearest million

55. 705,418 to its largest place

56. 96 to its largest place

57. 31,972 to its largest place

58. 4,913,440 to its largest place

Round each number as indicated.

59.

To the nearest	135,842	2,816,533
Hundred		
Thousand		
Ten thousand		
Hundred thousand		

60.

To the nearest	972,055	3,189,602
Thousand		
Ten thousand		
Hundred thousand		
Million		

Mixed Practice

Solve.

61. Write 12,051 in expanded form.

62. Identify the place occupied by the underlined digit in $\underline{2}$6,543,009.

63. Underline the digit that occupies the ten thousands place in 40,059.

64. Write five hundred forty-two thousand, sixty-seven in standard form.

65. Insert commas as needed, and then write 1 0 5 6 1 0 0 in words.

66. Round 26,255 to the nearest thousand.

Applications

Write each whole number in words.

67. Biologists have classified more than 900,000 species of insects. (*Source:* Smithsonian Institution)

68. Each pair of human lungs contains some 300,000,000 tiny air sacs.

69. The total land area of the Caribbean island of Puerto Rico is 8,959 square kilometers. (*Source: The World Fact-Book*, 2006)

70. Mercury is the closest planet to the Sun, a distance of approximately 36,000,000 miles.

71. There are 37,842 WiFi hotspots in the United States. (*Source: JiWire*, March 2006)

72. The Pyramid of Khufu in Egypt has a base of approximately 2,315,000 blocks. (*Source: The New Encyclopedia Britannica*)

Write each whole number in standard form.

73. Some one hundred billion nerve cells are part of the human brain.

74. Son of Beast, a roller coaster at Paramount's Kings Island in Ohio, has a track length of seven thousand thirty-two feet. (*Source:* American Coasters Network)

75. One of the largest giant sequoias in the United States is three thousand, two hundred eighty-eight inches tall. (*Source:* U.S. National Park Service)

76. The total land area of the United States is nine million, six hundred thirty-one thousand, four hundred eighteen square kilometers. (*Source: The World Factbook*, 2006)

77. The number of registered nurses employed in the United States is expected to grow to two million, nine hundred eight thousand by the year 2012. (*Source:* U.S. Bureau of the Census, *Statistical Abstract of the United States,* 2005)

78. George W. Bush received sixty-two million, thirty-nine thousand, seventy-three votes in the 2004 presidential election. (*Source:* U.S. National Archives and Records Administration)

Round to the indicated place.

79. The Statue of Liberty is 152 feet high. What is its height to the nearest 10 feet?

80. The Nile, with a length of 4,180 miles, is the longest river in the world. Find this length to the nearest thousand miles.

81. In 1949, Air Force Captain James Gallagher led the first team to make an around-the-world flight. The team flew 23,452 miles. What is this distance to the nearest ten thousand miles? (*Source:* Taylor and Mondey, *Milestones of Flight*)

82. The element copper changes from a liquid to a gas at the temperature 2,567 degrees Celsius (°C). Find this temperature to the nearest hundred degrees Celsius.

83. A weight of 454 grams is equivalent to 1 pound. How many grams is this to the nearest hundred?

84. The Rose Bowl stadium has a seating capacity of 92,542. Round this number to the nearest ten thousand. (*Source:* Rose Bowl Operating Company)

85. This chart displays the number of degrees awarded in the United States during a recent year.

Degree	Number Awarded
Associate	665,301
Bachelor's	1,399,542
Master's	558,940
Doctorate	48,378
Professional	83,041

(*Source: The Chronicle of Higher Education, Almanac,* 2006)

a. Write in words the number of bachelor's degrees awarded.

b. Round, to the nearest hundred thousand, the number of associate degrees awarded.

86. The following table lists the amount of gold produced during a recent year in six countries that are leading producers.

Country	Amount of Gold (in troy ounces)
South Africa	13,316,820
USA	10,108,180
Australia	10,050,308
China	7,552,199
Russia	6,462,290

(*Source:* Gold Fields Mineral Services Ltd., Gold Survey 2004)

a. Write in words the gold production for China.

b. Which country or countries produced ten million troy ounces, rounded to the nearest million troy ounces?

• *Check your answers on page A-1.*

MINDSTRETCHERS

Mathematical Reasoning

1. I am thinking of a certain whole number. My number, rounded to the nearest hundred, is 700. When it is rounded to the nearest ten, it is 750. What numbers could I be thinking of?

Writing

2. How does the number 10 play a special role in the way that we write whole numbers? Would it be possible to have the number 2 play this role? Explain.

Groupwork

3. Here are three ways of writing the number seven: 7 VII 𝍸||

Working with a partner, express each of the numbers 1, 2, … , 9 in these three ways.

1.2 Adding and Subtracting Whole Numbers

OBJECTIVES

- To add and subtract whole numbers
- To estimate the sum and difference of whole numbers
- To solve word problems involving the addition or subtraction of whole numbers

The Meaning and Properties of Addition and Subtraction

Addition is perhaps the most fundamental of all operations. One way to think about this operation is as *combining sets.* For example, suppose that we have two distinct sets of pens, with 5 pens in one set and 3 in the other. If we put the two sets together, we get a single set that has 8 pens.

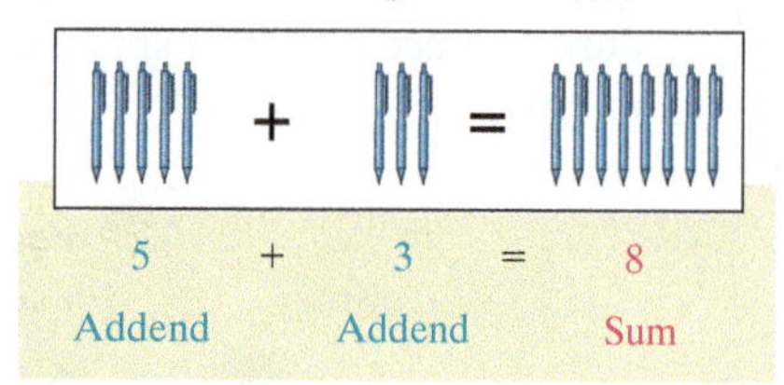

So we can say that 3 added to 5 is 8, or here, 5 pens plus 3 pens equals 8 pens. Numbers being added are called *addends*. The result is called the *sum*, or *total.*

In the above example, note that we are adding quantities of the same thing, or *like quantities.*

Another good way to think about the addition of whole numbers is as *moving to the right on a number line.* In this way, we start at the point on the line corresponding to the first number, 5. Then to add 3, we move 3 units to the right, ending on the point that corresponds to the answer, 8.

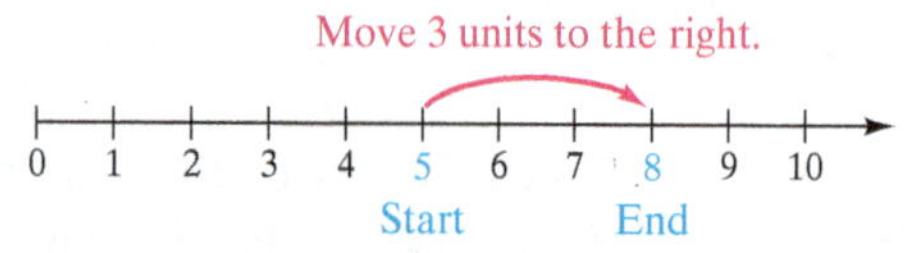

Now let's look at subtraction. One way to look at this operation is as *taking away.* For instance, when we subtract 5 pens from 8 pens, we take 5 pens away from 8 pens, leaving 3 pens.

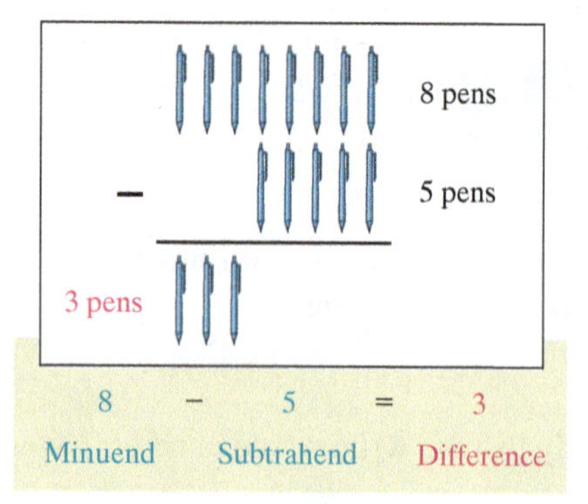

In a subtraction problem, the number from which we subtract is called the *minuend,* the number being subtracted is called the *subtrahend*, and the result is called the *difference*.

As in the preceding example, we can only subtract *like quantities*: we cannot subtract 5 pens from 8 scissors.

We can also think of subtraction as the *opposite of addition.*

$8 - 5 = 3$	because	$5 + 3 = 8$
Subtraction		**Related addition**

Note in this example that, if we add the 5 pens to the 3 pens, we get 8 pens.

Addition and subtraction problems can be written either horizontally or vertically.

$5 + 3 = 8$ $\quad 8 - 5 = 3$

Horizontal

$$\begin{array}{r} 5 \\ +3 \\ \hline 8 \end{array} \qquad \begin{array}{r} 8 \\ -5 \\ \hline 3 \end{array}$$

Vertical

Either format gives the correct answer. But it is generally easier to figure out the sum and difference of large numbers if the problems are written vertically.

Now let's briefly consider several special properties of addition that we use frequently. Examples appear to the right of each property.

The Identity Property of Addition

The sum of a number and zero is the original number.

$3 + 0 = 3$

$0 + 5 = 5$

The Commutative Property of Addition

Changing the order in which two numbers are added does not affect their sum.

$3 + 2 = 2 + 3$

↓ 5 ↓ 5

The Associative Property of Addition

When adding three numbers, regrouping addends gives the same sum. Note that the parentheses tell us which numbers to add first.

We add inside the parentheses first

$(4 + 7) + 2 = 4 + (7 + 2)$

$11 + 2 = 4 + 9$

$13 \quad 13$

Adding Whole Numbers

We add whole numbers by arranging the numbers vertically, keeping the digits with the same place value in the same column. Then we add the digits in each column.

Consider the sum 32 + 65. In the vertical format at the right, the sum of the digits in each column is 9 or less. The sum is 97. When the sum of the digits in a column is greater than 9, we must **regroup** and **carry**, because only a single digit can occupy a single place. Example 1 illustrates this process.

$$\begin{array}{r} 32 \\ +65 \\ \hline 97 \end{array}$$

EXAMPLE 1

Add 47 and 28.

Solution First, we write the addends in expanded form. Then, we add down the ones column.

$$\begin{array}{rcccl} 47 & = & 4 \text{ tens} + 7 \text{ ones} & = & \overset{1 \text{ ten}}{4} \text{ tens} + 7 \text{ ones} \\ +28 & = & 2 \text{ tens} + 8 \text{ ones} & = & 2 \text{ tens} + 8 \text{ ones} \\ \hline & & 15 \text{ ones} & & 5 \text{ ones} \end{array}$$

By regrouping, we express 15 ones as 1 ten + 5 ones. Then we carry the 1 ten to the tens place.

Next, we add down the tens column.

$$\begin{array}{l} \text{1 ten} \\ 4 \text{ tens} + 7 \text{ ones} \\ 2 \text{ tens} + 8 \text{ ones} \\ \hline 7 \text{ tens} + 5 \text{ ones} = 75 \end{array}$$

PRACTICE 1

Add: 178 + 207

The following rule tells how to add whole numbers without using expanded form.

To Add Whole Numbers

- Write the addends vertically, lining up the place values.
- Add the digits in the ones column, writing the rightmost digit of the sum on the bottom. If the sum has two digits, carry the left digit to the top of the next column on the left.
- Add the digits in the tens column, as in the preceding step.
- Repeat this process until you reach the last column on the left, writing the entire sum of that column on the bottom.

EXAMPLE 2

Add: 9,824 + 356 + 2,976

Solution We write the problem vertically, with the addends lined up on the right.

$$\begin{array}{r} 1 \\ 9,824 \\ 356 \\ +2,976 \\ \hline 6 \end{array}$$

The sum of the ones digits is 16 ones. We write the 6 and carry the 1 to the tens column.

$$\begin{array}{r} 11 \\ 9,824 \\ 356 \\ +2,976 \\ \hline 56 \end{array}$$

The sum of the tens digits is 15 tens. We write the 5 and carry the 1 to the hundreds column.

$$\begin{array}{r} 211 \\ 9,824 \\ 356 \\ +2,976 \\ \hline 156 \end{array}$$

The sum of the hundreds digits is 21 hundreds. We write the 1 and carry the 2 to the thousands column.

$$\begin{array}{r} 211 \\ 9,824 \\ 356 \\ +2,976 \\ \hline 13,156 \end{array}$$

The sum of the digits in the thousands column is 13, which we write completely—no need to carry here.

The sum is 13,156.

PRACTICE 2

Find the total: 838 + 96 + 9,502

In Example 3, let's apply the operation of addition to finding the geometric perimeter of a figure. The **perimeter** is the distance around a figure, which we can find by adding the lengths of its sides.

EXAMPLE 3

What is the perimeter of the region marked off for the construction of a swimming pool and an adjacent pool cabana?

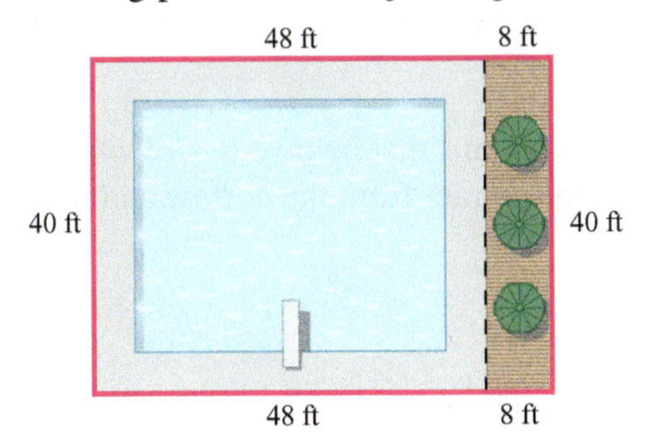

Solution This figure consists of two rectangles placed side by side. We note that the opposite sides of each rectangle are equal in length.

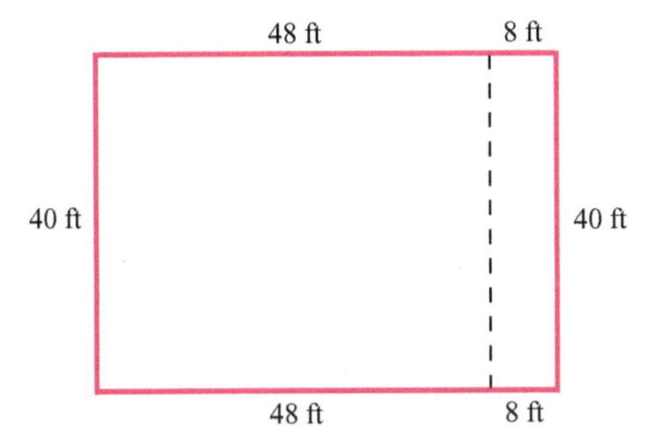

To compute the figure's perimeter, we need to add the lengths of all its sides.

$$\begin{array}{r} 40 \\ 48 \\ 8 \\ 40 \\ 8 \\ +\,48 \\ \hline 192 \end{array}$$

The figure's perimeter is 192 feet.

PRACTICE 3

How long a fence is needed to enclose the piece of land sketched?

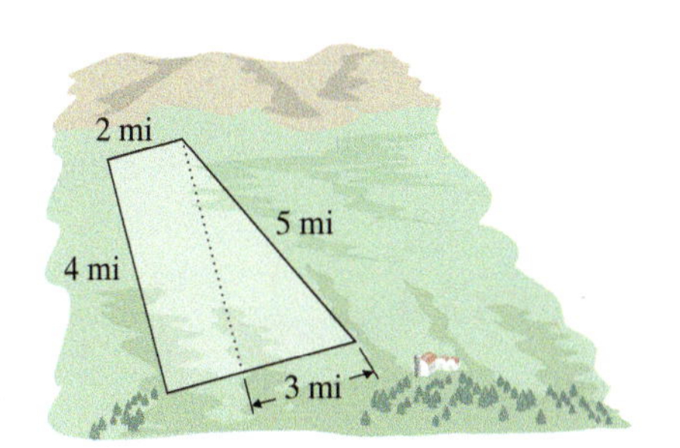

Subtracting Whole Numbers

Consider the subtraction problem 59 − 36, written vertically at the right. We write the whole numbers underneath one another, lined up on the right, so each column contains digits with the same place value. Subtracting the digits within each column, the bottom digit from the top, the result is a difference of 23.

$$\begin{array}{r} 59 \\ -36 \\ \hline 23 \end{array}$$

Keep in mind two useful properties of subtraction.

- When we subtract a number from itself, the result is 0: $6 - 6 = 0$
- When we subtract 0 from a number, the result is the original number: $25 - 0 = 25$

Tip When writing a subtraction problem vertically, be sure that

- the minuend—the number from which we are subtracting—goes on the top and that
- the subtrahend—the number being taken away—goes on the bottom.

Now we consider subtraction problems that involve *borrowing*. In these problems a digit on the bottom is too large to subtract from the corresponding digit on top.

EXAMPLE 4

Subtract: 329 − 87

Solution We first write these numbers vertically in expanded form.

$$\begin{array}{rr} 329 = & 3 \text{ hundreds} + 2 \text{ tens} + 9 \text{ ones} \\ \underline{-\ \ 87} = & \underline{-\qquad\qquad\qquad 8 \text{ tens} + 7 \text{ ones}} \end{array}$$

We then subtract the digits in the ones column: 7 ones from 9 ones gives 2 ones.

$$\begin{array}{rr} & 3 \text{ hundreds} + 2 \text{ tens} + 9 \text{ ones} \\ - & 8 \text{ tens} + 7 \text{ ones} \\ \hline & 2 \text{ ones} \end{array}$$

10 tens + 2 tens = 12 tens

We next go to the tens column. We cannot take 8 tens from 2 tens. But we can *borrow* 1 hundred from the 3 hundreds, leaving 2 in the hundreds place. We *exchange* this hundred for 10 tens (1 hundred = 10 tens). Then combining the 10 tens with the 2 tens gives 12 tens.

$$\begin{array}{rr} & \overset{2}{\not{3}} \text{ hundreds} + \overset{1}{\ }2 \text{ tens} + 9 \text{ ones} \\ - & 8 \text{ tens} + 7 \text{ ones} \\ \hline & 2 \text{ ones} \end{array}$$

We next take 8 from 12 in the tens column, giving 4 tens. Finally, we bring down the 2 hundreds. The difference is 242 in standard form.

$$\begin{array}{rr} & \overset{2}{\not{3}} \text{ hundreds} + \overset{1}{\ }2 \text{ tens} + 9 \text{ ones} \\ - & 8 \text{ tens} + 7 \text{ ones} \\ \hline & 2 \text{ hundreds} + 4 \text{ tens} + 2 \text{ ones} = 242 \end{array}$$

PRACTICE 4

Subtract: 748 − 97

Although we can always rewrite whole numbers in expanded form so as to subtract them, the following rule provides a shortcut.

To Subtract Whole Numbers

- On top, write the number *from which* we are subtracting. On the bottom, write the number that is being taken *away*, lining up the place values. Subtract in each column separately.
- Start with the ones column.
  - **a.** If the digit on top is *larger* than or *equal* to the digit on the bottom, subtract and write the difference below the bottom digit.
  - **b.** If the digit on top is *smaller* than the digit on the bottom, borrow from the digit to the left on top. Then subtract and write the difference below the bottom digit.
- Repeat this process until the last column on the left is finished.

Recall that for every subtraction problem there is a related addition problem. So we can use addition to check subtraction, as in the following example.

EXAMPLE 5

Find the difference between 500 and 293.

Solution We rewrite the problem vertically.

$$\begin{array}{r} 500 \\ -293 \\ \hline \end{array}$$

We cannot subtract 3 ones from 0 ones, and we cannot borrow from 0 tens. So we borrow from the 5 hundreds.

$$\begin{array}{r} \overset{4}{\cancel{5}}\ \overset{1}{0}\ 0 \\ -2\ 9\ 3 \\ \hline \end{array} \quad \leftarrow \text{5 hundreds} = \text{4 hundreds} + \text{10 tens}$$

We now borrow from the tens column.

$$\begin{array}{r} \overset{4}{\cancel{5}}\ \overset{\overset{9}{\cancel{10}}}{\cancel{0}}\ \overset{1}{0} \\ -2\ 9\ 3 \\ \hline 2\ 0\ 7 \end{array} \quad \leftarrow \text{10 tens} = \text{9 tens} + \text{10 ones}$$

Check We check the difference by adding it to the subtrahend. The sum turns out to be the original minuend, so our answer is correct.

$$\begin{array}{r} 207 \\ +293 \\ \hline 500 \end{array}$$

PRACTICE 5

Subtract 3,253 from 8,000.

EXAMPLE 6

There are a total of 116 chemical elements. Ninety-four of these occur naturally on Earth and the rest are synthetic. How many chemical elements are synthetic? (*Source:* *Wikipedia*)

Solution The total number of chemical elements equals the number of natural chemical elements plus the number of synthetic chemical elements.

Natural Chemical Elements (94)	Synthetic Chemical Elements (?)

All Chemical Elements (116)

To compute the number of synthetic chemical elements, we subtract the number of natural chemical elements from the total number of chemical elements.

Total number of chemical elements	116
Number of natural chemical elements	−94
Number of synthetic chemical elements	22

So 22 chemical elements are synthetic.

PRACTICE 6

Of the 1,300 endangered and threatened species (plants and animals) in the United States, 535 are animal species. How many plant species are endangered and threatened? (*Source:* U.S. Fish and Wildlife Service, March 2006)

EXAMPLE 7

The following graph shows the number of overseas visitors from various countries who came to the United States during a recent year.

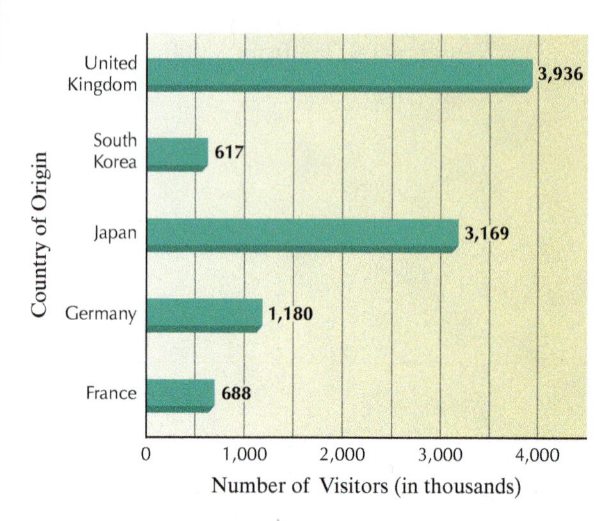

(*Source:* U.S. Department of Commerce, Office of Travel and Tourism Industries)

a. From which of these countries did the greatest number of visitors come?

b. How many visitors came from either the United Kingdom or Germany?

c. Was the number of German visitors greater or less than the total number of visitors who came from either South Korea or France?

Solution

a. More visitors came to the United States from the United Kingdom than from any other country.

b. To find how many visitors came to the United States from either the United Kingdom or Germany, we add the number of visitors from each country:

$$\begin{array}{rr} \text{United Kingdom} \rightarrow & 3{,}936{,}000 \\ \text{Germany} \rightarrow & \underline{+1{,}180{,}000} \\ \text{Sum} \rightarrow & 5{,}116{,}000 \end{array}$$

So 5,116,000 visitors came to the United States from either the United Kingdom or Germany.

c. The number of German visitors was 1,180,000. The total number of visitors who came from either South Korea or France is the sum of the number of visitors from each country:

$$\begin{array}{rr} \text{South Korea} \rightarrow & 617{,}000 \\ \text{France} \rightarrow & \underline{+688{,}000} \\ \text{Sum} \rightarrow & 1{,}305{,}000 \end{array}$$

Since 1,305,000 is greater than 1,180,000, the total number of visitors who came from either South Korea or France was greater than the number of German visitors.

PRACTICE 7

The following chart shows the projected employment change in the United States from 2004 to 2014 for some of the fastest-growing occupations.

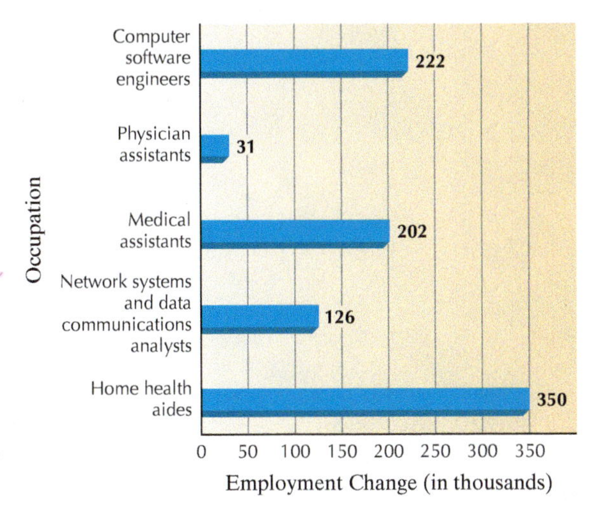

(*Source:* U.S. Bureau of Labor Statistics)

a. How much greater is the projected employment growth for home health aides than that for computer software engineers?

b. What is the combined increase in employment projected for physician assistants and medical assistants?

c. Is the combined change in employment projected for computer software engineers and network systems and data communications analysts greater or less than the projected change for home health aides?

Estimating Sums and Differences

Because everyone occasionally makes a mistake, we need to know how to check an answer so that we can correct it if it is wrong.

One method of checking addition and subtraction is by *estimation.* In this approach, we first compute and then estimate the answer. Then we compare the estimate and our "exact answer" to see if they are close. If they are, we can be confident that our answer is reasonable. If they are not close, we should redo the computation.

We can get different estimates for an answer, depending on how we round. For addition, one good way involves rounding each addend to the largest place value, as shown in Example 8. Similarly, for subtraction, we can round the minuend and subtrahend to the largest place value, as Example 9 illustrates.

EXAMPLE 8

Compute the sum of 1,923 + 898 + 754 + 2,873. Check by estimation.

Solution

$$\begin{array}{r} 1{,}923 \\ 898 \\ 754 \\ +2{,}873 \\ \hline 6{,}448 \end{array} \leftarrow \text{Exact sum}$$

Check We round each addend to the largest place value.

$$\begin{array}{rcr} 1{,}923 & \approx & 2{,}000 \\ 898 & \approx & 900 \\ 754 & \approx & 800 \\ +2{,}873 & \approx & +3{,}000 \\ \hline & & 6{,}700 \end{array} \leftarrow \text{Estimated sum}$$

Our exact answer (6,448) is reasonably close to the estimate, so we are done.

PRACTICE 8

Add: 3,945 + 849 + 4,001 + 682. Check by estimating the sum.

EXAMPLE 9

Subtract 1,994 from 8,253. Check by estimating the difference.

Solution

$$\begin{array}{r} 8{,}253 \\ -1{,}994 \\ \hline 6{,}259 \end{array} \leftarrow \text{Exact difference}$$

Check We round the minuend and subtrahend to the nearest thousand.

$$\begin{array}{rcr} 8{,}253 & \approx & 8{,}000 \\ -1{,}994 & \approx & -2{,}000 \\ \hline & & 6{,}000 \end{array} \leftarrow \text{Estimated difference}$$

Our exact answer (6,259) and the estimated difference (6,000) are fairly close.

With practice, we can mentally estimate and check differences and sums quickly and easily.

PRACTICE 9

Find the difference between 17,836 and 15,045. Then estimate this difference to check.

We have already seen how estimation helps us to check an exact answer that is computed. But sometimes an approximate answer is good enough.

EXAMPLE 10

The top two languages used online are English and Chinese, with 286,642,757 and 105,736,236 Internet users, respectively. Estimate how many more English users there are than Chinese.

Solution

$$\begin{array}{lrcr} \text{English users} \rightarrow & 286{,}642{,}775 & \approx & 300{,}000{,}000 \\ \text{Chinese users} \rightarrow & 105{,}736{,}236 & \approx & -100{,}000{,}000 \\ \hline & & & 200{,}000{,}000 \end{array}$$

So there are about 200,000,000 more English users than there are Chinese users. (*Source: The Top Ten of Everything 2006*)

PRACTICE 10

The diameters of Jupiter and Mercury are 88,863 miles and 3,032 miles, respectively. Estimate how much longer the diameter of Jupiter is than that of Mercury. (*Source:* www.montana.edu)

Adding and Subtracting Whole Numbers on a Calculator

Calculators are handy and powerful tools for carrying out complex computations. But it is easy to press a wrong key, so be sure to estimate the answer and compare this estimate to the displayed answer to see if it is reasonable.

EXAMPLE 11

On a calculator, compute the sum of 3,125 and 9,391.

Solution

Press	Display
3125 [+] 9391 [ENTER]	3125+9391 / 12516.

To check this answer, we mentally round the addends and then add.

$$\begin{aligned} 3125 &\approx 3{,}000 \\ 9391 &\approx \underline{9{,}000} \\ & \quad 12{,}000 \end{aligned}$$

This estimate is reasonably close to our answer 12,516.

PRACTICE 11

Use a calculator to add: 39,822 + 9,710

Pressing the clear key cancels the number in the display. Press this key after completing a computation to be sure that no number remains to affect the next problem. Note that calculator models vary as to how they work, so it may be necessary to consult the manual for a particular model.

EXAMPLE 12

Calculate: 39 + 48 + 277

Solution

A reasonable estimate is the sum of 40, 50, and 300, or 390—close to our calculated answer 364.

PRACTICE 12

Find the sum on a calculator:
23,801 + 7,116 + 982

When using a calculator to subtract,

- enter the numbers in the correct order—first enter the number **from which** we are subtracting and then the number **being** subtracted; and
- do not confuse the *negative sign key* [(−)] that some calculators have with the *subtraction key* [−].

EXAMPLE 13

Subtract on a calculator: 3,000 − 973

Solution

Press	Display
3000 [−] 973 [ENTER]	3000−973 / 2027.

A good estimate is 3,000 − 1,000, or 2,000, which is close to 2,027.

PRACTICE 13

Use a calculator to find the difference between 5,280 feet and 2,781 feet.

1.2 Exercises

FOR EXTRA HELP      

Mathematically Speaking

Fill in each blank with the most appropriate term or phrase from the given list.

subtrahend	addends	left	estimates
Commutative Property of Addition	right difference minuend	Identity Property of Addition sum	Associative Property of Addition

1. The operation of addition can be thought of as moving to the _______ on a number line.

2. The _______ states that the sum of a number and zero is the original number.

3. The result of addition is called the _______.

4. The _______ states that changing the order in which two numbers are added does not affect the sum.

5. The _______ states that when adding three numbers, regrouping addends gives the same sum.

6. In an addition problem, the numbers being added are called _______.

7. In a subtraction problem, the number being subtracted is called the _______.

8. The result of subtraction is called the _______.

Add and check by estimation.

9. $\begin{array}{r} 100{,}250 \\ +\ 77{,}528 \\ \hline \end{array}$

10. $\begin{array}{r} 3{,}505 \\ +\quad 11 \\ \hline \end{array}$

11. $\begin{array}{r} 8{,}132 \\ +6{,}578 \\ \hline \end{array}$

12. $\begin{array}{r} 60{,}725 \\ +38{,}928 \\ \hline \end{array}$

13. $\begin{array}{r} 7{,}481 \\ 702 \\ +5{,}819 \\ \hline \end{array}$

14. $\begin{array}{r} 99{,}103 \\ 33{,}450 \\ +\ 6{,}627 \\ \hline \end{array}$

15. $\begin{array}{r} 49{,}002 \\ 1{,}999 \\ +\ 5{,}187 \\ \hline \end{array}$

16. $\begin{array}{r} 55{,}998 \\ 40{,}003 \\ +17{,}827 \\ \hline \end{array}$

17. 1,903 + 5,075

18. 7,406 + 12,381

19. 800 + 20 + 4,000

20. 40,000 + 800 + 60

21. 31 + 93 + 277 + 12

22. 418 + 47 + 365 + 95

23. 3,911 + 2,947 + 8,007

24. 5,374 + 4,055 + 20,173

25. 6,482 meters + 9,027 meters

26. 17,812 miles + 4,283 miles

27. 35 hours + 47 hours

28. 225 square feet + 896 square feet

29. $92,258 + $7,447 + $5,126

30. $55,709 + $2,822 + $30,819

31. $1,863 + $1,089 + $9,772

32. 5,009 feet + 7,993 feet

33. 8,300 tons + 22,900 tons

34. 420,057 pounds + 900,808 pounds

35. 3,088,281
5,658,137
+4,550,239

36. 638,719
40,003
+984,035

37. 2,008,490
8,948,227
+11,956,174

38. 1,938,722
325,411
+ 517,827

In each addition table, fill in the empty spaces. Check that the sum in the shaded empty space is the same working both downward and across.

39.

+	400	200	1,200	300	Total
300					
800					
Total					

40.

+	4,000	300	3,000	2,000	Total
100					
900					
Total					

41.

+	389	172	1,155	324	Total
255					
799					
Total					

42.

+	3,749	279	2,880	1,998	Total
134					
896					
Total					

In each group of three sums, one is wrong. Use estimation to identify which sum is incorrect.

43. a. 814
9,106
+2,811
15,731

b. 30,812
47,045
+ 9,338
87,195

c. 183,066
78,911
+ 96,527
358,504

44. a. 1,035
5,210
+7,992
14,237

b. 5,801
3,882
+12,644
32,327

c. 801,716
78,001
+5,009,635
5,889,352

45. **a.** $\begin{array}{r} \$711{,}488 \\ 102{,}663 \\ +\ 95{,}003 \\ \hline \$809{,}154 \end{array}$ **b.** $\begin{array}{r} \$62{,}933 \\ 51{,}858 \\ +\ 49{,}612 \\ \hline \$164{,}403 \end{array}$ **c.** $\begin{array}{r} \$106{,}729 \\ 99{,}821 \\ +\ 103{,}277 \\ \hline \$309{,}827 \end{array}$

46. **a.** $\begin{array}{r} \$9{,}512{,}622 \\ 8{,}038{,}517 \\ +\ 2{,}615{,}334 \\ \hline \$20{,}166{,}473 \end{array}$ **b.** $\begin{array}{r} \$4{,}277{,}020 \\ 915{,}611 \\ +\ 3{,}688{,}402 \\ \hline \$8{,}881{,}033 \end{array}$ **c.** $\begin{array}{r} \$200{,}312 \\ 102{,}683 \\ +\ 504{,}113 \\ \hline \$707{,}108 \end{array}$

Subtract and check.

47. $\begin{array}{r} 379 \\ -162 \\ \hline \end{array}$ **48.** $\begin{array}{r} 362 \\ -110 \\ \hline \end{array}$ **49.** $\begin{array}{r} 200 \\ -110 \\ \hline \end{array}$ **50.** $\begin{array}{r} 210 \\ -100 \\ \hline \end{array}$

51. $\begin{array}{r} 401 \\ -\ 39 \\ \hline \end{array}$ **52.** $\begin{array}{r} 728 \\ -539 \\ \hline \end{array}$ **53.** $\begin{array}{r} 70{,}000 \\ -\ 1{,}759 \\ \hline \end{array}$ **54.** $\begin{array}{r} 8{,}000 \\ -1{,}691 \\ \hline \end{array}$

55. $\begin{array}{r} 5{,}062 \\ -2{,}777 \\ \hline \end{array}$ **56.** $\begin{array}{r} 3{,}005 \\ -1{,}666 \\ \hline \end{array}$ **57.** $\begin{array}{r} 72{,}000 \\ -19{,}001 \\ \hline \end{array}$ **58.** $\begin{array}{r} 2{,}001 \\ -\ \ 2 \\ \hline \end{array}$

59. $\begin{array}{r} 3{,}000 \\ -\ 57 \\ \hline \end{array}$ **60.** $\begin{array}{r} 52{,}947 \\ -27{,}997 \\ \hline \end{array}$ **61.** $\begin{array}{r} 261{,}406 \\ -57{,}941 \\ \hline \end{array}$ **62.** $\begin{array}{r} 729{,}888 \\ -192{,}889 \\ \hline \end{array}$

Find the difference and check.

63. 550 − 182

64. 1,448 − 962

65. 6,000 − 1,004

66. 8,602 − 907

67. 3,570 − 2,588

68. 2,182 − 899

69. 5,000 miles − 3,005 miles

70. 701 square feet − 206 square feet

71. $800 − $131

72. 622 hours − 137 hours

73. $4,812 − $1,203

74. 402 miles − 57 miles

75. 500 books − 227 books

76. $537 − $196

77. 527 meters − 318 meters

78. 1,266 tons − 597 tons

79. $\begin{array}{r} 30{,}000{,}000 \\ -27{,}999{,}000 \\ \hline \end{array}$ **80.** $\begin{array}{r} 1{,}973{,}000 \\ -\ 997{,}001 \\ \hline \end{array}$ **81.** $\begin{array}{r} 3{,}402{,}331 \\ -2{,}588{,}902 \\ \hline \end{array}$ **82.** $\begin{array}{r} 14{,}500{,}007 \\ -13{,}972{,}008 \\ \hline \end{array}$

In each group of three differences, one is wrong. Use estimation to identify which difference is incorrect.

83. **a.** $\begin{array}{r} 817{,}770 \\ -502{,}966 \\ \hline 314{,}804 \end{array}$ **b.** $\begin{array}{r} 11{,}172{,}055 \\ -\ 7{,}892{,}106 \\ \hline 3{,}279{,}949 \end{array}$ **c.** $\begin{array}{r} 71{,}384{,}612 \\ -32{,}016{,}594 \\ \hline 29{,}368{,}018 \end{array}$

84. **a.** $\begin{array}{r} 67{,}812 \\ -12{,}180 \\ \hline 55{,}632 \end{array}$ **b.** $\begin{array}{r} 3{,}997{,}401 \\ -1{,}125{,}166 \\ \hline 1{,}872{,}235 \end{array}$ **c.** $\begin{array}{r} 316{,}134 \\ -\ 89{,}164 \\ \hline 226{,}970 \end{array}$

85. a. $\begin{array}{r} \$381{,}882 \\ -\ 173{,}552 \\ \hline \$108{,}330 \end{array}$ **b.** $\begin{array}{r} \$479{,}116 \\ -\ 102{,}663 \\ \hline \$376{,}453 \end{array}$ **c.** $\begin{array}{r} \$200{,}072{,}639 \\ -\ 150{,}038{,}270 \\ \hline \$\ 50{,}034{,}369 \end{array}$

86. a. $\begin{array}{r} \$3{,}810{,}662 \\ -\quad 299{,}137 \\ \hline \$3{,}511{,}525 \end{array}$ **b.** $\begin{array}{r} \$4{,}718{,}287 \\ -\ 1{,}002{,}875 \\ \hline \$5{,}721{,}162 \end{array}$ **c.** $\begin{array}{r} \$381{,}975 \\ -\ 117{,}263 \\ \hline \$264{,}712 \end{array}$

Mixed Practice

Perform the indicated operation.

87. $\begin{array}{r} 7{,}415 \\ -\ 350 \\ \hline \end{array}$

88. $\begin{array}{r} 90{,}316 \\ 10{,}882 \\ +\ 5{,}281 \\ \hline \end{array}$

89. 281 + 758 + 104 + 533

90. \$5,233 + \$481 + \$82

91. 8,286 − 3,100

92. 410,700 miles − 280,900 miles

Applications

Solve and check.

93. In 1900, the population of the United States was approximately 76,000,000. During the next 100 years, the population grew by about 205,000,000 people. What was the population in 2000? (*Source:* U.S. Bureau of the Census)

94. It is recommended that a person drink 64 ounces of water each day. Will a person who drank 32 ounces so far today and then drinks another 24 ounces meet the recommended daily amount?

95. Of the 6,000,000 square miles of tropical rainforest that originally existed on earth, 3,400,000 square miles have been lost due to deforestation. How many square miles of rainforest still exist? (*Source:* The Nature Conservancy, 2006)

96. The Great Blue Norther of 1911 was the largest cold snap in U.S. history. On the day of the Norther, the temperature in Oklahoma City dropped from a record high of 83°F to a record low of 17°F in a 24-hour period. By how much did the temprature drop that day? (*Source:* National Weather Service)

97. The chart shows the 2006 Winter Olympic medal counts of selected countries.

Country	Gold	Silver	Bronze
Austria	9	7	7
Canada	7	10	7
Germany	11	12	6
Russia	8	6	8
United States	9	9	7

a. Calculate the total number of medals won by each country.

b. Which country won the most medals? (*Source:* http://www.olympic.org)

98. Consider the deposit slip shown.

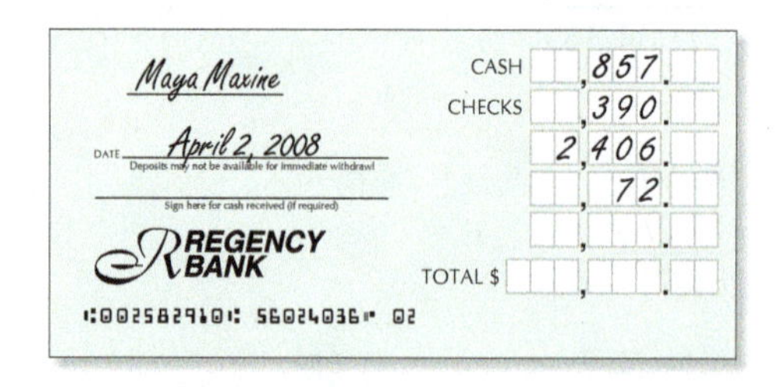

a. Estimate how much money is being deposited.

b. Fill in the exact total.

99. Blues singer Bessie Smith was born in 1894 and died in 1937. About how old was she when she died? (*Source: Encyclopedia of World Biography*)

100. The United States entered the First World War in 1917 and the Second World War in 1941. Approximately how many years apart were these two events?

101. A sign in an elevator reads: MAXIMUM CAPACITY: 1,000 POUNDS. The passengers in the elevator weigh 187 pounds, 147 pounds, 213 pounds, 162 pounds, 103 pounds, and 151 pounds. Will the elevator be overloaded?

102. A student would like to install a computer program that requires 128 megabytes (MB) of memory. The memory in his old computer is only 80 MB. If he increases the memory in his computer by 64 MB, will there be enough memory for him to run the program?

103. The thermometer at the right shows the boiling point and the freezing point of water in degrees Fahrenheit (°F). What is the difference between these two temperatures?

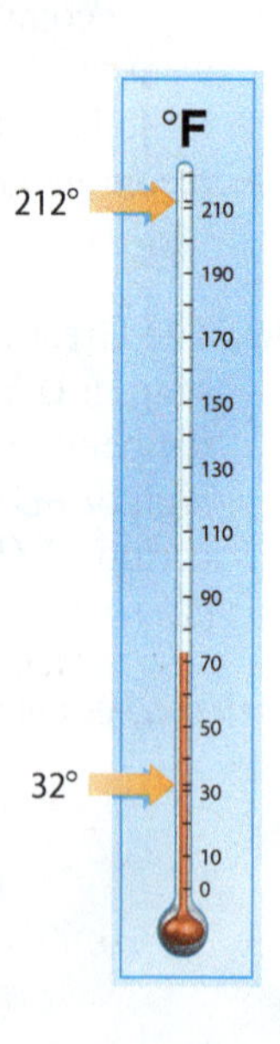

104. The following ad for a hybrid car was listed in a local newspaper. How much below the MSRP (manufacturer's suggested retail price) is the selling price?

105. Some friends cycle from town A to town B, to town C, to town D, and then back to A, as shown below. How far did they cycle in all?

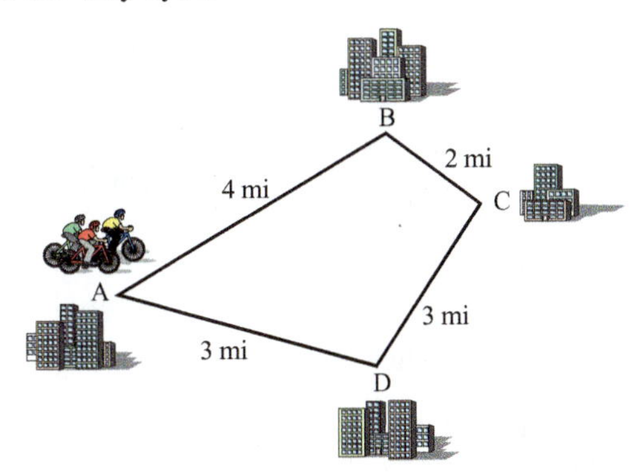

106. What is the length of the molding along the perimeter of the room pictured (yards = yd)?

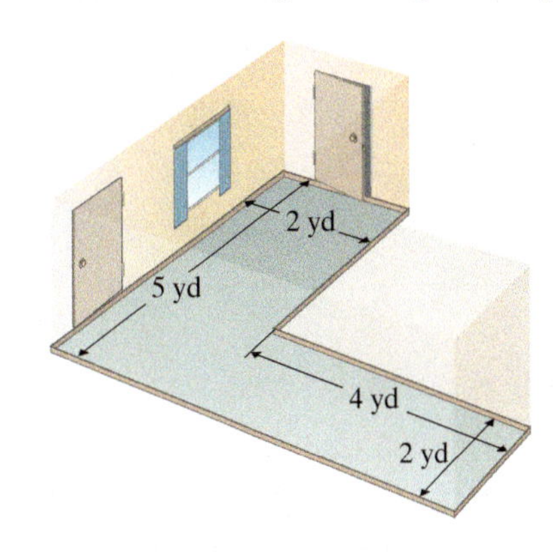

107. The total number of hours a typical person spends surfing the Web is expected to increase from 200 hours in 2005 to 236 hours in 2008. How big an increase is this? (*Source: Statistical Analysis*, Veronis Suhler Stevenson, 2006)

108. In a recent year, 4,822 merchant vessels were registered under the flag of Panama, in contrast to 412 under the U.S. flag. What is the difference between the number of merchant vessels in these fleets? (*Source:* Maritime Administration, U.S. Department of Commerce)

109. Find the total amount of money deposited in a checking account according to the following deposit slip.

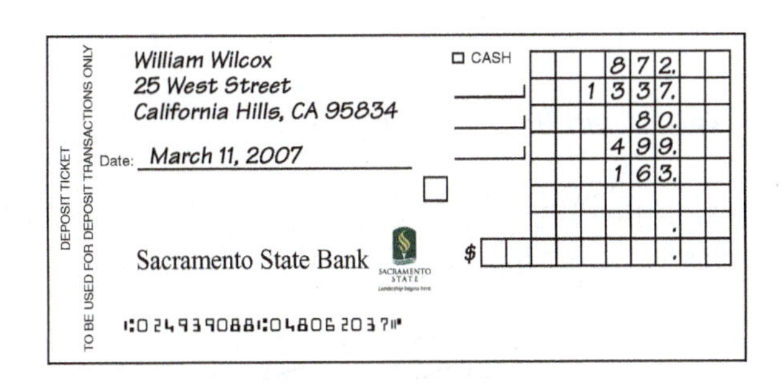

DEPOSIT TICKET

TO BE USED FOR DEPOSIT TRANSACTIONS ONLY

William Wilcox
25 West Street
California Hills, CA 95834

Date: *March 11, 2007*

CASH	
	872.
	1337.
	80.
	499.
	163.
	.
$	.

Sacramento State Bank

⑆02493908⑆04806 2037⑈

110. An oil tanker broke apart at sea. It spilled 150,000 gallons (gal) of crude oil the first day, 400,000 gal the second day, and 1,000,000 gal the third day. How much oil was spilled in all?

111. An airline limits the size of luggage that passengers can take with them. For each bag, the sum of the outside dimensions, that is, the length, width, and height, is limited to 62 inches for bags checked free of charge and 45 inches for carry-on bags. The the bag that a passanger wishes to take on a flight measures 21 inches by 10 inches by 19 inches. (*Source:* http://www.aa.com)

a. Will the passenger be allowed to carry the bag onto the plane?

b. To check the bag free of charge?

112. A particular credit card has a credit limit of \$3,500. On this card, there is a balance due of \$2,367.

a. Is there enough credit available to pay a tuition bill of \$1,295?

b. If, instead of the tuition bill, a DVD recorder for \$253 were purchased on the card, how much credit on the card would still be available?

113. The following graph gives the amount of red meat produced during a recent year in leading American states.

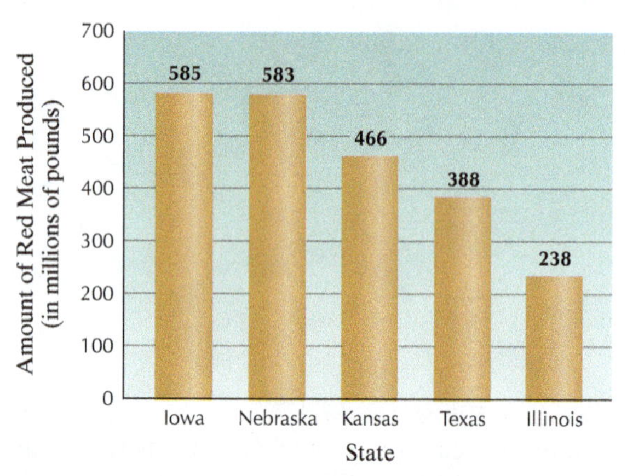

(*Source:* USDA)

a. How much more red meat did Iowa produce than Illinois?

b. Estimate the total amount of red meat produced in these five states.

c. Find the total amount of red meat produced in these five states.

114. A real estate broker based in Springfield, Illinois, covers the region shown below.

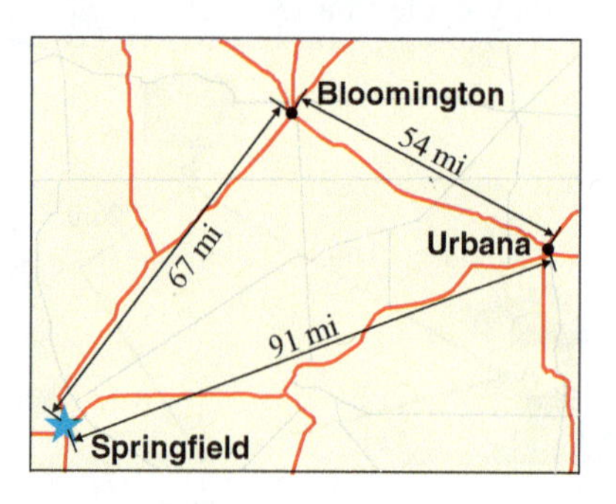

a. How much further from Springfield is Urbana than Bloomington?

b. How far does he drive going from Springfield to Bloomington by way of Urbana?

c. If the broker then drives back home to Springfield directly from Bloomington, how much shorter is the trip returning home than the earlier trip from Springfield?

Use a calculator to solve each problem, giving (a) the operation(s) carried out in the solution, (b) the exact answer, and (c) an estimate of the answer.

115. Is the total amount of the deposits shown on the following bank statement correct?

MBU Bank & Trust Co.

Your Account

Deposits	Date
$ 83	2/13
$ 59	2/14
$ 727	2/16
$ 183	2/17
$ 511	2/21

TOTAL $1,563

116. At its first eight games this season, a professional baseball team had the following paid attendance:

Game	Attendance
1	11,862
2	18,722
3	14,072
4	9,713
5	25,913
6	28,699
7	19,302
8	18,780

What was the combined attendance for these games?

• Check your answers on page A-1.

MINDSTRETCHERS

Writing

1. There are many different ways of putting numerical expressions into words.

a. For example, $3 + 2$ can be expressed as

the sum of 3 and 2, 2 more than 3, or 3 increased by 2

What are some other ways of reading this expression?

b. For example, $5 - 2$ can be expressed as

the difference between 5 and 2, 5 take away 2, or 5 decreased by 2

Write two other ways.

Critical Thinking

2. In a **magic square**, the sum of every row, column, and diagonal is the same number. Using the given information, complete the square at the right, which contains the whole numbers from 1 to 16. (*Hint:* The sum of every row, column, and diagonal is 34.)

16	3	2	
	10	11	
	6	7	

Groupwork

3. Two methods for borrowing in a subtraction problem are illustrated as follows. In method (a)—the method that we have already discussed—we borrow by taking 1 from the top, and in method (b) by adding 1 to the bottom.

a.
$$\begin{array}{r} \overset{7}{\cancel{8}}\ \overset{1}{}5\ 9 \\ -3\ 7\ 6 \\ \hline 4\ 8\ 3 \end{array}$$

b.
$$\begin{array}{r} 8\ \overset{1}{}5\ 9 \\ -\overset{4}{\cancel{3}}\ 7\ 6 \\ \hline 4\ 8\ 3 \end{array}$$

Note that we get the same answer with both methods. Working with a partner, discuss the advantages of each method.

1.3 Multiplying Whole Numbers

The Meaning and Properties of Multiplication

OBJECTIVES

- To multiply whole numbers
- To estimate the product of whole numbers
- To solve word problems involving the multiplication of whole numbers

What does it mean to multiply whole numbers? A good answer to this question is *repeated addition.*

For instance, suppose that you buy 4 packages of pens and each package contains 3 pens. How many pens are there altogether?

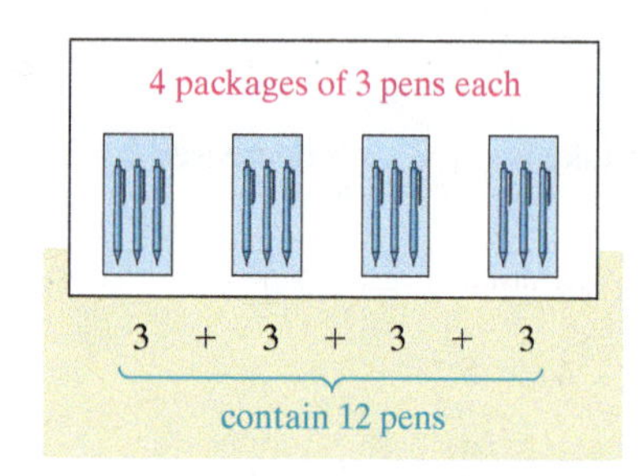

That is, $4 \times 3 = 3 + 3 + 3 + 3 = 12$. Generally, *multiplication means adding the same number repeatedly.*

We can also picture multiplication in terms of a rectangular figure, like this one, that represents 4×3.

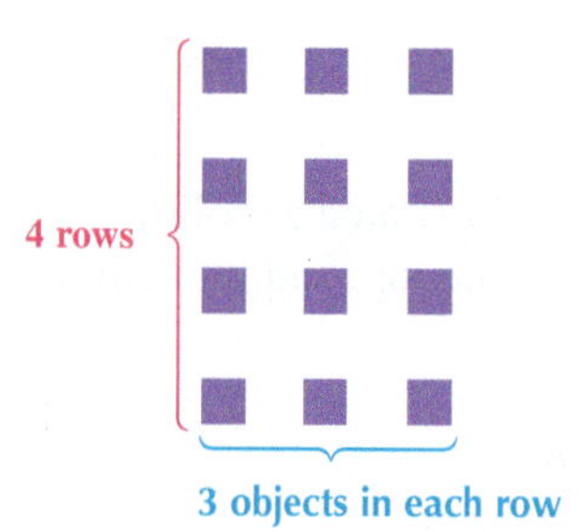

In a multiplication problem, the numbers being multiplied are called *factors*. The result is the *product*.

There are several ways to write a multiplication problem.

		Factor		Factor		Product
$\times$	the times sign	4	$\times$	3	=	12
$\cdot$	a multiplication dot	4	$\cdot$	3	=	12
( )( )	parentheses	(4)(3)			=	12

Like addition and subtraction, multiplication problems can be written either horizontally or vertically.

$8 \times 5 = 40$
Horizontal

$$\begin{array}{r} 8 \\ \times\ 5 \\ \hline 40 \end{array}$$
Vertical

The operation of multiplication has several important properties that we use frequently.

The Identity Property of Multiplication

The product of any number and 1 is that number.

$1 \times 12 = 12$
$5 \times 1 = 5$

The Multiplication Property of 0

The product of any number and 0 is 0.

$49 \times 0 = 0$
$0 \times 8 = 0$

The Commutative Property of Multiplication

Changing the order in which two numbers are multiplied does not affect their product.

$$2 \times 9 = 9 \times 2$$
$$18 = 18$$

The Associative Property of Multiplication

When multiplying three numbers, regrouping the factors gives the same product.

We multiply inside the parentheses first.

$$(3 \times 4) \times 5 = 3 \times (4 \times 5)$$
$$12 \times 5 = 3 \times 20$$
$$60 = 60$$

The next—and last—property of multiplication also involves addition.

The Distributive Property

Multiplying a factor by the sum of two numbers gives the same result as multiplying the factor by each of the two numbers and then adding.

$$2 \times (5 + 3) = (2 \times 5) + (2 \times 3)$$
$$2 \times 8 = 10 + 6$$
$$16 = 16$$

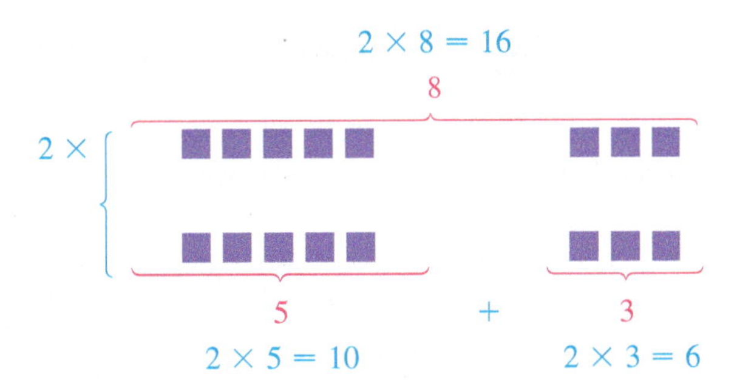

Before going on to the next section, study these properties of multiplication.

Multiplying Whole Numbers

Now let's consider problems in which we multiply any whole number by a single-digit whole number.

Note that, to multiply whole numbers with reasonable speed, you must commit to memory the products of all single-digit whole numbers.

EXAMPLE 1

Multiply: $98 \cdot 4$

Solution We recall that the dot means multiplication. We first write the problem vertically.

$$\begin{array}{r} \overset{3}{9}\,8 \\ \times\ 4 \\ \hline 2 \end{array}$$

← The product of 4 and 8 ones is 32 ones. We write the 2 and carry the 3 to the tens column.

We recall that the 9 in 98 means 9 tens.

$$\begin{array}{r} \overset{3}{9}\,8 \\ \times\ \ 4 \\ \hline 3\,9\,2 \end{array}$$

← The product of 4 and 9 tens is 36 tens. We add the 3 tens to the 36 tens to get 39 tens.

So the product of 98 and 4 is 392.

PRACTICE 1

Find the product of 76 and 8.

EXAMPLE 2

Calculate: (806) (7)

Solution We recall that parentheses side-by-side means to multiply. We write this problem vertically.

$$\begin{array}{r} 8\,\overset{4}{0}\,6 \\ \times\ \ \ \ 7 \\ \hline 5,\,6\,4\,2 \end{array}$$

Here, 7×0 tens $= 0$ tens. Add the carried 4 tens to the 0 tens to get 4 tens.

The product of 806 and 7 is 5,642.

PRACTICE 2

Find the product: (705)(6)

Now let's look at multiplying any two whole numbers.

Consider multiplying 32 by 48. We can write 32×48 as follows.

$$32 \times 48 = 32 \times (40 + 8)$$

We then use the distributive property to get the answer.

$$\begin{aligned} 32 \times (40 + 8) &= (32 \times 40) + (32 \times 8) \\ &= 1{,}280 + 256 \\ &= 1{,}536 \end{aligned}$$

Generally, we solve this problem vertically.

$$\begin{array}{r} \overset{1}{3}\,2 \\ \times\ \ 4\,8 \\ \hline 2\,5\,6 \\ 1\ 2\,8\,0 \\ \hline 1,5\,3\,6 \end{array}$$

256 ← Partial product (8×32)

1 280 ← Partial product (40×32)

1,536 ← Add the partial products.

Shortcut

$$\begin{array}{r} 32 \\ \times\ 48 \\ \hline 256 \\ 1\,28\ \ \\ \hline 1{,}536 \end{array}$$

256 ← (8×32)

128 ← (4×32)

If we use just the tens digit 4, we must write the product 128 leftward, starting at the tens column.

Example 2 suggests the following rule for multiplying whole numbers.

To Multiply Whole Numbers

- Multiply the top factor by the ones digit in the bottom factor, and write down this product.
- Multiply the top factor by the tens digit in the bottom factor, and write this product leftward, beginning with the tens column.
- Repeat this process until all the digits in the bottom factor are used.
- Add the partial products, writing down this sum.

EXAMPLE 3

Multiply: 300×50

Solution

```
    300
 ×   50
    000  ← 0 × 300 = 0
  15 00  ← 5 × 300 = 1,500
 15,000
```

PRACTICE 3

Find the product of 1,200 and 400.

In Example 3, note that the number of zeros in the product equals the total number of zeros in the factors. This result suggests a shortcut for multiplying factors that end in zeros.

```
    300  ← 2 zeros
 ×   50  ← 1 zero
 15,000  ← 2 + 1 = 3 zeros
```

Tip When multiplying two whole numbers that end in zeros, multiply the nonzero parts of the factors and then attach the total number of zeros to the product.

EXAMPLE 4

Simplify: $739 \cdot 305$

Solution

```
     739
 ×   305
   3 695  ← 5 × 739
   0 00   ← 0 × 739 = 0
  221 7   ← 3 × 739
 225,395
```

We don't have to write the row 000. Here is a shortcut.

```
     739
 ×   305
   3 695
  221 70  ← This one 0 represents the product of the tens digit 0
 225,395    and 739. This 0 lines up the products correctly.
```

PRACTICE 4

Find the product of 987 and 208.

Now let's apply the operation of multiplication to geometric area. Area means the number of square units that a figure contains.

In the rectangle at the right, each small square represents 1 square inch (sq in.). Finding the rectangle's area means finding the number of sq-in. units that it contains. A good strategy here is to find the number of units in each row and then multiply that number by the number of rows.

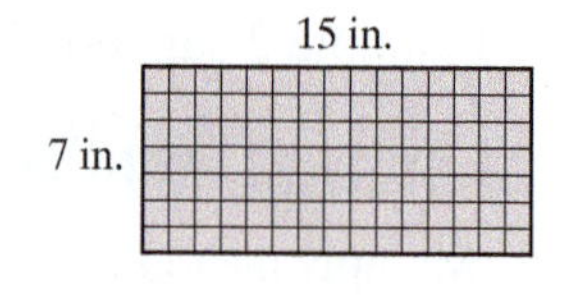

There are two ways to find that there are 15 squares in a row—either by directly counting the squares or by noting that the length of the figure is 15 in. Similarly, we find that the figure contains 7 rows. Therefore the area of the figure is 15 × 7, or 105 sq in.

In general, we can compute the *area of a rectangle* by finding the product of its length and its width.

EXAMPLE 5

Calculate the area of the home office shown in the diagram.

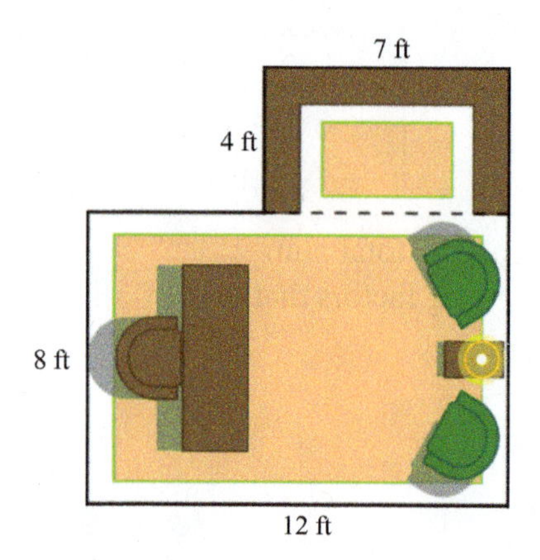

Solution The dashed line separates the office into two connected rectangles. The top retangle measures 7 feet by 4 feet, and so its area is 7 × 4, or 28 square feet. The bottom rectangle measures 12 feet by 8 feet, and its area is 12 × 8, or 96 square feet. The entire area of the office is the sum of two smaller areas: 28 + 96, or 124 square feet. So the area of the home office is 124 square feet.

PRACTICE 5

Find the area of the room pictured.

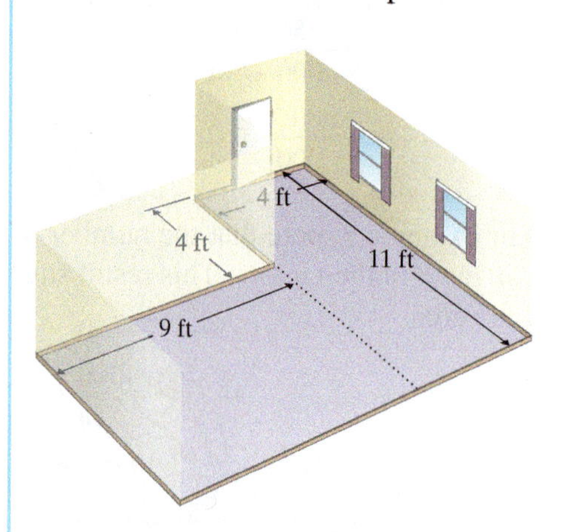

Estimating Products

As mentioned before, estimation is a valuable technique for checking an exact answer. When checking a product by estimation, round each factor to its largest place.

EXAMPLE 6

Multiply 328 by 179. Check the answer by estimation.

Solution

```
   328
 × 179
 2 952
22 96
32 8
58,712  ← Exact product
```

PRACTICE 6

Find the product of 455 and 248. Use estimation to check your answer.

Check

$$\begin{array}{rcrl} 328 & \approx & 300 & \leftarrow \text{The largest place is hundreds.} \\ \times\ 179 & \approx & \times\ 200 & \leftarrow \text{The largest place is hundreds.} \\ \hline 58{,}712 & & 60{,}000 & \leftarrow \text{Estimated product} \end{array}$$

Our exact product (58,712) and the estimated product (60,000) are fairly close.

When solving some multiplication problems, we are willing to settle for—or even prefer—an approximate answer.

EXAMPLE 7

A couple planning their wedding set aside \$2,000 for floral centerpieces. Each centerpiece costs \$72, and a total of 19 centerpieces are needed. By estimating, decide if the couple has set aside enough money for the centerpieces.

Solution To estimating a product, we first round each factor to its largest place value so that every digit after the first digit is 0.

$$\begin{array}{rcrl} 72 & \approx & 70 & \leftarrow \text{The largest place is tens.} \\ \times 19 & \approx & 20 & \leftarrow \text{The largest place is tens.} \end{array}$$

Then, we multiply the rounded factors.

$$70 \times 20 = 1{,}400$$

Since the centerpieces will cost about \$1,400 and since \$2,000 is greater than \$1,400, we conclude that the couple has set aside enough money for the centerpieces.

PRACTICE 7

Producing flyers for your college's registration requires 25,000 sheets of paper. If the college buys 38 reams of paper and there are 500 sheets in a ream, estimate to decide if there is enough paper to produce the flyers.

Multiplying Whole Numbers on a Calculator

Now let's use a calculator to find a product. When you are using a calculator to multiply large whole numbers, the answer may be too big to fit in the display. When this occurs the answer may be displayed in scientific notation (see Appendix xx).

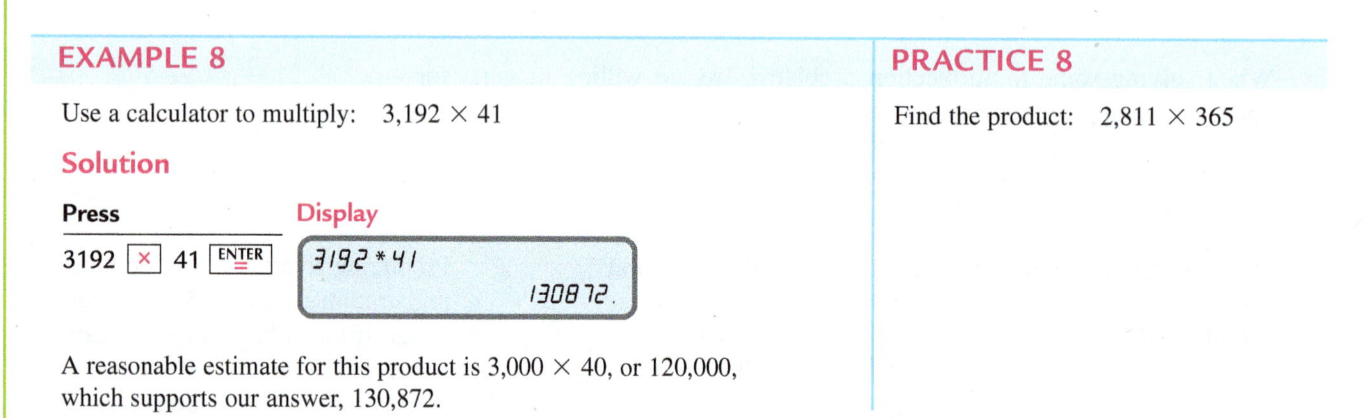

EXAMPLE 8

Use a calculator to multiply: $3{,}192 \times 41$

Solution

Press	Display
3192 [×] 41 [ENTER]	3192*41 130872.

A reasonable estimate for this product is $3{,}000 \times 40$, or 120,000, which supports our answer, 130,872.

PRACTICE 8

Find the product: $2{,}811 \times 365$

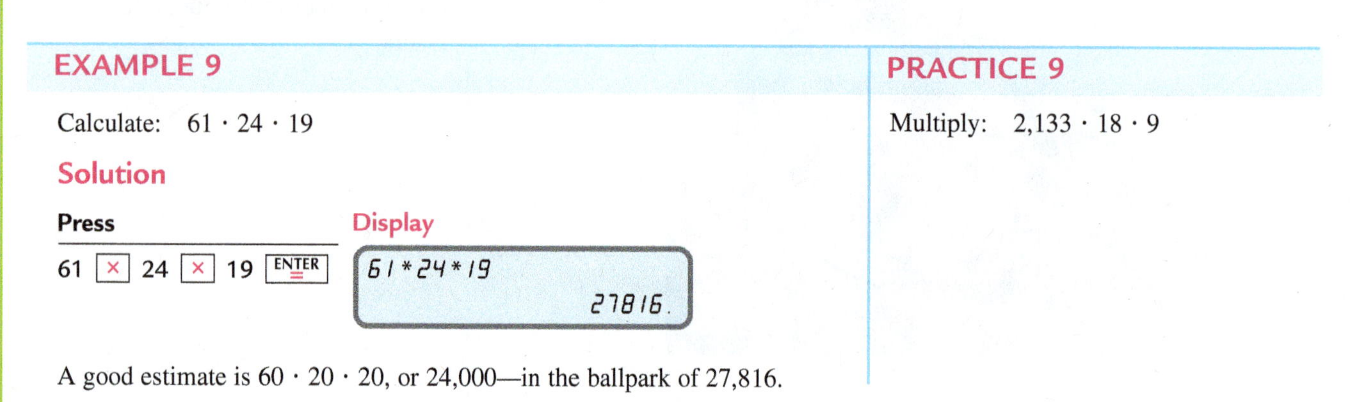

EXAMPLE 9

Calculate: $61 \cdot 24 \cdot 19$

Solution

Press	Display
61 [×] 24 [×] 19 [ENTER]	61*24*19 27816.

A good estimate is $60 \cdot 20 \cdot 20$, or 24,000—in the ballpark of 27,816.

PRACTICE 9

Multiply: $2{,}133 \cdot 18 \cdot 9$

1.3 Exercises

FOR EXTRA HELP

Mathematically Speaking

Fill in each blank with the most approppriate term or phrase from the given list.

Identity Property of Multiplication	addition	Distributive Property	product
subtraction	perimeter	sum	area
Multiplication Property of 0	Associative Property		

1. The result of multiplying two factors is called their _________.

2. The _________ is illustrated by $3 \times (7 + 2) = (3 \times 7) + (3 \times 2)$.

3. The _________ states that the product of any number and 1 is that number.

4. The _________ states that the product of any number and 0 is 0.

5. The multiplication of whole numbers can be thought of as repeated _________.

6. The _________ of a figure is the number of square units that it contains.

Compute.

7. 4×100

8. $1{,}000 \times 12$

9. 710×200

10. 270×50

11. $8{,}500 \times 20$

12. 680×300

13. $10{,}000 \times 700$

14. $1{,}000 \times 8{,}000$

Multiply and check by estimation.

15. $6{,}350 \times 2$

16. $8{,}864 \times 7$

17. 209×2

18. 703×9

19. $812{,}000 \times 4$

20. $19{,}250 \times 8$

21. 882×74

22. 881×28

23. $43 \cdot 19$

24. $85 \cdot 72$

25. $709 \cdot 48$

26. $602 \cdot 34$

27. 273×11

28. 607×65

29. 301×12

30. 513×34

31. $3{,}001 \times 19$

32. $4{,}005 \times 72$

33. $5{,}072 \times 48$

34. $8{,}801 \times 25$

35. $5{,}003 \times 40$

36. $2{,}881 \times 70$

Find the product and check by estimation.

37. (372)(403)

38. (699)(101)

39. $8{,}500 \times 17$

40. 700×207

41. 406×305

42. 702×509

43. $46 \cdot 8 \cdot 9$

44. $13 \cdot 11 \cdot 5$

45. $81 \times 2 \times 13$

46. $3 \times 5 \times 88$

47. (10)(10)(400)

48. (20)(80)(30)

49. $57 \times 81 \times 5$

50. $73 \times 4 \times 33$

51. $\begin{array}{r} 8{,}972 \\ \times \ 365 \\ \hline \end{array}$

52. $\begin{array}{r} 7{,}552 \\ \times \ 841 \\ \hline \end{array}$

53. $\begin{array}{r} 18{,}650 \\ \times \ 2{,}949 \\ \hline \end{array}$

54. $\begin{array}{r} 8{,}783 \\ \times \ 7{,}159 \\ \hline \end{array}$

In each group of three products, one is wrong. Use estimation to indentify which product is incorrect.

55. **a.** $802 \times 755 = 605{,}510$ **b.** $39 \times 4{,}722 = 184{,}158$ **c.** $77 \times 6{,}005 = 46{,}385$

56. **a.** $618 \times 555 = 342{,}990$ **b.** $86{,}331 \times 21 = 18{,}129{,}511$ **c.** $380 \times 772 = 293{,}360$

57. **a.** $9 \times 37{,}118 = 334{,}062$ **b.** $82 \times 961 = 7{,}882$ **c.** $13 \times 986 = 12{,}818$

58. **a.** $3{,}002 \times 9 = 2{,}718$ **b.** $58 \times 891 = 51{,}678$ **c.** $106 \times 68 = 7{,}208$

Mixed Practice

Multiply and check by estimation.

59. $48 \cdot 5 \cdot 12$

60. $89 \times 10{,}000$

61. $\begin{array}{r} 9{,}605 \\ \times \ 24 \\ \hline \end{array}$

62. (809)(201)

63. $357{,}000 \times 3$

64. $301 \cdot 34$

65. (50)(60)(100)

66. 495×21

Applications

Solve. Then check by estimation.

67. Underwater explorers in the eastern Mediterranean Sea found the wreck of an Egyptian ship that had sunk 33 centuries earlier. How long ago in years did the ship sink? (*Hint:* 1 century = 100 years)

68. Each day, an athlete in training takes two capsules. If each capsule contains 1,600 international units (IU) of vitamin A, how much vitamin A does he take daily?

69. The walls of a human heart are made of muscles that contract about 100,000 times a day.

a. How many contractions are there in 30 days? (*Source: American Heart Association's Your Heart: An Owner's Manual*)

b. How many more contractions are there in 40 days than in 30 days?

70. It is estimated that India would have to create 10 million new jobs per year to maintain the present unemployment rate.

a. How many new jobs altogether would have to be created during a 5-year period for the present unemployment rate to remain constant?

b. If 35 million new jobs are created during this time period, how short of the goal is this? (*Source:* Narayana Murthy, http://www.knowledgeplex.org/news/147232.html)

71. The 2006 Honda Civic gets about 40 miles per gallon of gasoline. If the fuel tank holds about 13 gallons of gasoline, can a person drive from San Francisco to Los Angeles, a distance of 276 miles, without refilling the car's fuel tank? (*Source:* American Honda Motor Company, Inc., 2006)

72. The area of a football field in the Canadian Football League (CFL) is 87,750 square feet. A football field in the National Football League (NFL) is 360 feet long and 160 feet wide. Is the area of a football field in the NFL larger than the area of a football field in the CFL?

73. Find the area of the countertop shown in the diagram.

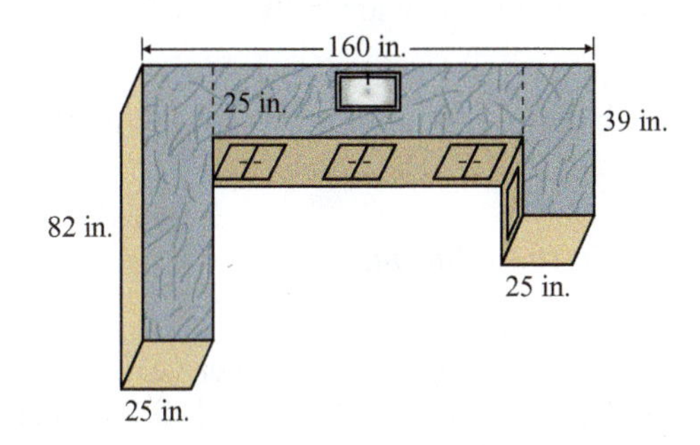

74. Calculate the area of the deck shown in the diagram.

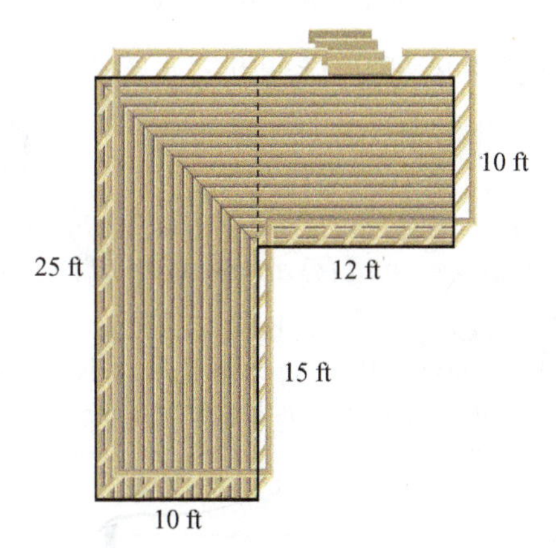

75. On the following map, 1 inch corresponds to 250 miles in the real world. How many miles actually separate towns A and B?

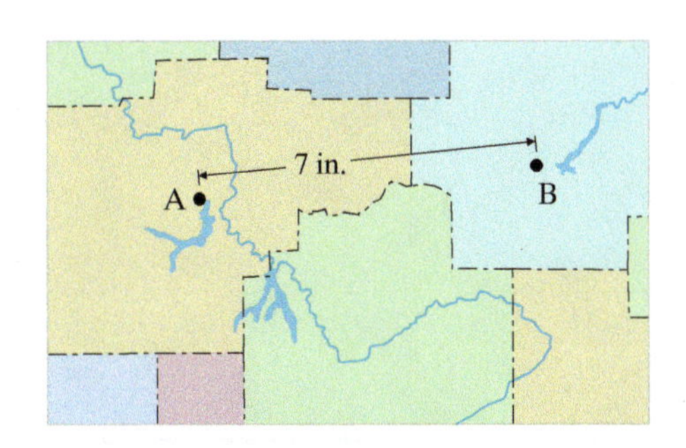

76. Angles are measured in either degrees (°) or radians (rad). A radian is about 57°. Express in degrees the measure of the angle shown.

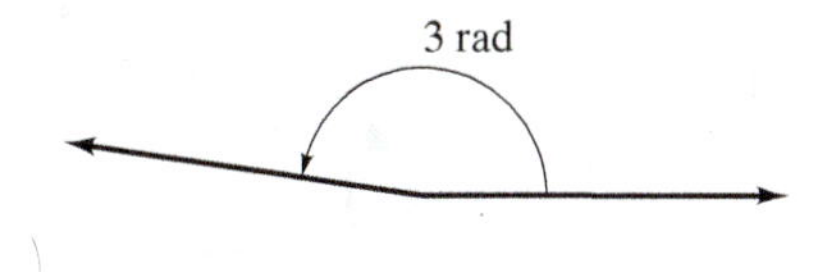

77. It costs $130 to join a local health club and $26 for each month of the membership. How much does a 1-year membership cost?

78. A customer bought a plasma television on the installment plan offered by the store. If the total cost of the television is $1,599 and the customer pays $134 each month, how much does he have left to pay off after 7 months of payments?

79. During a 5-day week, a truck driver daily drove 42 miles an hour for 7 hours.

a. How far did she drive in one day?

b. How far did she drive during the week?

80. A young couple took out a mortgage on a condo. They paid $790 per month for 15 years.

a. How much did they pay annually toward the mortgage?

b. How much did it cost them to pay off the mortgage altogether?

Use a calculator to solve each problem, giving (a) the operation(s) carried out in the solution, (b) the exact answer, and (c) an estimate of the answer.

81. The state of Colorado is approximately rectangular in shape, as shown. If the area of Kansas is about 82,000 sq mi, which state is larger? (*Source: The Columbia Gazeteer of the World*)

388 mi

275 mi

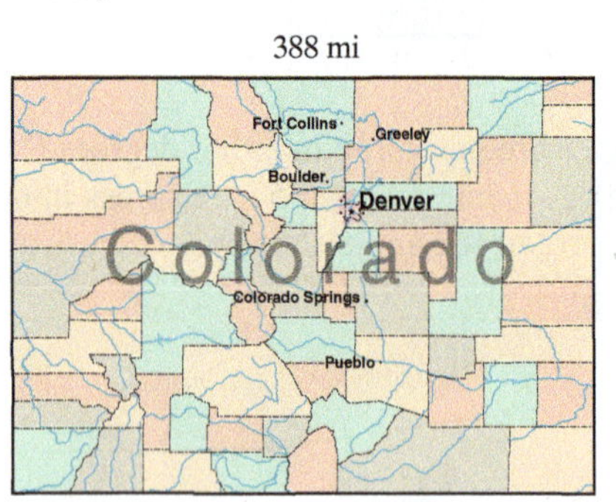

82. Tuition at a certain college is $2,125 per year for every full-time student. If there are 10,975 full-time students at the college, how much revenue is generated from student tuition?

• Check your answers on page A-2.

MINDSTRETCHERS

Writing

1. Study the following diagram. Explain how it justifies the Distributive Property.

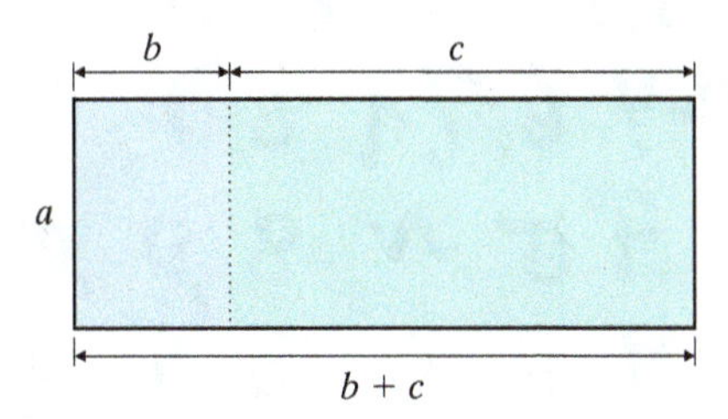

Mathematical Reasoning

2. Consider the six digits 1, 3, 5, 7, 8, and 9. Fill in the blanks with these digits, using each digit only once, so as to form the largest possible product.

9 5 1 × 8 7 3

Historical

3. Centuries ago in India and Persia, the **lattice method** of multiplication was popular. The following example, in which we multiply 57 by 43, illustrates this method. Explain how it works.

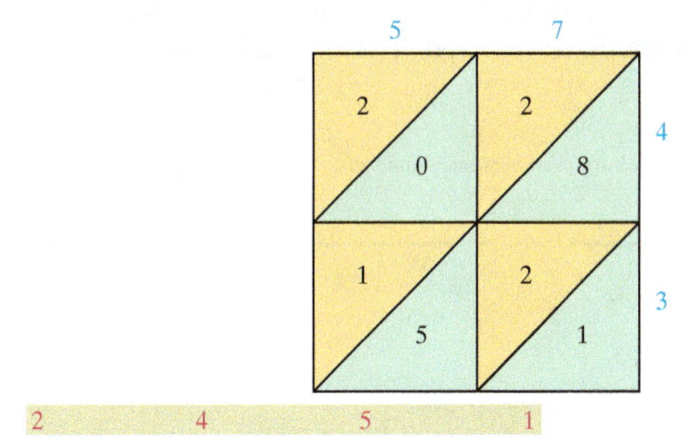

CULTURAL NOTE

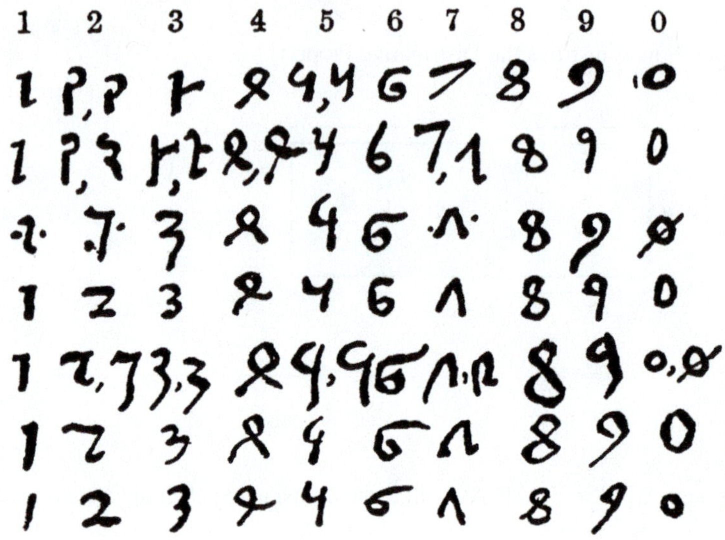

1 2 3 4 5 6 7 8 9 0

Twelfth century

A.D. 1197

A.D. 1275

c. A.D. 1294

c. A.D. 1303

c. A.D. 1360

c. A.D. 1442

The way the 10 digits are written has evolved over time. Early Hindu symbols found in a cave in India date from more than two thousand years ago. About twelve hundred years ago, an Indian manuscript on arithmetic, which had been translated into Arabic, was carried by merchants to Europe where it was later translated into Latin.

This table shows European examples of digit notation from the twelfth to the fifteenth century, when the printing press led to today's standardized notation. Through international trade, these symbols became known throughout the world.

Source: David Eugene Smith and Jekuthiel Ginsburg, *Numbers and Numerals, A Story Book for Young and Old* (New York: Bureau of Publications, Teachers College, Columbia University, 1937)

1.4 Dividing Whole Numbers

OBJECTIVES

- To divide whole numbers
- To estimate the quotient of whole numbers
- To solve word problems involving the division of whole numbers

The Meaning and Properties of Division

What does it mean to divide? One good answer is to think of division as *breaking up a set of objects* into a given number of equal smaller sets.

For instance, suppose that we want to split a set of 15 objects, say pens, evenly among 3 boxes.

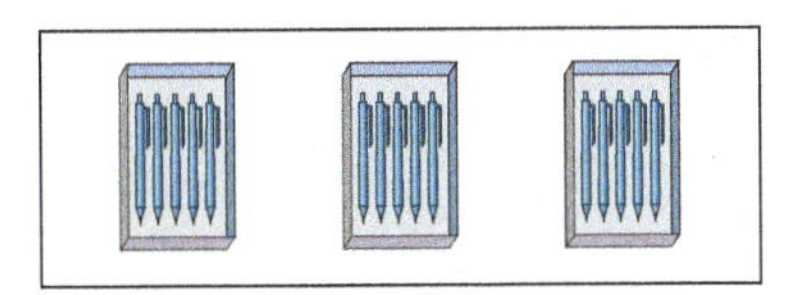

From the diagram we see that each box ends up with 5 pens. We therefore say that 15 divided by 3 is 5, which we can write as follows:

$$\text{Divisor} \quad 3\overline{)15} \quad \begin{matrix}5 & \text{Quotient} \\ & \text{Dividend}\end{matrix}$$

In a division problem, the number that is being used to divide another number is called the *divisor*. The number being divided is the *dividend*. The result is the *quotient*.

We can also think of division as the *opposite* (*inverse*) of multiplication. Consider the following pair of problems that illustrate this point.

$$\begin{array}{c} 5 \\ 3\overline{)15} \end{array} \quad \text{because} \quad 5 \times 3 = 15$$

Division — Related multiplication

The following relationship connects multiplication and division.

Quotient × Divisor = Dividend

Note that this relationship allows us to check our answer to a division problem by multiplying.

There are several common ways to write a division problem.

$$\begin{array}{c} 5 \\ 3\overline{)15} \end{array}, \quad \frac{15}{3} = 5, \quad \text{or} \quad 15 \div 3 = 5$$

Usually, we use the first of these to compute the answer. However, no matter which way we write this problem, 3 is the divisor, 15 is the dividend, and 5 is the quotient.

Tip When reading a division problem, we say that we are dividing either the divisor *into* the dividend or the dividend *by* the divisor. For instance, $3\overline{)15}$ is read either "3 divided into 15" or "15 divided by 3."

When calculating a quotient, we frequently use the following properties of division.

	Division	Related Multiplication
• Any whole number (except 0) divided by itself is 1.	$\begin{array}{r}1\\6\overline{)6}\end{array}$	$1 \times 6 = 6$
• Any whole number divided by 1 is the number itself.	$\begin{array}{r}12\\1\overline{)12}\end{array}$	$12 \times 1 = 12$
• Zero divided by any whole number (other than 0) is 0.	$\begin{array}{r}0\\8\overline{)0}\end{array}$	$0 \times 8 = 0$
• Division by 0 is not permitted.	$\begin{array}{r}?\\0\overline{)5}\end{array}$	$? \times 0 = 5$ **There is no number that when multiplied by 0 equals 5.**

Dividing Whole Numbers

Multiplication is the opposite of division. So in the simple division problem $3\overline{)15}$, we know that the answer is 5 because we have memorized that $5 \cdot 3$ is 15. But what should we do when the dividend is a larger number?

Consider the following problem: Divide 9 into 5,112 and check the answer.

- We start with the greatest place (thousands) in the dividend. We consider the dividend to be 5 thousands and think $9\overline{)5}$. Since $9 \cdot 1 = 9$ and 9 is larger than 5, there are no thousands in the quotient.

$$\begin{array}{r}0 \leftarrow \text{Thousands}\\9\overline{)5{,}112}\end{array}$$

- So we go to the hundreds place in the dividend. We consider the dividend to be 51 hundreds and think $9\overline{)51}$. Since $9 \cdot 5 = 45$, we position the 5 in the hundreds place of the quotient.

$$\begin{array}{rl}5 & \leftarrow \text{Hundreds}\\9\overline{)5{,}112} & \\-4{,}500 & \leftarrow 500 \cdot 9 = 4500\\\hline 612 & \leftarrow \text{Difference}\end{array}$$

- Next, we move to the tens place of the difference, 612. We consider the new dividend to be 61 tens and think $9\overline{)61}$. Since $9 \cdot 6 = 54$, we position the 6 in the tens place of the quotient.

- Finally, we go to the ones place of the difference, 72. We consider the new dividend to be 72 ones. So we think $9\overline{)72}$. Since $9 \cdot 8 = 72$, we position the 8 in the ones place of the quotient.

So 568 is our answer.

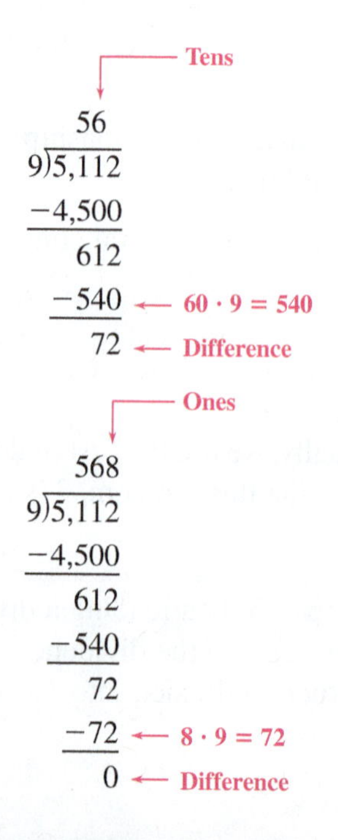

Instead of writing 0's as placeholders, we can use the following shortcut.

$$\begin{array}{r} 568 \\ 9\overline{)5{,}112} \\ -4\,5 \\ \hline 61 \\ -54 \\ \hline 72 \\ -72 \\ \hline 0 \end{array}$$

← These arrows help us to keep track of which digit we have brought down.

Check

$$\begin{array}{r} 568 \\ \times\ \ 9 \\ \hline 5{,}112 \end{array}$$

← The product equals the dividend, so our answer is correct.

Note that each time we subtract in a division problem, the difference is less than the divisor. Why must that be true?

EXAMPLE 1

Divide and check: 4,263 ÷ 7

Solution

$$\begin{array}{r} 609 \\ 7\overline{)4{,}263} \\ -4\,2 \\ \hline 06 \\ -0 \\ \hline 63 \\ -63 \\ \hline 0 \end{array}$$

- Think $7\overline{)42}$
- ← 6 × 7 = 42. Subtract.
- ← Think $7\overline{)6}$. There are zero 7's in 6.
- ← 0 × 7 = 0. Subtract.
- ← Think $7\overline{)6}$.
- ← 9 × 7 = 63. Subtract.

Check

$$\begin{array}{r} 609 \\ \times\ \ 7 \\ \hline 4{,}263 \end{array}$$

The product agrees with our dividend. Note the 0 in the quotient. Can you explain why the 0 is needed?

PRACTICE 1

Compute $9\overline{)7{,}263}$ and then check your answer.

Tip In writing your answer to a division problem, position the first digit of the quotient over the *right digit* of the number into which you are dividing (the 6 over the 2 in Example 1).

$$\begin{array}{r} 609 \\ 7\overline{)4{,}263} \end{array}$$

EXAMPLE 2

Compute $\frac{2{,}709}{9}$. Then check your answer.

Solution

$$\begin{array}{r} 301 \\ 9\overline{)2{,}709} \\ -2\,7 \\ \hline 00 \\ -0 \\ \hline 09 \\ -9 \\ \hline 0 \end{array}$$

Check

$$\begin{array}{r} 301 \\ \times\ \ 9 \\ \hline 2{,}709 \end{array}$$

PRACTICE 2

Carry out the following division and check your answer.

$$8\overline{)56{,}016}$$

In Examples 1 and 2, note that the remainder is 0; that is, the divisor goes evenly into the dividend. However, in some division problems, that is not the case. Consider, for instance, the problem of dividing 16 pens *equally* among 3 boxes.

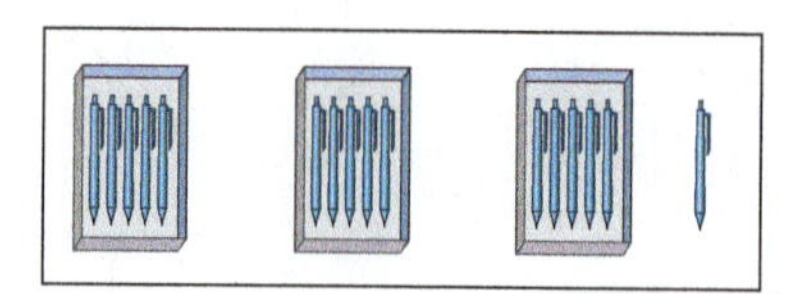

From the diagram, we see that each box contains 5 pens *but* that 1 pen—the *remainder*—is left over.

$$\begin{array}{rrl} & 5 & \leftarrow \text{Number of pens in each box} \\ \text{Number of boxes} \rightarrow & 3\overline{)16} & \leftarrow \text{Total number of pens} \\ & \underline{-15} & \leftarrow \text{Total number of pens in the boxes} \\ & 1 & \leftarrow \text{Number of pens remaining} \end{array}$$

We write the answer to this problem as 5 R1 (read "5 Remainder 1"). Note that $(3 \times 5) + 1 = 16$. The following relationship is always true.

> (Quotient × Divisor) + Remainder = Dividend

When a division problem results in a remainder as well as a quotient, we use this relationship for checking.

EXAMPLE 3

Find the quotient of 55,811 and 6. Then check.

Solution

$$\begin{array}{r} 9{,}301 \text{ R5} \\ 6\overline{)55{,}811} \\ \underline{-54} \\ 1\,8 \\ \underline{-1\,8} \\ 01 \\ \underline{-0} \\ 11 \\ \underline{-6} \\ 5 \end{array}$$

Our answer is therefore 9,301 R5.

(Quotient × Divisor) + Remainder × Dividend

Check $(9{,}301 \times 6) + 5 =$

$55{,}806 + 5 = 55{,}811$

Since this matches the dividend, our answer checks.

PRACTICE 3

Compute $8\overline{)42{,}329}$ and check.

Now let's consider division problems in which a divisor has more than one digit. Notice that such problems involve rounding.

EXAMPLE 4

Compute $\dfrac{2{,}574}{34}$ and check.

Solution In order to estimate the first digit of the quotient, we round 34 to 30 and 257 to 260.

$$\begin{array}{r} 8 \\ 34\overline{)2{,}574} \\ \underline{-2\ 72} \end{array}$$

8 ← Think 260 ÷ 30, or 26 ÷ 3. The quotient 8 goes over the 7 because we are dividing 34 into 257.

−2 72 ← 8 × 34 = 272. Try to subtract.

Because 272 is too large, we reduce our estimate in the quotient by 1 and try 7.

$$\begin{array}{r} 76 \\ 34\overline{)2{,}574} \\ \underline{-2\ 38} \\ 194 \\ \underline{-204} \end{array}$$

−2 38 ← 7 × 34 = 238. Subtract.

194 ← Think 190 ÷ 30 or 19 ÷ 3.

−204 ← 6 × 34 = 204. Try to subtract.

Because 204 is too large, we reduce our estimate in the quotient by 1 and try 5.

$$\begin{array}{r} 75\ \text{R24} \\ 34\overline{)2{,}574} \\ \underline{-2\ 38} \\ 194 \\ \underline{-170} \\ 24 \end{array}$$

So our answer is 75 R24.

Check $(75 \times 34) + 24 = 2{,}574$

Since 2,574 is the dividend, our answer checks.

PRACTICE 4

Divide 23 into 1,818. Then check.

EXAMPLE 5

Divide $26\overline{)1{,}849}$ and then check.

Solution First we round 26 to 30 and 184 to 180. Think $180 \div 30 = 6$.

$$\begin{array}{r} 6 \\ 26\overline{)1{,}849} \\ \underline{-1\ 56} \\ 28 \end{array}$$

28 ← This difference is larger than the divisor, so we increase the 6 in the quotient by 1.

$$\begin{array}{r} 71 \\ 26\overline{)1{,}849} \\ \underline{-182} \\ 29 \\ \underline{-26} \\ 3 \end{array}$$

Our answer is therefore 71 R3.

Check $(71 \times 26) + 3 = 1{,}849$

PRACTICE 5

Compute and check: $15\overline{)1{,}420}$

Tip If the divisor has more than one digit, estimate each digit in the quotient by rounding and then dividing. If the product is too large or too small, adjust it up or down by 1 and then try again.

EXAMPLE 6

Find the quotient of 13,559 and 44. Then check.

Solution

```
       308 R7
  44)13,559
    -13 2
       35   <-- This number is smaller than the divisor, so the next
       -0       digit in the quotient is 0.
       359
      -352
         7
```

Check $(308 \times 44) + 7 = 13,559$

PRACTICE 6

Divide 16,999 by 28. Then check your answer.

EXAMPLE 7

Divide and check: $6,000 \div 20$

Solution We set up the problem as before.

```
      300
  20)6,000
     -60
      00
     -00
       00
      -00
        0
```

Check

```
   300
 × 20
 6,000
```

Because the divisor and dividend both end in zero, a quicker way to do Example 7 is by dropping zeros.

20̸)6,000̸ ← Drop one 0 from both the divisor and the dividend.

```
   300
 2)600  <-- Then divide.
```

PRACTICE 7

Compute 40)8,000 and then check.

Tip Dropping the same number of zeros at the right end of both the divisor and the dividend does not change the quotient.

Estimating Quotients

As for other operations, estimating is an important skill for division. Checking a quotient by estimation is faster than checking it by multiplication, although less exact. And in some division problems, we need only an approximate answer.

How do we estimate a quotient? A good way is to round the divisor to its greatest place. The new divisor then contains only one nonzero digit and so is relatively easy to divide by mentally. Then we round the dividend to the place of our choice.

Finally, we compute the estimated quotient by calculating its first digit and then attaching the appropriate number of zeros.

EXAMPLE 8

Calculate $\frac{7{,}004}{34}$ and then check by estimation.

Solution

$$\begin{array}{r} 206 \\ 34\overline{)7{,}004} \\ -6\,8 \\ \hline 204 \\ -204 \\ \hline \end{array}$$

206 ← Exact quotient

Check $34\overline{)7{,}004}$ Round 34 to 30 and round 7,004 to 7,000.

$30\overline{)7{,}000}$ Think 70 ÷ 30 or 7 ÷ 3.

$$\begin{array}{r} 200 \\ 30\overline{)7{,}000} \end{array}$$

200 ← Estimated quotient

Note that, to the right of the 2 in the estimated quotient, we added a 0 over each of the digits in the dividend. Our answer (206) is close to our estimate (200), and so our answer is reasonable.

PRACTICE 8

Compute 100,568 ÷ 104 and use estimation to check.

EXAMPLE 9

Sound travels at about 340 meters per second, whereas light travels at 299,792,458 meters per second. Estimate how many times as fast as the speed of sound is the speed of light.

Solution To estimate a quotient, we first round the divisor and the dividend to their largest place value.

$340\overline{)299{,}792{,}458}$

$300\overline{)300{,}000{,}000}$ Round 340 to 300 and 299,792,458 to 300,000,000.

Then we divide.

$$\begin{array}{r} 1{,}000{,}000 \\ 300\overline{)300{,}000{,}000} \end{array}$$

So the speed of light is about 1,000,000 times faster than the speed of sound.

PRACTICE 9

Based on population projections, China will have a population of 1,366,205,049 in the year 2012. In that same year, Brazil will have a population of 199,083,155. Estimate how many times the population of Brazil the population of China will be in 2012. (*Source:* U.S. Bureau of the Census, International Database)

Dividing Whole Numbers on a Calculator

When using a calculator to divide, we must enter the numbers in the correct order to get the correct answer. We first enter the number *into* which we are dividing (the dividend) and then the number *by* which we are dividing (the divisor).

EXAMPLE 10

Use a calculator to divide $18\overline{)11{,}718}$.

Solution

A reasonable estimate is 10,000 ÷ 20, or 500, which is fairly close to 651.

PRACTICE 10

Find the following quotient with a calculator:

$$\frac{47{,}034}{78}$$

1.4 Exercises

FOR EXTRA HELP

PRACTICE

WATCH

DOWNLOAD

READ

REVIEW

Mathematically Speaking

Fill in each blank with the most appropriate term or phrase from the given list.

subtraction	product	increased
quotient	divisor	multiplication
divided		

1. When dividing, the dividend is divided by the __________.
2. The result of dividing is called the __________.
3. The opposite operation of division is __________.
4. Any whole number __________ by 1 is equal to the number itself.

Divide and check.

5. $5\overline{)2,000}$ **6.** $5\overline{)10,000}$ **7.** $5\overline{)2,800}$ **8.** $8\overline{)12,504}$

9. $9\overline{)2,709}$ **10.** $2\overline{)5,780}$ **11.** $7\overline{)21,021}$ **12.** $5\overline{)27,450}$

13. $3\overline{)606}$ **14.** $2\overline{)30,534}$ **15.** $9\overline{)4,500}$ **16.** $3\overline{)4,512}$

Find the quotient and check.

17. $300 \div 10$ **18.** $400 \div 20$ **19.** $700 \div 50$ **20.** $6,000 \div 20$

21. $\frac{8,400}{200}$ **22.** $\frac{7,500}{300}$ **23.** $\frac{16,000}{40}$ **24.** $\frac{48,000}{20}$

25. $6,996 \div 44$ **26.** $9,660 \div 92$ **27.** $80,295 \div 15$ **28.** $936 \div 72$

29. $39,078 \div 39$ **30.** $49,497 \div 21$ **31.** $249,984 \div 36$ **32.** $499,992 \div 24$

33. $52\overline{)52,052}$ **34.** $24\overline{)48,072}$ **35.** $12\overline{)36,600}$ **36.** $36\overline{)25,560}$

37. $6,512 \div 10$ **38.** $8,922 \div 25$ **39.** $304 \div 27$ **40.** $206 \div 45$

41. $\frac{10,175}{87}$ **42.** $\frac{21,109}{25}$ **43.** $\frac{63,002}{90}$ **44.** $\frac{12,509}{61}$

45. $47\overline{)34,000}$ **46.** $66\overline{)99,980}$ **47.** $14\overline{)6,000}$ **48.** $32\overline{)3,007}$

49. $537\overline{)387,177}$ **50.** $265\overline{)197,160}$ **51.** $638\overline{)98,890}$ **52.** $152\overline{)34,048}$

In each group of three quotients, one is wrong. Use estimation to identify which quotient is incorrect.

53. **a.** $455,260 \div 65 = 704$ **b.** $11,457 \div 57 = 201$ **c.** $10,044 \div 93 = 108$

54. **a.** $18,473 \div 91 = 203$ **b.** $43,364 \div 74 = 586$ **c.** $14,562 \div 18 = 8,009$

55. **a.** $43,710 \div 93 = 47$ **b.** $71,048 \div 107 = 664$ **c.** $11,501 \div 31 = 371$

56. **a.** $178,267 \div 89 = 2,003$ **b.** $350,007 \div 21 = 1,667$ **c.** $37,185 \div 37 = 1,005$

Mixed Practice

Divide and check.

57. $38,095 \div 42$ **58.** $\frac{63,147}{21}$ **59.** $6\overline{)12,000}$ **60.** $4,907 \div 7$

61. $\frac{48,000}{20}$ **62.** $36\overline{)249,986}$ **63.** $\frac{3,330}{9}$ **64.** $4,090 \div 91$

Applications

Solve and check.

65. A part-time student is taking 9 credit-hours this semester at a local community college. If her tuition bill is \$1,215, how much does each credit-hour cost?

66. A car used 15 gallons of gas on a 300-mile trip. How many miles per gallon (mpg) of gas did the car get?

67. The area of the Pacific Ocean is about 64 million square miles, and the area of the Atlantic Ocean is approximately 32 million square miles. The Pacific is how many times as large as the Atlantic? (*Source: The New Encyclopedia Britannica*)

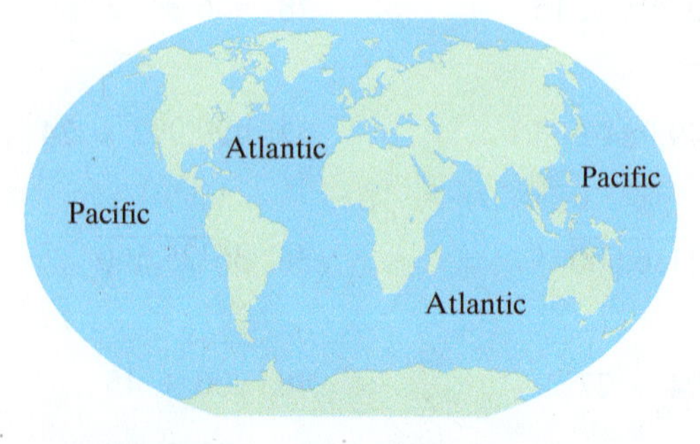

68. The diameter of Earth is about 8,000 miles, whereas the diameter of the Moon is about 2,000 miles. How many times the Moon's diameter is Earth's? (*Source: The New Encyclopedia Britannica*)

8,000 mi
2,000 mi
Moon
Earth

69. In the year 2030, Ohio is projected to have a population of about 12,300,000 people. If Ohio has a total land area of about 41,000 square miles, how many people per square mile will there be in 2030? (*Source:* Ohio Department of Development)

70. A certified medical assistant has an annual salary of \$26,472. What is her gross monthly income?

71. A 150-pound person can burn about 360 calories in 1 hour doing yoga. How many calories are burned in 1 minute? (*Source:* American Cancer Society)

72. Derek Jeter signed a 10-year contract for $189,000,000 with the New York Yankees in 2001. What is his pay per year from the contract? (*Source: The New York Times*)

73. A homeowner is remodeling a bathroom with dimensions 96 inches and 114 inches. For the floor, she has selected tiles that measure 6 inches by 6 inches.

a. How many tiles must she purchase?

b. The tiles come in boxes of 12. How many boxes of tiles must she purchase?

c. If each box of tiles costs $18, how much will she spend on the tiles for the floor?

74. The group admission rate for 15 or more people at Six Flags Great Adventure amusement park is $30 per person. A student group hosted a field trip to the park and charged $46 per ticket, covering both the cost of admission to the park and the bus transportation. (*Source:* Six Flags Great Adventure, 2006)

a. If the total amount the group collected for tickets was $1,656, how many students went on the field trip?

b. Calculate the total cost of admissions for the students on the field trip.

c. What was the cost of the bus transportation?

Use a calculator to solve each problem, giving (a) the operation(s) carried out in the solution, (b) the exact answer, and (c) an estimate of the answer.

75. Although the areas of the United States and of China are roughly the same, the United States has a much smaller population. The population of China is 1,306,313,800, whereas that of the United States is only 295,734,100. Is the population of China more or less than 4 times that of the United States? Explain. (*Source: The World Factbook*, 2005)

76. A couple set aside $3,300 for mortgage payments. If they pay $281 per month toward the mortgage, for how many months can they make full payments?

• *Check your answers on page A-2.*

MINDSTRETCHERS

Writing

1. Use the problem $10 \div 2 = 5$ to help explain why division can be thought of as repeated subtraction.

Mathematical Reasoning

2. Consider the following pair of problems.

a. $2\overline{)7}$ **b.** $4\overline{)13}$

Are the answers the same? Explain.

Groupwork

3. In the following division problem, A, B, and C each stand for a different digit. Working with a partner, identify all the digits. (*Hint:* There are two answers.)

$$\begin{array}{r} \text{ABA} \\ \text{AB}\overline{)\text{CACAB}} \\ \underline{-\text{CAB}\phantom{\text{AB}}} \\ \text{CA}\phantom{\text{B}} \\ \underline{-\text{B}\phantom{\text{B}}} \\ \text{CAB} \\ \underline{-\text{CAB}} \end{array}$$

1.5 Exponents, Order of Operations, and Averages

OBJECTIVES

- To evaluate expressions involving exponents
- To evaluate expressions using the rule for order of operations
- To compute averages

Exponents

There are many mathematical situations in which we multiply a number by itself repeatedly. Writing such expressions in **exponential form** provides a shorthand method for representing this repeated multiplication of the same factor.

For instance, we can write $5 \cdot 5 \cdot 5 \cdot 5$ in exponential form as

$$5^4$$

(4 ← Exponent; 5 ← Base)

This expression is read "5 to the fourth *power*" or simply "5 to the fourth."

Definition

An **exponent** (or **power**) is a number that indicates how many times another number (called the **base**) is used as a factor.

We read the power 2 or the power 3 in a special way. For instance, 5^2 is usually read "5 *squared*" rather than "5 to the second power." Similarly, we usually read 5^3 as "5 *cubed*" instead of "5 to the third power."

Let's look at a number written in exponential form—namely, 2^4. To evaluate this expression, we multiply 4 factors of 2.

$$\begin{aligned} 2^4 &= \underbrace{2 \cdot 2} \cdot 2 \cdot 2 \\ &= \underbrace{4 \cdot 2} \cdot 2 \\ &= \underbrace{8 \cdot 2} \\ &= 16 \end{aligned}$$

In short, $2^4 = 16$. Do you see the difference between 2^4 and $2 \cdot 4$?

Sometimes we prefer to shorten expressions by writing them in exponential form. For instance, we can write $3 \cdot 3 \cdot 4 \cdot 4 \cdot 4$ in terms of powers of 3 and 4.

$$\underbrace{3 \cdot 3}_{\text{2 factors of 3}} \cdot \underbrace{4 \cdot 4 \cdot 4}_{\text{3 factors of 4}} = 3^2 \cdot 4^3$$

EXAMPLE 1

Rewrite

$$6 \cdot 6 \cdot 6 \cdot 10 \cdot 10 \cdot 10 \cdot 10$$

in exponential form.

Solution

$$\underbrace{6 \cdot 6 \cdot 6}_{\text{3 factors of 6}} \cdot \underbrace{10 \cdot 10 \cdot 10 \cdot 10}_{\text{4 factors of 10}} = 6^3 \cdot 10^4$$

PRACTICE 1

Write

$$5 \cdot 5 \cdot 5 \cdot 5 \cdot 5 \cdot 2 \cdot 2$$

in terms of powers.

EXAMPLE 2

Compute:

a. 1^5

b. 22^2

Solution

a. $1^5 = \underbrace{1 \cdot 1 \cdot 1 \cdot 1 \cdot 1}_{} = 1$

Note that 1 raised to any power is 1.

b. $22^2 = 22 \cdot 22 = 484$

After considering this example, can you explain the difference between squaring and doubling a number?

PRACTICE 2

Calculate:

a. 1^8

b. 11^3

EXAMPLE 3

Write $4^3 \cdot 5^3$ in standard form.

Solution

$$4^3 \cdot 5^3 = (4 \cdot 4 \cdot 4) \cdot (5 \cdot 5 \cdot 5) = 64 \cdot 125 = 8{,}000$$

From this example, do you see the difference between cubing and tripling a number?

PRACTICE 3

Express $7^2 \cdot 2^4$ in standard form.

It is especially easy to compute powers of 10.

$$10^2 = 10 \cdot 10 = 1\underbrace{00}_{\text{2 zeros}}, \qquad 10^3 = 10 \cdot 10 \cdot 10 = 1{,}\underbrace{000}_{\text{3 zeros}}$$

$$10^4 = 10 \cdot 10 \cdot 10 \cdot 10 = 1\underbrace{0{,}000}_{\text{4 zeros}}$$

and so on.

Do you see the pattern?

EXAMPLE 4

Astronomical distances are commonly expressed in terms of light-years. Our galaxy is approximately 100,000 light-years in diameter. Express this number in terms of a power of 10. (*Source: The Time Almanac 2006*)

Solution

$$1\underbrace{00{,}000}_{\text{5 zeros}} = 10^5$$

So the diameter of our galaxy is 10^5 light-years.

PRACTICE 4

In 1850, the world population was approximately 1,000,000,000. Represent this number as a power of 10. (*Source: World Almanac and Book of Facts 2006*)

Order of Operations

Some mathematical expressions involve more than one mathematical operation. For instance, consider $5 + 3 \cdot 2$. This expression seems to have two different values, depending on the order in which we perform the given operations.

Adding first	**Multiplying first**
$\underbrace{5 + 3} \cdot 2$	$5 + \underbrace{3 \cdot 2}$
$= \underbrace{8 \cdot 2}$	$= \underbrace{5 + 6}$
$= 16$	$= 11$

How are we to know which operation to carry out first? By consensus we agree to follow the rule called the **order of operations** so that everyone always gets the same value for an answer.

Order of Operations Rule

To evaluate mathematical expressions, carry out the operations *in the following order.*

1. First, perform the operations within any grouping symbols, such as parentheses ( ) or brackets [ ].
2. Then, raise any number to its power $\square^{\square}$.
3. Next, perform all multiplications and divisions as they appear from left to right.
4. Finally, do all additions and subtractions as they appear from left to right.

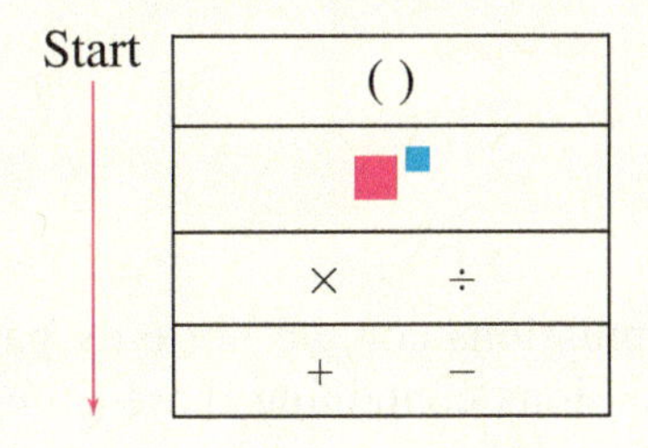

Applying this rule to the preceding example gives us the following result.

$$5 + \underbrace{3 \cdot 2} \quad \text{Multiply first.}$$
$$= \underbrace{5 + 6} \quad \text{Then add.}$$
$$= 11$$

So 11 is the correct answer.

Let's consider more examples that depend on the order of operations rule.

EXAMPLE 5

Simplify: $18 - 7 \cdot 2$

Solution Applying the rule, we multiply first, and then subtract.

$$18 - \underbrace{7 \cdot 2} =$$
$$18 - 14 = 4$$

PRACTICE 5

Evaluate: $2 \cdot 8 + 4 \cdot 3$

EXAMPLE 6

Find the value of $3 + 2 \cdot (8 + 3^2)$.

Solution

$$\begin{aligned} 3 + 2 \cdot (8 + 3^2) &= 3 + 2 \cdot (8 + 9) && \text{First, perform the operations in parentheses: square the 3.} \\ &= 3 + 2 \cdot 17 && \text{Then, add 8 and 9.} \\ &= 3 + 34 && \text{Next, multiply 2 by 17.} \\ &= 37 && \text{Finally, add 3 and 34.} \end{aligned}$$

PRACTICE 6

Simplify: $(4 + 1)^2 \times 6 - 4$

Tip When a division problem is written in the format $\frac{\square}{\square}$, parentheses are understood to be around both the dividend and the divisor. For instance,

$$\frac{10 - 2}{3 + 1} \text{ means } \frac{(10 - 2)}{(3 + 1)}.$$

EXAMPLE 7

Evaluate: $6 \cdot 2^3 - \frac{21 - 11}{2}$

Solution

$$\begin{aligned} 6 \cdot 2^3 - \frac{21 - 11}{2} &= 6 \cdot 2^3 - \frac{10}{2} && \text{First, simplify the dividend by subtracting.} \\ &= 6 \cdot 8 - \frac{10}{2} && \text{Then, cube.} \\ &= 48 - 5 && \text{Next, multiply and divide.} \\ &= 43 && \text{Finally, subtract.} \end{aligned}$$

PRACTICE 7

Simplify: $10 + \frac{24}{12 - 8} - 3 \times 4$

Some arithmetic expressions contain not only parentheses but also brackets. When simplifying expressions containing these grouping symbols, first perform the operations within the innermost grouping symbols and then continue to work outward.

EXAMPLE 8

Simplify: $5 + [4(10 - 3^2) - 2]$

Solution

$$\begin{aligned} 5 + [4(10 - 3^2) - 2] &= 5 + [4(10 - 9) - 2] && \text{Perform the operation in parentheses: square the 3.} \\ &= 5 + [4 \cdot 1 - 2] && \text{Subtract 9 from 10.} \\ &= 5 + [4 - 2] && \text{Multiply.} \\ &= 5 + 2 && \text{Subtract.} \\ &= 7 && \text{Add.} \end{aligned}$$

PRACTICE 8

Evaluate: $[4 + 3(2^3 - 5)] + 10$

EXAMPLE 9

Young's Rule is a rule of thumb for calculating the dose of medicine recommended for a child of a given age. According to this rule, the dose of acetaminophen in milligrams (mg) for a child who is eight years old can be calculated using the expression.

$$\frac{8 \times 500}{8 + 12}$$

What is the recommended dose?

Solution

$$\frac{8 \times 500}{8 + 12} = \frac{4{,}000}{20}$$ First, simplify the dividend and the divisor.

$$= 200$$ Then, divide.

So the recommended dose is 200 milligrams.

PRACTICE 9

The minimum distance (in feet) that it takes a car to stop if it is traveling on a particular road surface at a speed of 30 miles per hour is given by the expression.

$$\frac{10 \times 30^2}{30 \times 5}$$

What is this minimum stopping distance?

Averages

We use an **average** to represent a set of numbers. Averages allow us to compare two or more sets. (For example, do the men or the women in your class spend more time studying?) Averages also allow us to compare an individual with a set. (For example, is the amount of time you spend studying above or below the class average?) The following definition shows how to compute an average.

Definition

The **average** (or **mean**) of a set of numbers is the sum of those numbers divided by however many numbers are in the set.

EXAMPLE 10

What is the average of 100, 94, and 100?

Solution The average equals the sum of these three numbers divided by 3.

$$\frac{100 + 94 + 100}{3} = \frac{294}{3} = 98$$

PRACTICE 10

Find the average of \$30, \$0, and \$90.

EXAMPLE 11

The following map shows the five Great Lakes. The maximum depth of each of these lakes is given in the table. (*Source:* U.S. Environmental Protection Agency)

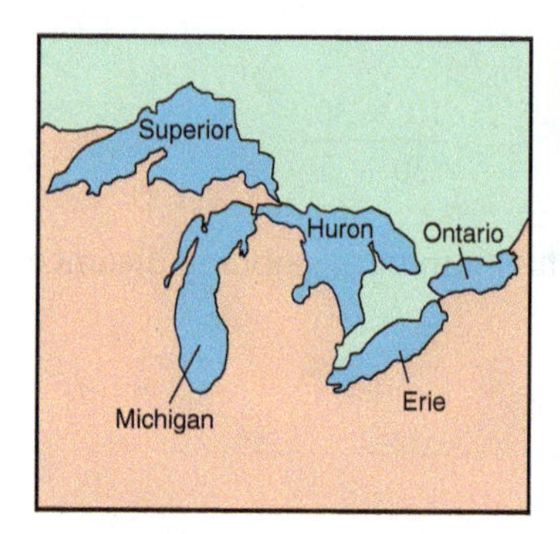

Lake	Maximum Depth (in meters)
Erie	64
Huron	229
Michigan	282
Ontario	244
Superior	406

a. What is the average maximum depth of the Great Lakes?

b. Which of the Great Lakes has a maximum depth that is above the average?

Solution

a. $\frac{\text{The sum of the depths}}{\text{The number of lakes}} = \frac{64 + 229 + 282 + 244 + 406}{5}$

$= \frac{1,225}{5}$

$= 245$

So the average maximum depth is 245 meters.

b. Lake Michigan and Lake Superior are deeper than the average.

PRACTICE 11

The table shown gives the number of fatalities due to tornadoes in the United States in each year from 2002 through 2005. (*Source:* NOAA/National Weather Service, Storm Prediction Center)

Year	Number of Fatalities
2002	55
2003	54
2004	36
2005	39

a. What was the average annual number of fatalities for these years?

b. In which years was the number of fatalities below the average?

Powers and Order of Operations on a Calculator

Let's use a calculator to carry out computations that involve either powers or the order of operations rule.

EXAMPLE 12

Calculate 23^3.

Solution

Press	Display
23 [^] 3 [ENTER]	23^3 12167.

PRACTICE 12

Use a calculator to compute 375^2.

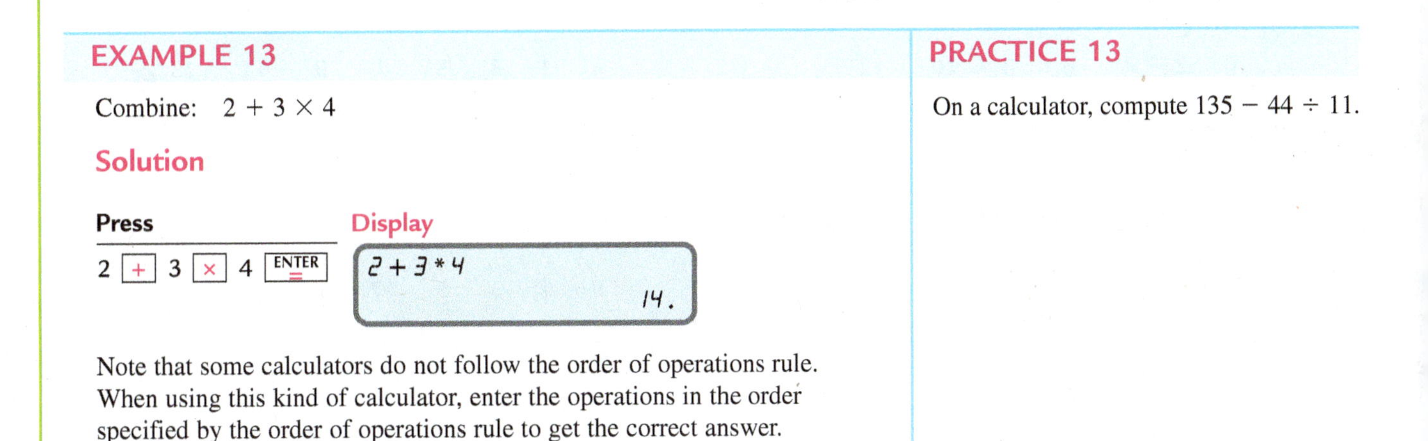

EXAMPLE 13

Combine: $2 + 3 \times 4$

Solution

Press	Display
2 [+] 3 [×] 4 [ENTER]	2+3*4 14.

Note that some calculators do not follow the order of operations rule. When using this kind of calculator, enter the operations in the order specified by the order of operations rule to get the correct answer.

PRACTICE 13

On a calculator, compute $135 - 44 \div 11$.

1.5 Exercises

FOR EXTRA HELP

Mathematically Speaking

Fill in each blank with the most appropriate term or phrase from the given list.

product	sum	adding	listing
subtracting	grouping	power	base

1. An exponent indicates how many times the _______ is used as a factor.

2. Parentheses and brackets are examples of _______ symbols.

3. An average of numbers on a list is found by _______ the numbers and then dividing by how many numbers there are on the list.

4. In evaluating an expression involving both a sum and a product, the _______ is evaluated first.

Complete each table by squaring the numbers given.

5.

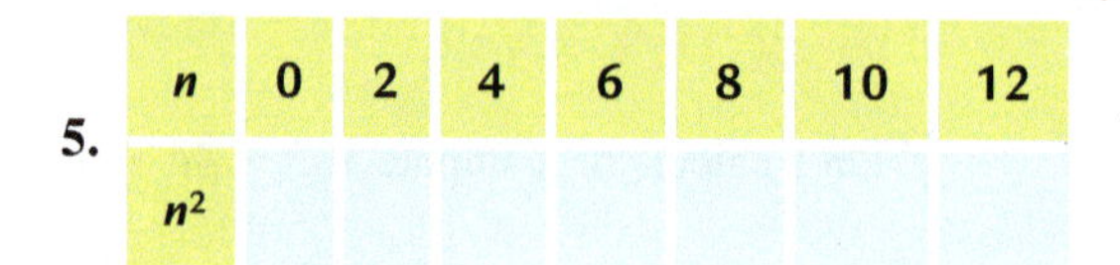

n	0	2	4	6	8	10	12
n^2							

6.

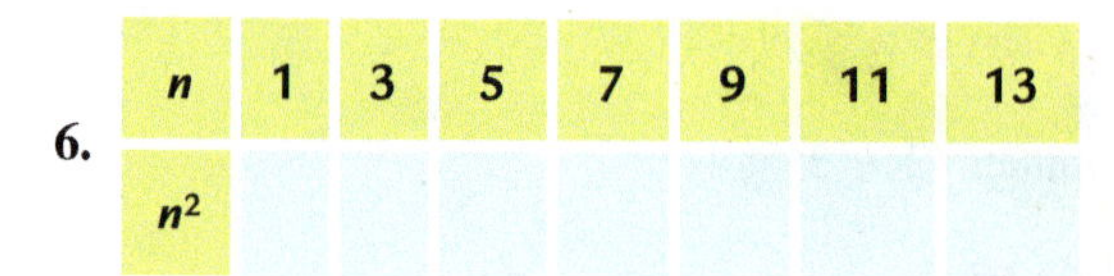

n	1	3	5	7	9	11	13
n^2							

Complete each table by cubing the numbers given.

7.

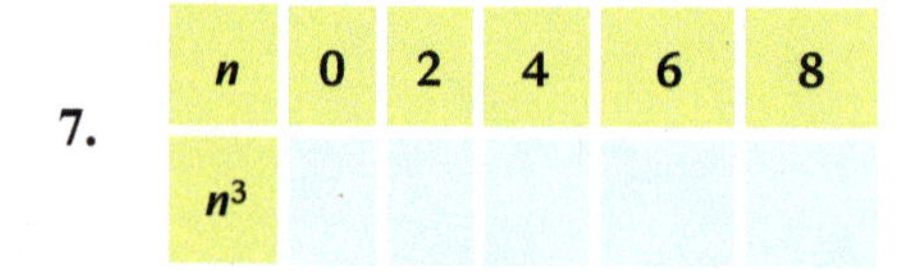

n	0	2	4	6	8
n^3					

8.

n	1	3	5	7	9
n^3					

Express each number as a power of 10.

9. $100 = 10^{\square}$

10. $1{,}000 = 10^{\square}$

11. $10{,}000 = 10^{\square}$

12. $100{,}000 = 10^{\square}$

13. $1{,}000{,}000 = 10^{\square}$

14. $10{,}000{,}000 = 10^{\square}$

Write each number in terms of powers.

15. $2 \cdot 2 \cdot 3 \cdot 3 = 2^{\square} \cdot 3^{\square}$

16. $2 \cdot 2 \cdot 5 \cdot 2 \cdot 5 = 2^{\square} \cdot 5^{\square}$

17. $5 \cdot 4 \cdot 4 \cdot 4 = 4^{\square} \cdot 5^{\square}$

18. $6 \cdot 7 \cdot 6 \cdot 7 \cdot 6 \cdot 7 = 6^{\square} \cdot 7^{\square}$

Write each number in standard form.

19. $6^2 \cdot 5^2$

20. $10^3 \cdot 9^2$

21. $2^5 \cdot 7^2$

22. $3^4 \cdot 4^3$

Evaluate.

23. $8 + 5 \cdot 2$

24. $9 + 10 \div 2$

25. $8 - 2 \times 3$

26. $12 - 6 \div 2$

27. $10 + 5^2$

28. $9 - 2^3$

29. $(9 - 2)^3$

30. $(10 + 5)^2$

31. 10×5^2

32. $12 \div 2^2$

33. $(12 \div 2)^2$

34. $(10 \times 5)^2$

35. $(24 \div 4) + 2$

36. $(15 \cdot 6) - 2$

37. $15 \cdot 6 + 2$

38. $24 \div 4 - 2$

39. $15 \cdot (6 - 2)$

40. $24 \div (4 + 2)$

41. $2^6 - 6^2$

42. $3^5 - 5^3$

43. $8 + 5 - 3 - 2 \times 2$

44. $7 - 1 + 2 + 3 \cdot 2$

45. $(10 - 1)(10 + 1)$

46. $(8 - 1)(8 + 1)$

47. $10^2 - 1$

48. $8^2 - 1$

49. $\left(\dfrac{8 + 2}{7 - 2}\right)^2$

50. $\left(\dfrac{9 - 1}{3 + 5}\right)^3$

51. $\dfrac{5^3 - 2^3}{3}$

52. $\dfrac{3^2 + 5^2}{2}$

53. $(2 + 14) \div 4(9 - 5)$

54. $3 + 10(20 - 2^3)$

55. $10 \cdot 3^2 + \dfrac{10 - 4}{2}$

56. $\dfrac{3^3 + 1^3 + 2^3}{4}$

57. $[9 + 2(3^2 - 8)] + 7$

58. $15 + [3(8 - 2^2) - 6]$

59. $32 + 9 \cdot 215 \div 5$

60. $84 \cdot 27 + 32 \cdot 27^2 \div 2$

61. $48(48 - 31)(48 - 24)(48 - 41)$

62. $137^2 - 4(36)(22)$

In each exercise, the three squares stand for the numbers 4, 6, and 8 in some order. Fill in the squares to make true statements.

63. $\square \cdot 3 + \square \cdot 5 + \square \cdot 7 = 98$

64. $\square + 10 \times \square - \dfrac{\square}{2} = 42$

65. $(\square)(3 + \square) - 2 \cdot \square = 44$

66. $\square \cdot 3 + \square \cdot 5 + \square \cdot 7 = 82$

67. $\square + 10 \times \square - \square \div 2 = 45$

68. $\dfrac{48}{\square} - \dfrac{\square}{2} + (3 + \square)^2 = 127$

Insert parentheses, if needed, to make the expression on the left equal to the number on the right.

69. $5 + 2 \cdot 4^2 = 112$

70. $5 + 2 \cdot 4^2 = 69$

71. $5 + 2 \cdot 4^2 = 169$

72. $5 + 2 \cdot 4^2 = 37$

73. $8 - 4 \div 2^2 = 1$

74. $8 - 4 \div 2^2 = 7$

Find the area of each shaded region.

75.

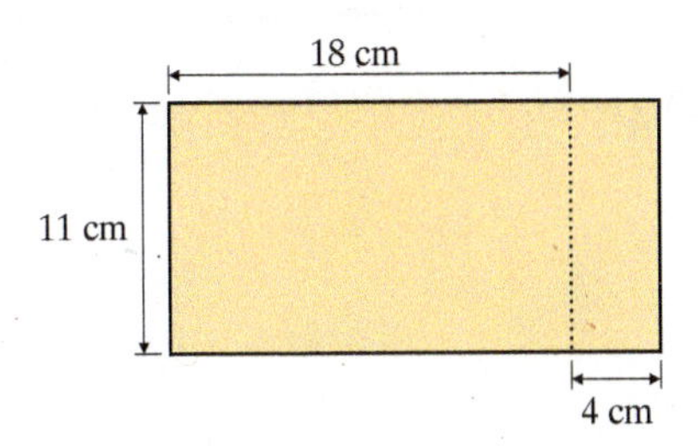

76.

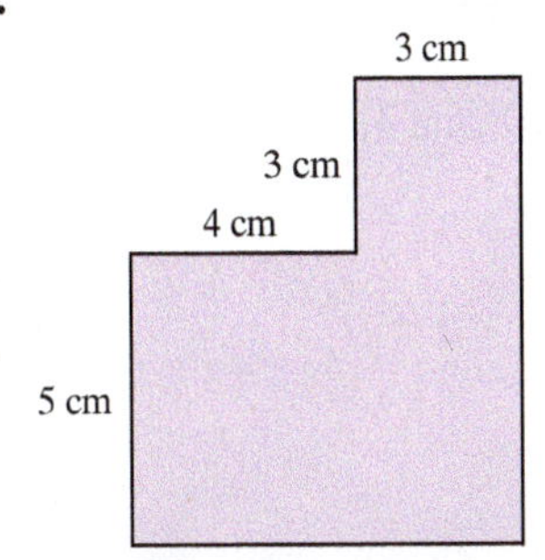

77.

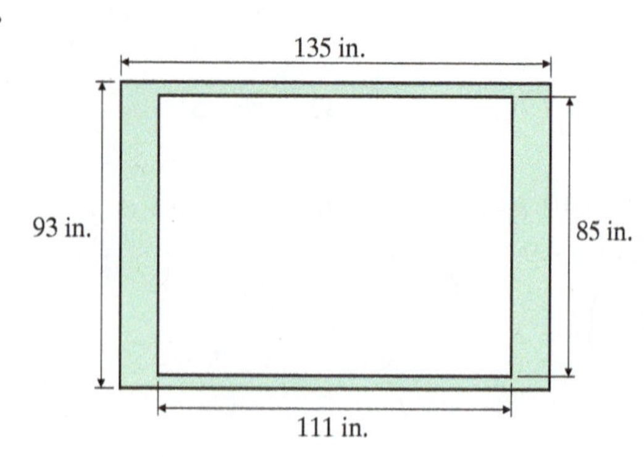

78.

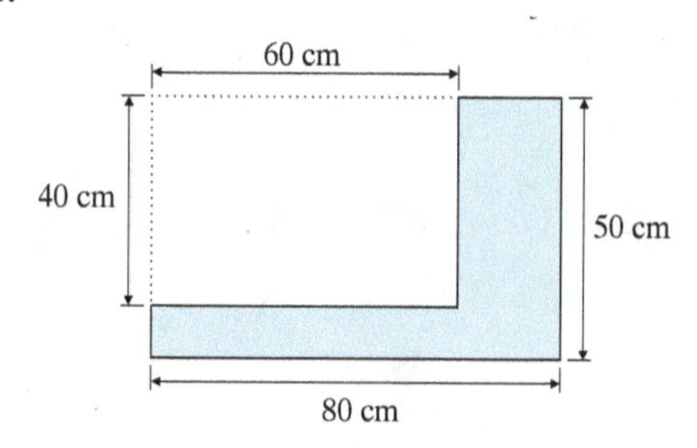

Complete each table.

79.

Input	Output
0	$21 + 3 \times \mathbf{0} =$
1	$21 + 3 \times \mathbf{1} =$
2	$21 + 3 \times \mathbf{2} =$

80.

Input	Output
0	$14 - 5 \times \mathbf{0} =$
1	$14 - 5 \times \mathbf{1} =$
2	$14 - 5 \times \mathbf{2} =$

Find the average of each set of numbers.

81. 20 and 30

82. 10 and 50

83. 30, 60, and 30

84. 17, 17, and 26

85. 10, 0, 3, and 3

86. 5, 7, 7, and 17

87. 3,527 miles, 1,788 miles, and 1,921 miles

88. 7 hours, 6 hours, 10 hours, 9 hours, and 8 hours

89. Six 10's and four 5's

90. Sixteen 5's and four 0's

Mixed Practice

Solve.

91. Express 100,000,000 as a power of 10.

92. Find the area of the shaded region.

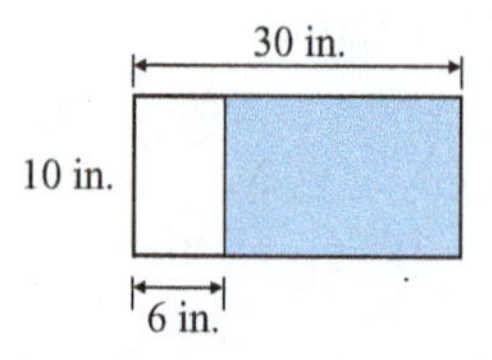

93. Square 17.

94. Rewrite in terms of powers of 2 and 7:
$2 \cdot 2 \cdot 2 \cdot 7 \cdot 7 = 2^{} \cdot 7^{}$

95. Simplify: $50 - 2(10 - 3^2)$.

96. Cube 10.

97. Find the average of 10, 10, and 4.

98. Evaluate: 6×4^2.

Applications

Solve and check.

99. A 40-story office building has 25,000 square feet of space to rent. What is the average rental space on a floor?

100. The total area of the 50 states in the United States is about 3,700,000 square miles. If the state of Georgia's area is about 60,000 square miles, is its size above the average of all the states? Explain. (*Source: The New Encyclopedia Britannica*)

101. In a branch of mathematics called number theory, the numbers 3, 4, and 5 are called a *Pythagorean triple* because $3^2 + 4^2 = 5^2$ (that is, $9 + 16 = 25$). Show that 5, 12, and 13 are a Pythagorean triple.

102. If an object is dropped off a cliff, after 10 seconds it will have fallen $\frac{32 \cdot 10^2}{2}$ feet, ignoring air resistance. Express this distance in standard form, without exponents.

103. The solar wind streams off the Sun at speeds of about 1,000,000 miles per hour. Express this number as a power of 10. (*Source:* NASA)

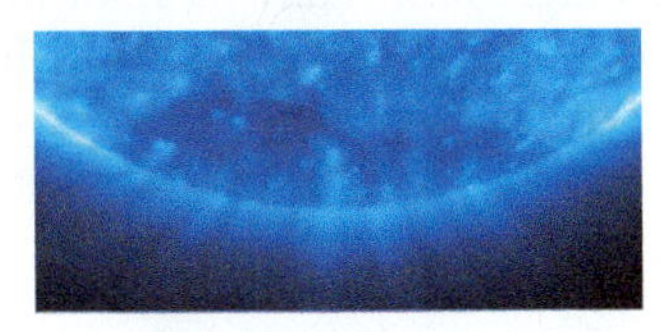

104. In 2005, about $100,000,000,000 worth of twenty-dollar bills were in circulation. Represent this number as a power of 10. (*Source:* Financial Management Service *Treasury Bulletin*, March 2006)

105. The following table shows a lab assistant's salary in various years.

Year	1	2	3
Salary	$19,400	$21,400	$23,700

a. Find the average salary for the three years.

b. How much greater was her average salary for the last two years than for all three years?

106. The following grade book shows a student's math test scores.

Test 1	Test 2	Test 3	Test 4
85	63	98	82

a. What is the average of his math scores?

b. If he were to get a 92 on the next math test, by how much would his average score increase?

107. In the last four home games, a college basketball team had scores of 68, 79, 57, and 72.

a. What was the average score for these games?

b. The average score for the last four away games was 64. On average, did the team score more at home or away? Explain.

108. A small theater company's production of *Romeo and Juliet* had 10 performances over two weekends. The attendance for each performance during the second weekend was 171, 297, 183, 347, and 232.

a. What was the average attendance for the performances during the second weekend?

b. If the average attendance at a performance during the first weekend was 272, was this average greater in the first or second weekend? Explain.

109. The number of hours the typical American spent watching broadcast television and cable and satellite television each year is given in the following table. (*Source:* Veronis Suhler Stevenson)

Year	Broadcast Television	Cable and Satellite Television
2001	833	843
2002	787	918
2003	769	975
2004	782	1,010
2005	785	1,042
2006	790	1,068

a. Find the average annual number of hours spent watching broadcast television and the average annual number of hours spent watching cable and satellite television.

b. Which average is higher? By how much?

110. The following table shows how many one-family homes were bought in the Northeast and Midwest regions of the United States each year from 2001 through 2005. (*Source:* National Association of Realtors)

Year	Northeast	Midwest
2001	710,000	1,155,000
2002	730,000	1,217,000
2003	769,000	1,322,000
2004	821,000	1,389,000
2005	839,000	1,410,000

a. Find the average annual number of one-family homes bought in each region for the 5-year period.

b. On the average, how many more of these homes were bought per year in the Midwest as compared to the Northeast?

Use a calculator to solve each problem giving (a) the operation(s) carried out in the solution, (b) the exact answer, and (c) an estimate of the answer.

111. Last week, newspaper A had an average daily circulation of 72,073. The daily circulation for newspaper B was as follows.

Day	B's Circulation
M	85,774
Tu	72,503
W	68,513
Th	74,812
F	89,002
Sa	92,331
Su	102,447

Which newspaper had a greater daily average circulation last week? By how many newspapers?

112. The hospital chart shown is a record of a patient's temperature for two days.

Time	Temp. (°F)	Time	Temp. (°F)
6 A.M.	98	6 A.M.	101
10 A.M.	100	10 A.M.	102
2 P.M.	98	2 P.M.	101
6 P.M.	100	6 P.M.	102
10 P.M.	98	10 P.M.	100
2 A.M.	100	2 A.M.	100

What was her average temperature for this period of time?

• *Check your answers on page A-2.*

MINDSTRETCHERS

Writing

1. Evaluate the expressions in parts (a) and (b).

a. $7^2 + 4^2$ ____

b. $(7 + 4)^2$ ____

c. Are the answers to parts (a) and (b) the same? ____ If not, explain why not.

Mathematical Reasoning

2. The square of any whole number (called a **perfect square**) can be represented as a geometric square, as follows:

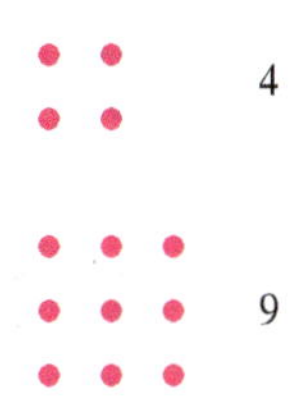

Try to represent the numbers 16, 25, 5, and 8 the same way.

16 25 5 8

Critical Thinking

3. Find the average of the whole numbers from 1 through 999.

1.6 More on Solving Word Problems

OBJECTIVE

- To solve word problems involving the addition, subtraction, multiplication, or division of whole numbers using various problem-solving strategies

What Word Problems Are and Why They Are Important

In this section, we consider some general tips to help solve word problems.

Word problems can deal with any subject—from shopping to physics and geography to business. Each problem is a brief story that describes a particular situation and ends with a question. Our job, after reading and thinking about the problem, is to answer that question by using the given information.

Although there is no magic formula for solving word problems, you should keep the following problem-solving steps in mind.

To Solve Word Problems

- Read the problem carefully.
- Choose a strategy (such as drawing a picture, breaking up the question, substituting simpler numbers, or making a table).
- Decide which basic operation(s) are relevant and then translate the words into mathematical symbols.
- Perform the operations.
- Check the solution to see if the answer is reasonable. If it is not, start again by rereading the problem.

Reading the Problem

In a math problem, each word counts. So it is important to read the problem slowly and carefully, and not to scan it as if it were a magazine or newspaper article.

When reading a problem, we need to understand the problem's key points: *What information is given* and *what question is posed*. Once these points are clear, jot them down so as to help keep them in mind.

After taking notes, decide on a plan of action that will lead to the answer. For many problems, just thinking back to the meaning of the four basic operations will be helpful.

Operation	Meaning
$+$	Combining
$-$	Taking away
$\times$	Adding repeatedly
$\div$	Splitting up

Many word problems contain *clue words* that suggest performing particular operations. If we spot a clue word in a problem, we should consider whether the operation indicated in the following table will lead us to a solution.

+	−	×	÷
• add	• subtract	• multiply	• divide
• sum	• difference	• product	• quotient
• total	• take away	• times	• over
• plus	• minus	• double	• split up
• more	• less	• twice	• fit into
• increase	• decrease	• triple	• per
• gain	• loss	• of	• goes into

However, be on guard—a clue word can be misleading. For instance, in the problem *What number increased by 2 is 6?*, we solve by subtracting, not adding.

Consider the following "translations" of these clues.

The patient's fever **increased by 5°**.	$+ 5$
The number of unemployed people **tripled**.	$\times 3$
The length of the bedroom is **8 feet less** than the kitchen's.	$- 8$
The company's earnings were **split** among the **four** partners.	$\div 4$

Choosing a Strategy

If no method of solution comes to mind after reading problem, there are a number of problem-solving strategies that may help. Here we discuss four of these strategies: drawing a picture, breaking up the question, substituting simpler numbers, and making a table.

Drawing a Picture

Sketching even a rough representation of a problem—say, a diagram or a map—can provide insight into its solution, if the sketch accurately reflects the given information.

EXAMPLE 1

In an election, everyone voted for one of three candidates. The winner received 188,000 votes, and the second-place candidate got 177,000 votes. If 380,000 people voted in the election, how many people voted for the third candidate?

Solution To help us understand the given information, let's draw a diagram to represent the situation.

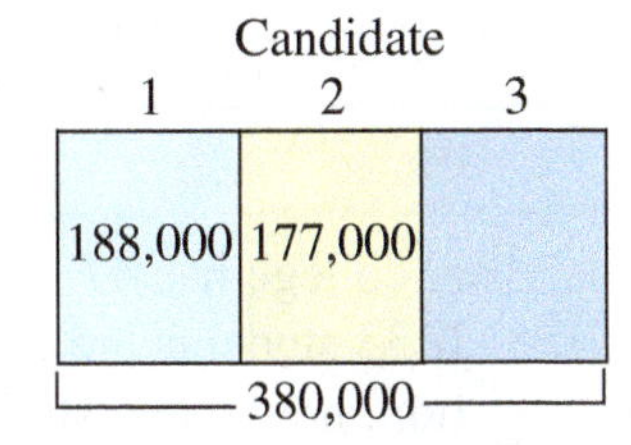

We see from this diagram that to find the answer we need to do two things.

PRACTICE 1

A company slashed its workforce by laying off 1,150 employees during one month and laying off 2,235 employees during another month. Afterward, 7,285 employees remained. How many employees worked for the company before the layoffs began?

- First, we need to add 188,000 to 177,000.

$$\begin{array}{r} 188{,}000 \\ +\ 177{,}000 \\ \hline 365{,}000 \end{array}$$

- Then, we need to subtract this sum from 380,000.

$$\begin{array}{r} 380{,}000 \\ -365{,}000 \\ \hline 15{,}000 \end{array}$$

A good way to check our answer here is by adding.

$$188{,}000 + 177{,}000 + 15{,}000 = 380{,}000$$

Our answer checks, so 15,000 people voted for the third candidate.

Breaking Up the Question

Another effective problem-solving strategy is to break up the given question into a chain of simpler questions.

EXAMPLE 2

Suppose that a student took a math test consisting of 20 questions, and that she answered 3 questions wrong. How many more questions did she get right than wrong?

Solution Say that we do not know how to answer this question directly. Try to split it into several easier questions that lead to a solution.

- How many questions did the student get *right*? $20 - 3 = 17$
- How many questions did she get *wrong*? 3
- How many *more* questions did she get right than wrong? $17 - 3 = 14$

So the student had 14 more questions right than wrong.

This answer seems reasonable, because it must be less than 17, the number of questions that the student answered correctly.

PRACTICE 2

Teddy and Franklin Roosevelt were both U.S. presidents. Teddy was born in 1858 and died in 1919. Franklin was born in 1882 and died in 1945. How much longer did Franklin live than Teddy?

(*Source:* Foner and Garraty, *The Reader's Companion to American History*)

Substituting Simpler Numbers

A word problem involving large numbers often seems difficult just because of these numbers. A good problem-solving strategy here is to consider first the identical problem but with simpler numbers. Solve the revised problem and then return to the original problem.

EXAMPLE 3

Raffle tickets cost $4 each. How many tickets must be sold for the raffle to break even if the prizes total $4,736?

Solution Suppose that we are not sure which operation to perform to solve this problem. Let's try substituting a simpler amount (say, $8) for the break-even amount of $4,736 and see if we can solve the resulting problem.

PRACTICE 3

A college has 47 sections of Math 110. If 33 students are enrolled in each section, how many students are taking Math 110?

The question would then become: How many $4 tickets must be sold to make back $8? Because it is a "fit-in" question, we must *divide* the $8 by the $4. Going back to the original problem, we see that we must divide $4,736 by 4.

$$\$4{,}736 \div 4 = 1{,}184 \text{ tickets}$$

Is this answer reasonable? We can check either by estimating ($5{,}000 \div 4 = 1{,}250$, which is close to our answer) or by multiplying ($1{,}184 \times 4 = 4{,}736$), which also checks).

Making a Table

Finally, let's consider a strategy for solving word problems that involve many numbers. Organizing these numbers into a table often leads to a solution.

EXAMPLE 4

A borrower promises to pay back $50 per month until a $1,000 loan is settled. What is the remaining loan balance at the end of 5 months?

Solution We can solve by organizing the information in a table.

After Month	Remaining Balance
1	$1{,}000 - 50 = 950$
2	$950 - 50 = 900$
3	$900 - 50 = 850$
4	$850 - 50 = 800$
5	$800 - 50 = 750$

From the table, we see that the remaining balance after 5 months is $750.

We can also solve this problem by breaking up the question into simpler questions.

- How much money did the borrower pay after 5 months? $5 \cdot 50 = \$250$
- How much money did the borrower still owe after 5 months? $1{,}000 - 250 = \$750$

Again, the remaining balance after 5 months is $750.

PRACTICE 4

An athlete weighs 210 pounds and decides to go on a diet. If he loses 2 pounds a week while on the diet, how much will he weigh after 15 weeks?

1.6 Exercises

Choose a strategy. Solve and check.

1. In retailing, the difference between the gross sales and customer returns and allowances is called the net sales. If a store's gross sales were $2,538 and customer returns and allowances amounted to $388, what was the store's net sales?

2. The population of the United States in 1800 was 5,308,483. Ten years later, the population had grown to 7,239,881. During this period of time, did the country's population double? Justify your answer. (***Source:*** *The Time Almanac 2000*)

3. Suppose that you drive 27 miles north, 31 miles east, 45 miles west, and 14 miles east. How far are you from your starting point?

4. In your office, there are 19 reams of paper. If you need 7,280 sheets of paper for a printing job, do you have enough reams? (***Hint:*** A ream is 500 sheets of paper.)

5. A blue whale weighs about 300,000 pounds, and a great white shark weighs about 4,000 pounds. How many times the weight of a great white shark is the weight of a blue whale? (***Source:*** wikipedia.com)

6. A movie fan installed shelves for his collection of 400 DVDs. If 36 DVDs fit on each shelf, how many shelves did he need to house his entire collection?

7. A sales representative flew from Los Angeles to Miami (2,339 miles), then to New York (1,092 miles), and finally back to LA (2,451 miles). How many total miles did he fly?

8. Two major naval disasters of the twentieth century involved the sinking of British ships—the *Titanic* and the *Lusitania.* The *Titanic,* which weighed about 93,000,000 pounds, was the most luxurious liner of its time; it struck an iceberg on its maiden voyage in 1912. The *Lusitania,* which weighed about 63,000,000 pounds, was sunk by a German submarine in 1915. How much heavier was the *Titanic* than the *Lusitania?* (***Source:*** *The Oxford Companion to Ships and the Sea*)

The New York Times. EXTRA

LUSITANIA SUNK BY A SUBMARINE, PROBABLY 1,260 DEAD; TWICE TORPEDOED OFF IRISH COAST; SINKS IN 15 MINUTES; CAPT. TURNER SAVED, FROHMAN AND VANDERBILT MISSING; WASHINGTON BELIEVES THAT A GRAVE CRISIS IS AT HAND

9. Immigrants from all over the world came to the United States between 1931 and 1940 in the following numbers: 348,289 (Europe), 15,872 (Asia), 160,037 (Americas, outside the United States), 1,750 (Africa), and 2,231 (Australia/New Zealand). What was the total number of immigrants? (***Source:*** George Thomas Kurian, *Datepedia of the United States*)

10. In 2004, there were 36,652 movie screens in the United States. A year later, there were 37,740 movie screens. What was the increase in the number of movie screens from 2004 to 2005? (***Source:*** National Association of Theater Owners, 2006)

11. For each 4×6 print of a digital photo, a lab usually charges 10¢. During a promotion, the first 20 prints of each order are free. How much does the lab charge for 50 4×6 prints during the promotion?

12. Because of a noisy neighbor, a young man decided to put acoustical tiles on the living room ceiling, which measures 21 feet by 18 feet. The tiles are square, with a side length of 1 foot. If the tiles cost $3 apiece, what is the total cost to cover the ceiling?

13. After it was on the market for 4 months, the sellers of a home reduced the asking price by $14,000. After 6 months, they reduced the asking price a second and final time. If the original asking price had been $229,000 and the final asking price was $198,000, by how much did the sellers reduce the price the final time?

14. A nurse sets the drip rate for an IV medication at 25 drops each minute. How many drops does a patient receive in 2 hours?

15. A car dealer offered to lease a car for $1,500 down and $189 per month. If a customer accepted a lease contract of 2 years, how much did the customer have to pay over the lease period?

16. A doctor instructed a patient to take 100 milligrams of a medication daily for 4 weeks. The local pharmacy dispensed 120 tablets, each containing 25 milligrams of the medication. After taking the tablets for 4 weeks, how many remained?

17. The part-time tuition rates per credit-hour at a community college were $95 for in-state residents and $257 for out-of-state residents. To take 9 credit-hours, how much more than an in-state resident does an out-of-state resident pay?

18. A shirt placed on sale is marked down by $16. At the register, the customer receives an additional discount of $6. If the final sale price of the shirt was $18, what was the original price?

19. An office manager needs to order 1,000 pens from an office supply catalog. If the catalog sells pens by the gross (that is, in sets of 144) and 7 gross were ordered, how many extra pens did the manager order?

20. While shopping, a mother decides to buy three shirts costing $39 apiece and two pairs of shoes at $62 per pair. If she has $300 with her, is that enough money to pay for these items? Explain.

21. Eisenhower beat Stevenson in the 1952 and 1956 presidential elections. In 1952, Eisenhower received 442 electoral votes and Stevenson 89. In 1956, Eisenhower got 457 electoral votes and Stevenson 73. Which election was closer? By how many electoral votes? (*Source: World Almanac*)

22. A garden is rectangular in shape—26 feet in length and 14 feet in width. If fencing costs $13 a foot, how much will it cost to enclose the garden with this fencing?

Use a calculator to solve each problem, giving (a) the operation(s) carried out in the solution, (b) the exact answer, and (c) an estimate of the answer.

23. A couple agrees to pay the seller of the house of their dreams $165,000. They put down $23,448 and promise to pay the balance in 144 equal installments. How much money will each installment be?

24. Earth revolves around the Sun in 365 days, but the planet Mercury does so in only 88 days. Compared to Earth, how many more complete revolutions will Mercury make in 1,000 days?

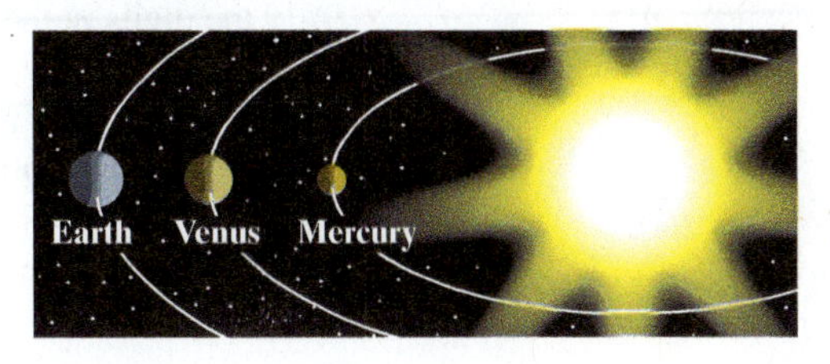

• Check your answers on page A-3.

KEY CONCEPTS AND SKILLS

CONCEPT SKILL

Concept/Skill	Description	Example
[1.1] Place value	**Thousands**: Hundreds Tens Ones | **Ones**: Hundreds Tens Ones	846,120 4 is in the ten thousands place.
[1.1] To read a whole number	Working from left to right, • read the number in each period, and then • name the period in place of the comma.	71,400 is read "seventy-one thousand, four hundred".
[1.1] To write a whole number	Working from left to right, • write the number named in each period, and • replace each period name with a comma.	"Five thousand, twelve" is written 5,012.
[1.1] To round a whole number	• Underline the place to which you are rounding. • Look at the digit to the right of the underlined digit, called the *critical digit.* If this digit is 5 or more, add 1 to the underlined digit; if it is less than 5, leave the underlined digit unchanged. • Replace all the digits to the right of the underlined digit with zeros.	$3\underline{8}6 \approx 390$ $4,\underline{8}17 \approx 4,800$
[1.2] Addend, Sum	In an addition problem, the numbers being added are called *addends.* The result is called their *sum.*	6 + 4 = 10 6 = Addend, 4 = Addend, 10 = Sum
[1.2] The Identity Property of Addition	The sum of a number and zero is the original number.	$4 + 0 = 4$ $0 + 7 = 7$
[1.2] The Commutative Property of Addition	Changing the order in which two numbers are added does not affect their sum.	$7 + 8 = 8 + 7$
[1.2] The Associative Property of Addition	When adding three numbers, regrouping addends gives the same sum.	$(5 + 4) + 1 =$ $5 + (4 + 1)$
[1.2] To add whole numbers	• Write the addends vertically, lining up the place values. • Add the digits in the ones column, writing the rightmost digit of the sum on the bottom. If the sum has two digits, carry the left digit to the top of the next column on the left. • Add the digits in the tens column as in the preceding step. • Repeat this process until you reach the last column on the left, writing the entire sum of that column on the bottom.	1 1 1 (carries) 7,385 92,551 + 2,007 101,943

continued

Concept/Skill	Description	Example
[1.2] **Minuend, Subtrahend, Difference**	In a subtraction problem, the number that is being subtracted from is called the *minuend.* The number that is being subtracted is called the *subtrahend.* The answer is called the *difference.*	Difference ↓ 10 − 6 = 4 ↑ Minuend ↑ Subtrahend
[1.2] **To subtract whole numbers**	• On top, write the number *from which* we are subtracting. On the bottom, write the number that is being *taken away,* lining up the place values. Subtract in each column separately. • Start with the ones column. **a.** If the digit on top is *larger* than or *equal* to the digit on the bottom, subtract and write the difference below. **b.** If the digit on top is *smaller* than the digit on the bottom, borrow from the digit to the left on top. Then subtract and write the difference below the bottom digit. • Repeat this process until the last column on the left is finished, subtracting and writing its difference below.	8 14 1 7, 9 5 2 −1, 8 8 3 6, 0 6 9
[1.3] **Factor, Product**	In a multiplication problem, the numbers being multiplied are called *factors*. The result is called their *product*.	Factor Product $4 \times 5 = 20$
[1.3] **The Identity Property of Multiplication**	The product of any number and 1 is that number.	$1 \times 6 = 6$ $7 \times 1 = 7$
[1.3] **The Multiplication Property of 0**	The product of any number and 0 is 0.	$51 \times 0 = 0$ $0 \times 9 = 0$
[1.3] **The Commutative Property of Multiplication**	Changing the order in which two numbers are multiplied does not affect their product.	$3 \times 2 = 2 \times 3$
[1.3] **The Associative Property of Multiplication**	When multiplying three numbers, regrouping the factors gives the same product.	$(4 \times 5) \times 6$ $= 4 \times (5 \times 6)$
[1.3] **The Distributive Property**	Multiplying a factor by the sum of two numbers gives the same result as multiplying the factor by each of the two numbers and then adding.	$2 \times (4 + 3)$ $= (2 \times 4) + (2 \times 3)$
[1.3] **To multiply whole numbers**	• Multiply the top factor by the ones digit in the bottom factor and write this product. • Multiply the top factor by the tens digit in the bottom factor and write this product leftward, beginning with the tens column. • Repeat this process until all the digits in the bottom factor are used. • Add the partial products, writing this sum.	693 × 71 693 48 51 49,203

continued

CONCEPT SKILL

Concept/Skill	Description	Example
[1.4] **Divisor, Dividend, Quotient**	In a division problem, the number that is being used to divide another number is called the *divisor*. The number into which it is being divided is called the *dividend*. The result is called the *quotient*.	Quotient: 3 $4\overline{)12}$ Divisor: 4, Dividend: 12
[1.4] **To divide whole numbers**	• Divide 17 into 39, which gives 2. Multiply the 17 by 2 and subtract the result (34) from 39. Beside the difference (5), bring down the next digit (3) of the dividend. • Repeat this process, dividing the divisor (17) into 53. • At the end, there is a remainder of 2. Write it beside the quotient on top.	23 R2 $17\overline{)393}$ 34 53 51 2
[1.5] **Exponent (or Power), Base**	An *exponent* (or *power*) is a number that indicates how many times another number (called the *base*) is used as a factor.	Exponent: 3 $5^3 = 5 \times 5 \times 5$ Base: 5
[1.5] **Order of Operations Rule**	To evaluate mathematical expressions, carry out the operations *in the following order.* **1.** First, perform the operations within any grouping symbols, such as parentheses ( ) or brackets [ ]. **2.** Then, raise any number to its power. **3.** Next, perform all multiplications and divisions as they appear from left to right. **4.** Finally, do all additions and subtractions as they appear from left to right. ( ) × ÷ + −	$8 + 5 \cdot (3 + 1)^2 = 8 + 5 \cdot 4^2$ $= 8 + 5 \cdot 16$ $= 8 + 80$ $= 88$
[1.5] **Average (or Mean)**	The *average* (or *mean*) of a set of numbers is the sum of those numbers divided by however many numbers are in the set.	The average of 3, 4, 10, and 3 is 5 because $\frac{3 + 4 + 10 + 3}{4} = \frac{20}{4} = 5$
[1.6] **To solve word problems**	• Read the problem carefully. • Choose a strategy (such as drawing a picture, breaking up the question, substituting simpler numbers, or making a table). • Decide which basic operation(s) are relevant and then translate the words into mathematical symbols. • Perform the operations. • Check the solution to see if the answer is reasonable. If it is not, start again by rereading the problem.	

Chapter 1 Review Exercises

To help you review this chapter, solve these problems.

[1.1] *In each whole number, identify the place that the digit 3 occupies.*

1. 23 **2.** 30,802 **3.** 385,000,000 **4.** 30,000,000,000

Write each number in words.

5. 497 **6.** 2,050 **7.** 3,000,007 **8.** 85,000,000,000

Write each number in standard form.

9. Two hundred fifty-one

10. Nine thousand, two

11. Fourteen million, twenty-five

12. Three billion, three thousand

Express each number in expanded form.

13. 2,500,000

14. 42,707

Round each number to the place indicated.

15. 571 to the nearest hundred

16. 938 to the nearest thousand

17. 384,056 to the nearest ten thousand

18. 68,332 to its largest place

[1.2] *Find the sum and check.*

19. $\begin{array}{r} 102 \\ 4{,}251 \\ +\ 5{,}133 \\ \hline \end{array}$

20. $\begin{array}{r} 53{,}569 \\ 10{,}000 \\ +\ \ 2{,}123 \\ \hline \end{array}$

21. $\begin{array}{r} 48{,}758 \\ 37{,}226 \\ +\ 87{,}559 \\ \hline \end{array}$

22. $\begin{array}{r} 95{,}000 \\ 25{,}895 \\ +\ 30{,}000 \\ \hline \end{array}$

Add and check.

23. 972,558 + 87,055 + 36,488 + 861,724

24. \$138,865 + \$729 + \$8,002 + \$75,471

Find the difference and check.

25. $\begin{array}{r} 876 \\ -\ 431 \\ \hline \end{array}$

26. $\begin{array}{r} 56{,}000 \\ -\ 45{,}984 \\ \hline \end{array}$

27. $\begin{array}{r} 98{,}118 \\ -\ 87{,}009 \\ \hline \end{array}$

28. $\begin{array}{r} 7{,}100 \\ -\ 1{,}590 \\ \hline \end{array}$

29. 60,000,000 − 48,957,777

30. \$5,000,000 − \$2,937,148

31. From 67,502 subtract 56,496.

32. Subtract 89,724 from 92,713.

[1.3] *Find the product and check.*

33. $\begin{array}{r} 72 \\ \times\ 6 \\ \hline \end{array}$

34. $\begin{array}{r} 400 \\ \times\ 3 \\ \hline \end{array}$

35. $\begin{array}{r} 2,923 \\ \times\ 51 \\ \hline \end{array}$

36. $\begin{array}{r} 6,000 \\ \times\ 2,000 \\ \hline \end{array}$

37. $\begin{array}{r} 14,921 \\ \times\ 32 \\ \hline \end{array}$

38. $\begin{array}{r} 8,152 \\ \times\ 125 \\ \hline \end{array}$

Multiply and check.

39. $2,751 \cdot 508$

40. $(681)(498)(555)$

[1.4] *Divide and check.*

41. $\dfrac{975}{25}$

42. $21\overline{)6,450}$

43. $13\overline{)491}$

44. $7,488 \div 11$

Find the quotient and check.

45. $8\overline{)205,000}$

46. $347\overline{)332,079}$

[1.5] *Compute.*

47. 7^3

48. 1^{10}

49. $2^3 \cdot 3^2$

50. $3 \cdot 10^5$

51. $20 - 3 \times 5$

52. $(9 + 4)^2$

53. $10 - \dfrac{6 + 4}{2}$

54. $3 + (5 - 1)^2$

55. $5 + [4^2 - 3(2 + 1)]$

56. $17 + [2(3^2 - 6) - 5]$

57. $98(50 - 1)(50 - 2)(50 - 3)$

58. $\dfrac{28^3 + 29^3 + 37^3 - 10}{(7 - 1)^2}$

Rewrite each expression, using exponents.

59. $7 \cdot 7 \cdot 5 \cdot 5 = 7^{■} \cdot 5^{■}$

60. $5 \cdot 2 \cdot 5 \cdot 2 \cdot 5 = 2^{■} \cdot 5^{■}$

Find the average.

61. 34 and 44

62. 20, 0, and 1

63. 5, 8, and 5

64. 4, 6, 3, and 7

Mixed Applications

Solve and check.

65. Beetles about the size of a pinhead destroyed 2,400,000 acres of forest. Express this number in words.

66. Scientists in Utah found a dinosaur egg one hundred fifty million years old. Write this number in standard form.

67. In a part-time job, a graduate student earned $15,964 a year. How much money did she earn per week? (*Hint:* 1 year equals 52 weeks.)

68. Halley's Comet was sighted in 1682. If it reappeared 76 years later, in what year did it reappear? (*Source: The Time Almanac 2000*)

69. What is the land area of Texas to the nearest hundred thousand square miles? (*Source: Time Almanac 2006*)

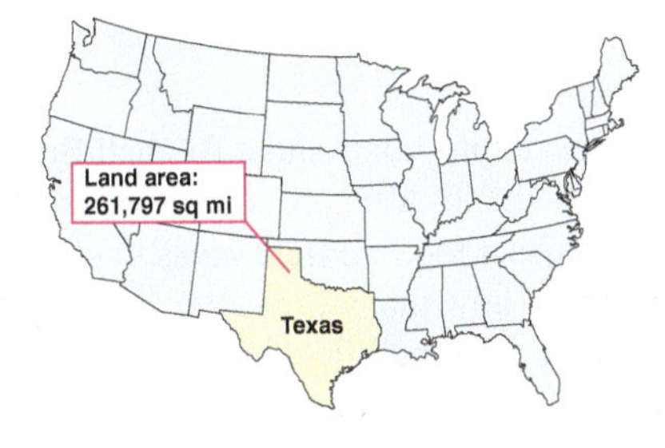

70. Apple Computer sold 22,497,000 iPods in 2005. How many iPods is this to the nearest million? (*Source:* Apple Computer, 2005 10-K Annual Report)

71. The Empire State Building is 1,250 feet high, and the Statue of Liberty is 152 feet in height. What is the minimum number of Statues of Liberty that would have to be stacked to be taller than the Empire State Building?

72. A millipede—a small insect with 68 body segments—has 4 legs per segment. How many legs does a millipede have?

73. Taipei 101 in Taiwan, the tallest building in the world as of 2006, is 67 meters taller than the Sears Tower in Chicago. If the Sears Tower is 442 meters tall, what is the height of Taipei 101? (*Source:* Emporis Buildings)

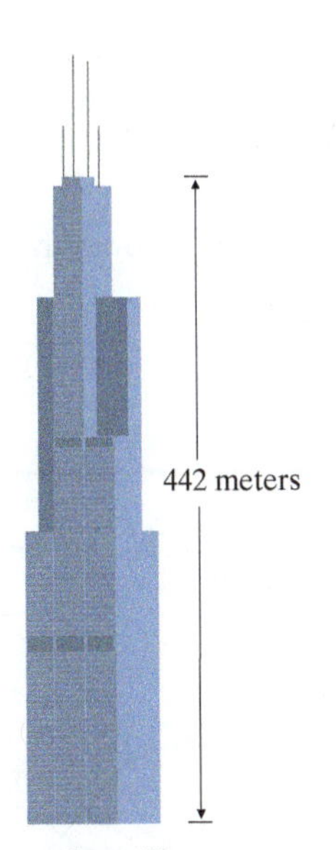

Sears Tower

?

Taipei 101

74. A landscaper needs 550 flower plants for a landscaping project. If a local garden center sells flats containing 24 plants, how many flats should the landscaper buy?

75. The following graph shows the consolidated assets of the six largest banks in the United States (in millions of U.S. dollars). Find the combined assets of Wells Fargo and Wachovia.

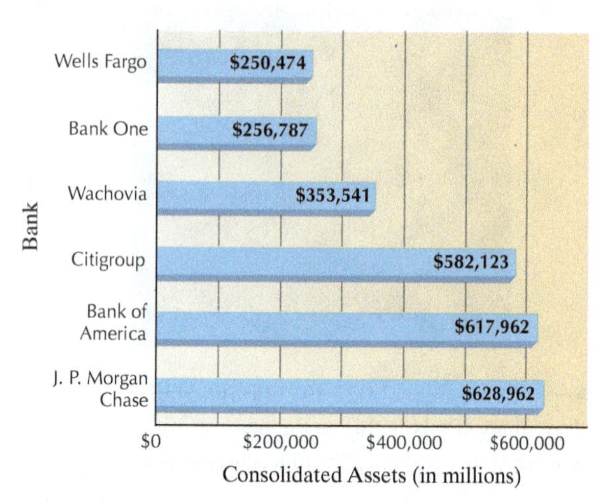

(*Source: Time Almanac 2006*)

76. Compute a company's net profit by completing the following business *skeletal profit and loss statement.*

Net sales	$430,000
− Cost of merchandise sold	− 175,000
Gross margin	$
− Operating expenses	− 135,000
Net profit	$

77. In Giza, Egypt, the pyramid of Khufu has a base that measures 230 meters by 230 meters, whereas the pyramid of Khafre has a base that measures 215 meters by 215 meters. In area, how much larger than the base of the pyramid of Khafre is that of Khufu? (*Source:* pbs.org)

78. Both a singles tennis court and a football field are rectangular in shape. A tennis court measures 78 feet by 27 feet, whereas a football field measures 360 feet by 160 feet. About how many times the area of a tennis court is that of a football field?

79. Richard Nixon ran for the U.S. presidency three times. According to the table below, which was greater—the increase from 1960 to 1972 in the number of votes he got or the increase from 1968 to 1972? (*Source: The World Almanac & Book of Facts 2000*)

Year	Number of Votes for Nixon
1960	34,108,546
1968	31,785,480
1972	47,165,234

80. On a business trip, a sales representative flew from Chicago to Los Angeles to Boston. The chart below shows the air distances in miles between these cities.

Air Distance	Chicago	Los Angeles	Boston
Chicago	—	1,745	1,042
Los Angeles	1,745	—	2,596
Boston	1,042	2,596	—

If the sales rep earned a frequent flier point for each mile flown, how many points did he earn?

81. The Tour de France is a 20-stage bicycle race held in France annually. The chart shows the distances for the first 10 stages of the 2006 Tour de France. (*Source:* Le Tour de France)

Stage	Distance (in kilometers)
1	183
2	223
3	216
4	215
5	219
6	184
7	52
8	177
9	170
10	193

a. What was the total distance covered in the first 10 stages?

b. The entire race covered a distance of 3,632 kilometers. How many kilometers were covered in the last 10 stages?

82. The number of students enrolled annually in public colleges in the United States from 2000 through 2005 is given in the table. (*Source:* National Center for Education Statistics)

Year	Enrollment
2000	11,753,000
2001	12,233,000
2002	12,752,000
2003	12,952,000
2004	13,092,000
2005	13,283,000

a. What was the average annual enrollment in the years 2000 through 2005?

b. The enrollments in 2006 and 2007 were 13,518,800 and 13,752,000, respectively. How would the average annual enrollment change if the enrollments for 2006 and 2007 were included?

83. Find the area of the figure.

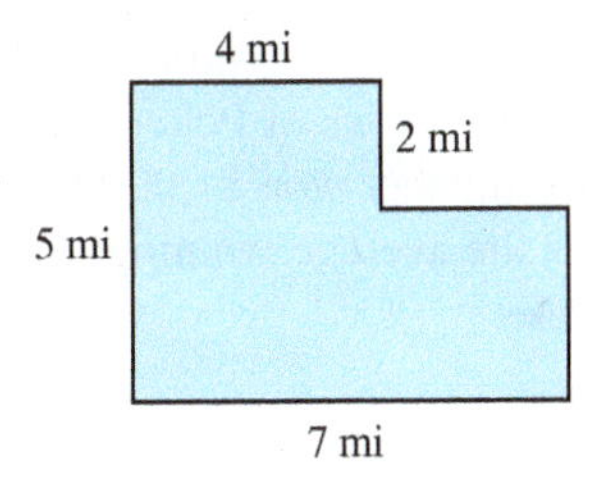

84. Find the perimeter of the figure.

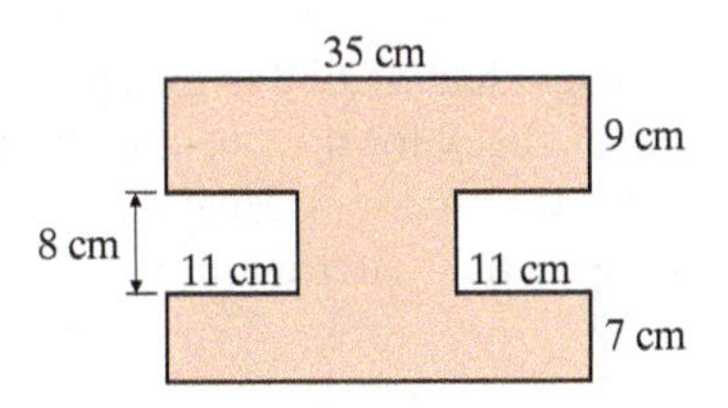

• *Check your answers on page A-3.*

Chapter 1 POSTTEST

FOR EXTRA HELP  Test solutions are found on the enclosed CD.

To see if you have already mastered the topics in this chapter, take this test.

1. Write two hundred twenty-five thousand, sixty-seven in standard form.

2. Underline the digit that occupies the ten thousands place in 1,768,405.

3. Write 1,205,007 in words.

4. Round 196,593 to the nearest hundred thousand.

5. Find the sum of 398 and 1,496.

6. Subtract 398 from 1,005.

7. Subtract: $2{,}000 - 1{,}853$

8. Multiply: 328×907

9. Compute: $\dfrac{23{,}923}{47}$

10. Find the quotient: $59\overline{)36{,}717}$

11. Evaluate: 5^4

12. Write $5 \cdot 5 \cdot 4 \cdot 4 \cdot 4$ using exponents.

Simplify.

13. $4 \cdot 9 + 3 \cdot 4^2$

14. $29 - 3^3 \cdot (10 - 9)$

Solve and check.

15. The two largest continents in the world are Asia and Africa. To the nearest hundred thousand square miles, Asia's area is 17,400,000 and Africa's is 11,700,000. How much larger is Asia than Africa? (*Source: The Time Almanac 2006*)

16. In the year 2005, the state of Kansas had about 64,000 farms with an average size of 732 acres. How many acres of land in Kansas were devoted to farming? (*Source:* U.S. Department of Agriculture, *Farms and Land in Farms*, 2006)

17. A part-time student had $1,679 in his checking account. He wrote a $625 check for tuition, a $546 check for rent, and a $39 check for groceries. How much money remained in the account after these checks cleared?

18. A total of $27,609,360 was paid out to the top twenty places at the World Series of Poker main event in 2005. The ninth-place finisher won $1,000,000. Was this amount above or below the average winnings? Explain. (*Source:* World Series of Poker)

19. A homeowner wishes to carpet the hallway shown below. If the cost of carpeting is about $10 per square foot, approximately how much will the carpeting cost?

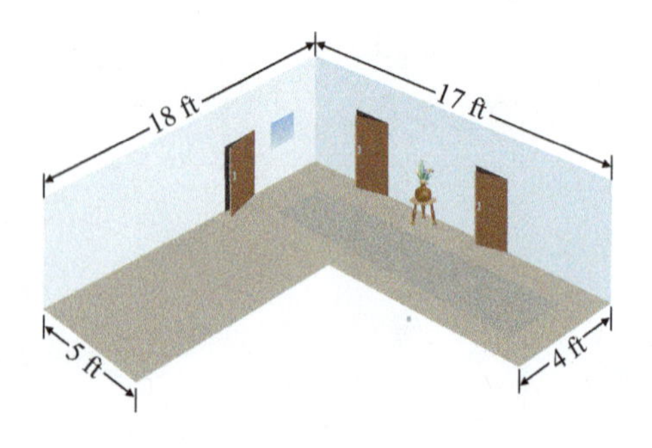

20. For breakfast, you have an 8-ounce (oz) serving of yogurt, a cup of black coffee, and 2 cups of pineapple juice. Based on the following table, how much more vitamin C do you need to reach the recommended 60 milligrams (mg)?

Food	Quantity	Vitamin C Content
Pineapple juice	1 cup	23 mg
Yogurt	8 oz	2 mg
Black coffee	1 cup	0 mg

• *Check your answers on page A-3.*

CHAPTER 2

Fractions

Fractions and Cooking

Using recipes in cooking has a very long and complicated history that goes back thousands of years. For instance, archaeologists have found Mesopotamian recipes written on clay tablets dating from 1700 B.C. Until the Industrial Revolution, at which time standard measurements and precise cooking directions were introduced, recipes gave just a list of ingredients and a general description for cooking a dish.

Today, it is much easier to follow recipes and to cook satisfying dishes because standard measurements and detailed instructions are given in recipes. Many recipes give fractional amounts of ingredients. As a result, cooks need to know how to use fractions.

For example, consider the ingredients given in a recipe for Italian-style meatloaf. The recipe, which serves 8 people, calls for $1\frac{1}{2}$ pounds of ground meat, of which $\frac{1}{3}$ is pork and $\frac{2}{3}$ is beef. To follow this recipe, a cook uses multiplying fractions to determine that for $1\frac{1}{2}$ pounds of ground meat, the amount of ground pork needed is $\frac{1}{3} \times 1\frac{1}{2}$, or $\frac{1}{2}$ pound, and the amount of ground beef needed is $\frac{2}{3} \times 1\frac{1}{2}$, or 1 pound. Multiplying fractions is also used if the cook wants to increase or decrease the number of servings that the recipe yields. (*Source:* http://www.foodtimeline.org)

Chapter 2 PRETEST

To see if you have already mastered the topics in this chapter, take this test.

1. Find all the factors of 20.

2. Express 72 as the product of prime factors.

3. What fraction does the shaded part of the diagram represent?

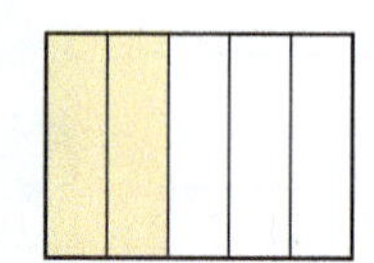

4. Write $20\frac{1}{3}$ as an improper fraction.

5. Express $\frac{31}{30}$ as a mixed number.

6. Write $\frac{9}{12}$ in simplest form.

7. What is the least common multiple of 10 and 4?

8. Which is greater, $\frac{1}{8}$ or $\frac{1}{9}$?

Add.

9. $\frac{1}{2} + \frac{7}{10}$

10. $7\frac{1}{3} + 5\frac{1}{2}$

Subtract.

11. $8\frac{1}{4} - 6$

12. $12\frac{1}{2} - 7\frac{7}{8}$

Multiply.

13. $2\frac{1}{3} \times 1\frac{1}{2}$

14. $\frac{5}{8} \times 96$

15. Divide: $3\frac{1}{3} \div 5$

16. Calculate: $2 + 1\frac{1}{3} \div \frac{4}{5}$

Solve. Write your answer in simplest form.

17. In 2006, the Pittsburgh Steelers won their fifth Super Bowl championship game. If 40 Super Bowl championship games have been played, what fraction of these games has this team won? (*Source:* National Football League)

18. In a biology class, three-fourths of the students received a passing grade. If there are 24 students in the class, how many students received failing grades?

19. Find the perimeter of the traffic island shown.

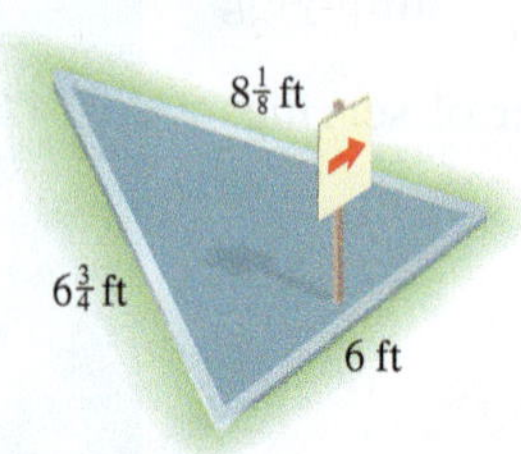

20. According to the nutrition information given, one serving of Honey Nut Cheerios® contains 22 grams of carbohydrates. If one serving is $\frac{3}{4}$ cup, what amount of carbohydrates is contained in $2\frac{1}{4}$ cups of Honey Nut Cheerios? (*Source:* General Mills)

• *Check your answers on page A-3.*

2.1 Factors and Prime Numbers

OBJECTIVES

- To identify prime and composite numbers
- To find the factors and prime factorization of a whole number
- To find the least common multiple of two or more numbers
- To solve word problems using factoring or the LCM

What Factors Mean and Why They Are Important

Recall that in a multiplication problem, the whole numbers that we are multiplying are called **factors** of the product. For instance, 2 is said to be a factor of 8 because $2 \cdot 4 = 8$. Likewise, 4 is a factor of 8.

Another way of expressing the same idea is in terms of division: We say that 8 is *divisible* by 2, meaning that there is a remainder 0 when we divide 8 by 2.

$$\frac{8}{2} = 4 \text{ R0}$$

Note that 1, 2, 4, and 8 are all factors of 8.

Although we factor whole numbers, a major application of factoring involves working with fractions, as we demonstrate in the next section.

Finding Factors

To identify the factors of a whole number, we divide the whole number by the numbers 1, 2, 3, 4, 5, 6, and so on, looking for remainders of 0.

EXAMPLE 1

Find all the factors of 6.

Solution

Starting with 1, we divide each whole number into 6.

$\frac{6}{1} = 6\text{ R}0$	$\frac{6}{2} = 3\text{ R}0$	$\frac{6}{3} = 2\text{ R}0$	$\frac{6}{4} = 1\text{ R}2$	$\frac{6}{5} = 1\text{ R}1$	$\frac{6}{6} = 1\text{ R}0$
A factor	A factor	A factor	Not a factor	Not a factor	A factor

In finding the factors of 6, we do not need to divide 6 by the numbers 7 or greater. The reason is that no number larger than 6 could divide evenly into 6, that is, divide into 6 with no remainder.

So the factors of 6 are 1, 2, 3, and 6. Note that

- 1 is a factor of 6 and
- 6 is a factor of 6.

PRACTICE 1

What are the factors of 7?

Tip For any whole number, both *the number itself* and *1* are always factors. Therefore, all whole numbers (except 1) have at least two factors.

When checking to see if one number is a factor of another, it is generally faster to use the following **divisibility tests** than to divide.

The number is divisible by	if
2	the ones digit is 0, 2, 4, 6, or 8, that is, if the number is even.
3	the sum of the digits is divisible by 3.
4	the number named by the last two digits is divisible by 4.
5	the ones digit is either 0 or 5.
6	the number is even and the sum of the digits is divisible by 3.
9	the sum of the digits is divisible by 9.
10	the ones digit is 0.

EXAMPLE 2

What are the factors of 45?

Solution Let's see if 45 is divisible by 1, 2, 3, and so on, using the divisibility tests wherever they apply.

Is 45 divisible by	Answer
1?	Yes, because 1 is a factor of any number; $\frac{45}{1} = 45$, so 45 is also a factor.
2?	No, because the ones digit is not even.
3?	Yes, because the sum of the digits, $4 + 5 = 9$, is divisible by 3; $\frac{45}{3} = 15$, so 15 is also a factor.
4?	No, because 4 will not divide into 45 evenly.
5?	Yes, because the ones digit is 5; $\frac{45}{5} = 9$, so 9 is also a factor.
6?	No, because 45 is not even.
7?	No, because $45 \div 7$ has remainder 3.
8?	No, because $45 \div 8$ has remainder 5.
9?	We already know that 9 is a factor.
10?	No, because the ones digit is not 0.

The factors of 45 are, therefore, 1, 3, 5, 9, 15, and 45.

Note that we really didn't have to check to see if 9 was a factor—we learned that it was when we checked for divisibility by 5. Also, because the factors were beginning to repeat with 9, there was no need to check numbers greater than 9.

PRACTICE 2

Find all the factors of 75.

EXAMPLE 3

Identify all the factors of 60.

Solution Let's check to see if 60 is divisible by 1, 2, 3, 4, and so on.

Is 60 divisible by	Answer
1?	Yes, because 1 is a factor of all numbers; $\frac{60}{1} = 60$, so 60 is also a factor.
2?	Yes, because the ones digit is even; $\frac{60}{2} = 30$, so 30 is also a factor.
3?	Yes, because the sum of the digits, $6 + 0 = 6$, is divisible by 3; $\frac{60}{3} = 20$, so 20 is also a factor.
4?	Yes, because 4 will divide into 60 evenly; $\frac{60}{4} = 15$, so 15 is also a factor.
5?	Yes, because the ones digit is 0; $\frac{60}{5} = 12$, so 12 is also a factor.
6?	Yes, because the number is even, and the sum of the digits is divisible by 3; $\frac{60}{6} = 10$, so 10 is also a factor.
7?	No, because $60 \div 7$ has remainder 4.
8?	No, because $60 \div 8$ has remainder 4.
9?	No, because the sum of the digits, $6 + 0 = 6$, is not divisible by 9.
10?	We already know that 10 is a factor.

The factors of 60 are, therefore, 1, 2, 3, 4, 5, 6, 10, 12, 15, 20, 30, and 60.

PRACTICE 3

What are the factors of 90?

EXAMPLE 4

A presidential election takes place in the United States every year that is a multiple of 4. Was there a presidential election in 1866? Explain.

Solution The question is: Does 4 divide into 1866 evenly? Using the divisibility test for 4, we check whether 66 is a multiple of 4.

$$\frac{66}{4} = 16 \text{ R2}$$

Because $\frac{66}{4}$ has remainder 2, 4 is not a factor of 1866. So there was no presidential election in 1866.

PRACTICE 4

The doctor instructs a patient to take a pill every 3 hours. If the patient took a pill at 8:00 this morning, should she take one tomorrow at the same time? Explain.

Identifying Prime and Composite Numbers

Now let's discuss the difference between prime numbers and composite numbers.

Definitions

A **prime number** is a whole number that has exactly two different factors: itself and 1.

A **composite number** is a whole number that has more than two factors.

Note that the numbers 0 and 1 are neither prime nor composite. But every whole number greater than 1 is either prime or composite, depending on its factors.

For instance, 5 is prime because its only factors are 1 and 5. But 8 is composite because it has more than two factors (it has four factors: 1, 2, 4, and 8).

Let's practice distinguishing between primes and composites.

EXAMPLE 5

Indicate whether each number is prime or composite.

a. 2 **b.** 78 **c.** 51 **d.** 19 **e.** 31

Solution

a. The only factors of 2 are 1 and 2. Therefore, 2 is prime.

b. Because 78 is even, it is divisible by 2. Having 2 as an "extra" factor—in addition to 1 and 78—means that 78 is composite. Do you see why all even numbers, except for 2, are composite?

c. Using the divisibility test for 3, we see that 51 is divisible by 3, because the sum of the digits 5 and 1, or 6, is divisible by 3. Because 51 has more than two factors, it is composite.

d. The only factors of 19 are itself and 1. Therefore, 19 is prime.

e. Because 31 has no factors other than itself and 1, it is prime.

PRACTICE 5

Decide whether each number is prime or composite.

a. 3 **b.** 57 **c.** 29

d. 34 **e.** 17

Finding the Prime Factorization of a Number

Every composite number can be written as the product of prime factors. This product is called its **prime factorization.** For instance, the prime factorization of 12 is $2 \cdot 2 \cdot 3$.

Definition

The **prime factorization** of a whole number is the number written as the product of its prime factors.

Being able to find the prime factorization of a number is an important skill to have for working with fractions, as we show later in this chapter. A good way to find the prime factorization of a number is by making a **factor tree,** as illustrated in Example 6.

EXAMPLE 6

Write the prime factorization of 72.

Solution We start building a factor tree for 72 by dividing 72 by the smallest prime, 2. Because 72 is $2 \cdot 36$, we write both 2 and 36 underneath the 72. Then we circle the 2 because it is prime.

72
2 36 ← Continue to divide by 2.
2 18 ← Continue to divide by 2.
2 9 ← 9 cannot be divided by 2, so we try 3.
3 3 ← 3 is prime, so we stop dividing.

Next we divide 36 by 2, writing both 2 and 18, and circling 2 because it is prime. Below the 18, we write 2 and 9, again circling the 2. Because 9 is not divisible by 2, we divide it by the next smallest prime, 3. We continue this process until all the factors in the bottom row are prime. The prime factorization of 72 is the product of the circled factors.

$$72 = 2 \times 2 \times 2 \times 3 \times 3$$

We can also write this prime factorization as $2^3 \times 3^2$.

PRACTICE 6

Write the prime factorization of 56, using exponents.

EXAMPLE 7

Express 80 as the product of prime factors.

Solution The factor tree method for 80 is as shown.

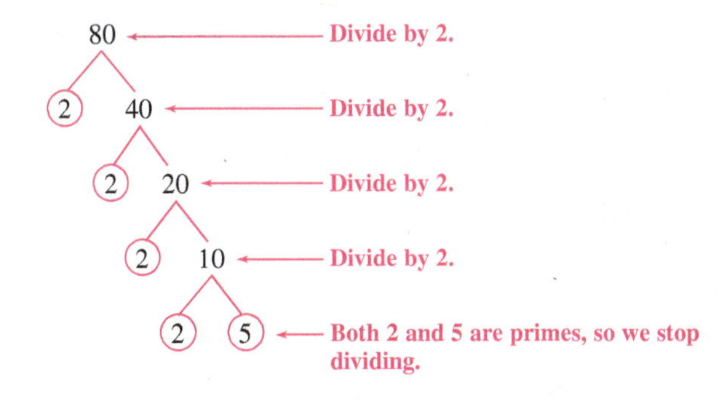

The prime factorization of 80 is $2 \times 2 \times 2 \times 2 \times 5$, or $2^4 \times 5$.

PRACTICE 7

What is the prime factorization of 75?

Finding the Least Common Multiples

The *multiples* of a number are the products of that number and the whole numbers. For instance, some multiples of 5 are the following:

$$\underbrace{0}_{0 \times 5} \quad \underbrace{5}_{1 \times 5} \quad \underbrace{10}_{2 \times 5} \quad \underbrace{15}_{3 \times 5}$$

A number that is a multiple of two or more numbers is called a *common multiple* of these numbers. To find the common multiples of 6 and 8, we first list the multiples of 6 and the multiples of 8 separately.

- The multiples of 6 are 0, 6, 12, 18, 24, 30, 36, 42, 48, 54, 60, ...
- The multiples of 8 are 0, 8, 16, 24, 32, 40, 48, 56, 64, ...

So the common multiples of 6 and 8 are 0, 24, 48, ... Of the nonzero common multiples, the *least* common multiple of 6 and 8 is 24.

Definition

The **least common multiple (LCM)** of two or more whole numbers is the smallest nonzero whole number that is a multiple of each number.

A shortcut for finding the LCM—often faster than listing multiples—involves prime factorization.

To Compute the Least Common Multiple (LCM)

- Find the prime factorization of each number.
- Identify the prime factors that appear in each factorization.
- Multiply these prime factors, using each factor the greatest number of times that it occurs in any of the factorizations.

EXAMPLE 8

Find the LCM of 8 and 12.

Solution We first find the prime factorization of each number.

$$8 = 2 \times 2 \times 2 = 2^3 \qquad 12 = 2 \times 2 \times 3 = 2^2 \times 3$$

The factor 2 appears *three times* in the factorization of 8 and *twice* in the factorization of 12, so it must be included three times in forming the least common multiple. Also, the factor 3 appears once in the prime factorization of 12.

The highest power of 2

$$\text{LCM} = 2^3 \times 3 = 8 \times 3 = 24$$

As always, it is a good idea to check that our answer makes sense. We do so by verifying that 8 and 12 really are factors of 24.

PRACTICE 8

What is the LCM of 9 and 6?

EXAMPLE 9

Find the LCM of 5 and 9.

Solution First, we write each number as the product of primes.

$$5 = 5 \qquad 9 = 3 \times 3 = 3^2$$

To find the LCM, we multiply the highest power of each prime.

$$\text{LCM} = 5 \times 3^2 = 5 \times 9 = 45$$

So the LCM of 5 and 9 is 45. Note that 45 is also the product of 5 and 9.

PRACTICE 9

Find the LCM for 3 and 22.

Tip If two or more numbers have no common factor (other than 1), the LCM is their product. If one number is a multiple of another number, then their LCM is the larger of the two numbers.

Now let's find the LCM of three numbers.

EXAMPLE 10

Find the LCM of 3, 5, and 6.

Solution First, we find the prime factorizations of these three numbers.

$$3 = 3 \qquad 5 = 5 \qquad 6 = 2 \times 3$$

The LCM is therefore the product $2 \times 3 \times 5$, which is 30. Note that 30 is a multiple of 3, 5, and 6, which supports our answer.

PRACTICE 10

Find the LCM of 2, 3, and 4.

EXAMPLE 11

A gym that is open every day of the week offers aerobics classes every third day and yoga classes every fourth day. A student took both classes this morning. In how many days will the gym offer both classes on the same day?

Solution To answer this question, we ask: What is the LCM of 3 and 4? As usual, we begin by finding prime factorizations.

$$3 = 3 \qquad 4 = 2 \times 2 = 2^2$$

To find the LCM, we multiply 3 by 2^2.

$$\text{LCM} = 2^2 \times 3 = 12$$

So both classes will be offered again on the same day in 12 days.

PRACTICE 11

Suppose that a Senate seat and a House of Representatives seat were both filled this year. If the Senate seat is filled every 6 years and the House seat every 2 years, in how many years will both seats be up for election?

2.1 Exercises

FOR EXTRA HELP

Mathematically Speaking

Fill in each blank with the most appropriate term or phrase from the given list.

division	least common multiple	composite	factor tree
divisibility	prime	remainders	common multiple
prime factorization	factors	multiples	

1. 1, 2, 3, 5, 6, 10, 15, and 30 are _______ of 30.

2. A(n) _______ number is a whole number that has more than two factors.

3. A(n) _______ number has exactly two different factors: itself and 1.

4. The _______ of two or more numbers is the smallest nonzero number that is a multiple of each number.

5. A number written as the product of its prime factors is called its _______.

6. The _______ test for 10 is to check if the ones digit is 0.

List all the factors of each number.

7. 21 **8.** 10 **9.** 17 **10.** 9

11. 12 **12.** 15 **13.** 31 **14.** 47

15. 36 **16.** 35 **17.** 29 **18.** 73

19. 100 **20.** 98 **21.** 28 **22.** 48

Indicate whether each number is prime or composite. If it is composite, identify a factor other than the number itself and 1.

23. 13 **24.** 7 **25.** 16 **26.** 24 **27.** 49

28. 75 **29.** 11 **30.** 31 **31.** 81 **32.** 45

Write the prime factorization of each number.

33. 8 **34.** 10 **35.** 49 **36.** 14

37. 24 **38.** 18 **39.** 50 **40.** 40

41. 77 **42.** 63 **43.** 51 **44.** 57

45. 25 **46.** 49 **47.** 32 **48.** 64

49. 21 **50.** 22 **51.** 104 **52.** 105

53. 121 **54.** 169 **55.** 142 **56.** 62

57. 100

58. 200

59. 125

60. 90

61. 135

62. 400

Find the LCM in each case.

63. 3 and 15

64. 9 and 12

65. 8 and 10

66. 4 and 6

67. 9 and 30

68. 20 and 21

69. 10 and 11

70. 15 and 60

71. 18 and 24

72. 30 and 150

73. 40 and 180

74. 100 and 90

75. 12, 5, and 50

76. 2, 8, and 10

77. 4, 7, and 12

78. 2, 3, and 5

79. 3, 5, and 7

80. 6, 8, and 12

81. 5, 15, and 20

82. 8, 24, and 56

Mixed Practice

Solve.

83. Write the prime factorization of 75.

84. Is 63 prime or composite? If it is composite, identify a factor other than the number itself and 1.

85. List all the factors of 72.

86. Find the LCM of 5, 10, and 12.

Applications

Solve.

87. The federal government conducts a census every year that is a multiple of 10. Explain whether there was a census in

a. 1995.

b. 1990.

88. Because of production considerations, the number of pages in a book that you are writing must be a multiple of 4. Can the book be

a. 196 pages long?

b. 198 pages long?

89. In 2006, the men's World Cup soccer tournament was held in Munich, Germany. If the tournament is held every 4 years, will there be a tournament in 2036? (*Source:* FIFA World Cup; Soccer Hall of Fame)

90. A car manufacturer recommends changing the oil every 3,000 miles. Would an oil change be recommended at 21,000 miles? Explain.

91. There are 9 players on a baseball team and 11 players on a soccer team. What is the smallest number of students in a college that can be split evenly into either baseball or soccer teams?

92. The Fields Medal, the highest scientific award for mathematicians, is awarded every 4 years. The Dantzig Prize, an achievement award in the field of mathematical programming, is awarded every 3 years. If both were given in 2006, in what year will both be given again? (*Source:* Eric W. Weisstein et al., "Fields Medals," Math World: A Wolfram Web Resource; Mathematical Programming Society)

93. Two friends work in a hospital. One gets a day off every 5 days, and the other every 6 days. If they were both off today, in how many days will they again both be off?

94. A family must budget for life insurance premiums every 6 months, car insurance premiums every 3 months, and payments for a home security system every 4 months. If all these bills were due this month, in how many months will they again all fall due?

MINDSTRETCHERS

Historical

1. The eighteenth-century mathematician Christian Goldbach made several famous conjectures (guesses) about prime numbers. One of these conjectures states: Every odd number greater than 7 can be expressed as the sum of three odd prime numbers. For instance, 11 can be expressed as $3 + 3 + 5$. Write the following odd numbers as the sum of three odd primes.

   a. $57 = \square + \square + \square$ **b.** $81 = \square + \square + \square$

Mathematical Reasoning

2. What is the smallest whole number divisible by every whole number from 1 to 10?

Critical Thinking

3. Choose a three-digit number, say, 715. Find three prime numbers so that, when 715 is multiplied by the product of the three prime numbers, the product of all four numbers is 715,715.

2.2 Introduction to Fractions

OBJECTIVES

- To read and write fractions and mixed numbers
- To write improper fractions as mixed numbers and mixed numbers as improper fractions
- To find equivalent fractions and to write fractions in simplest form
- To compare fractions
- To solve word problems with fractions

What Fractions Are and Why They Are Important

A fraction can mean *part of a whole*. Just as a whole number answers the question How many?, a fraction answers the question What part of? Every day we use fractions in this sense. For example, we can speak of *two-thirds* of a class (meaning two of every three students) or *three-fourths* of a dollar (indicating that we have split a dollar into four equal parts and have taken three of these parts).

A fraction can also mean *the quotient of two whole numbers*. In this sense, the fraction $\frac{3}{4}$ tells us what we get when we divide the whole number 3 by the whole number 4.

Definition

A **fraction** is any number that can be written in the form $\frac{a}{b}$, where a and b are whole numbers and b is nonzero.

From this definition, $\frac{1}{2}, \frac{3}{9}, \frac{6}{5}, \frac{8}{2}$, and $\frac{0}{1}$ are all fractions.

When written as $\frac{a}{b}$, a fraction has three components: $\frac{\text{Numerator}}{\text{Denominator}}$ ← Fraction line

- The **denominator** (on the bottom) stands for the number of parts into which the whole is divided.
- The **numerator** (on top) tells us how many parts of the whole the fraction contains.
- The **fraction line** separates the numerator from the denominator and stands for "out of" or "divided by."

Alternatively, a fraction can be represented as either a decimal or a percent. We discuss decimals and percents in Chapters 3 and 6.

Fraction Diagrams and Proper Fractions

Diagrams help us work with fractions. The fraction three-fourths, or $\frac{3}{4}$, is represented by the shaded part in each of the following diagrams:

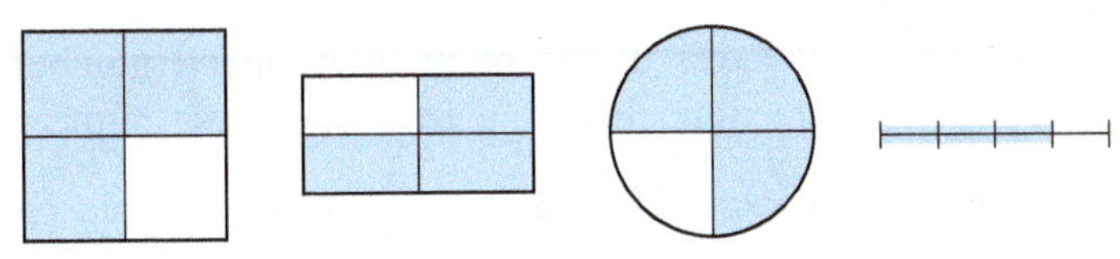

Note that in each diagram the whole has been divided into 4 *equal* parts, with 3 of the parts shaded.

The number $\frac{3}{4}$ is an example of a **proper fraction** because its numerator is smaller than its denominator. Let's consider some other examples of proper fractions.

EXAMPLE 1

In the diagram, what does the shaded portion represent?

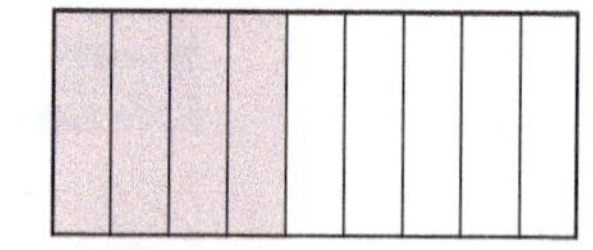

Solution In this diagram, the whole is divided into 9 equal parts, so the denominator of the fraction shown is 9. Four of these parts are shaded, so the numerator is 4. The diagram represents the fraction $\frac{4}{9}$.

PRACTICE 1

The diagram illustrates what fraction?

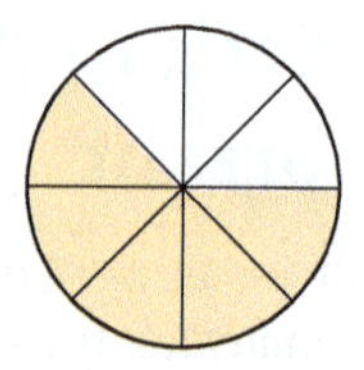

EXAMPLE 2

A college accepted 147 out of 341 applicants for admission into the nursing program. What fraction of the applicants were accepted into this program?

Solution Since there was a total of 341 applicants, the denominator of our fraction is 341. Because 147 of the applicants were accepted, 147 is the numerator. So the college accepted $\frac{147}{341}$ of the applicants into the nursing program.

PRACTICE 2

During a 30-minute television program, 7 minutes were devoted to commercials. What fraction of the time was for commercials?

EXAMPLE 3

The U.S. Senate approved a foreign-aid spending bill by a vote of 83 to 17. What fraction of the senators voted against the bill?

Solution First, we find the total number of senators. Because 83 senators voted for the bill and 17 voted against it, the total number of senators is 83 + 17, or 100. So $\frac{17}{100}$ of the senators voted against the bill.

PRACTICE 3

Through the 2005 baseball season, the American League had won 60 World Series championships, whereas the National League had won 41. What fraction of the World Series championships were *not* won by the American League? (*Source:* http://mlb.com)

Mixed Numbers and Improper Fractions

On many jobs, if you work overtime, the rate of pay increases to one-and-a-half times the regular rate. A number such as $1\frac{1}{2}$, with a whole number part and a proper fraction part, is called a mixed number. A mixed number can also be expressed as an improper fraction, that is, a fraction whose numerator is greater than or equal to its denominator. The number $\frac{3}{2}$ is an example of an improper fraction.

Diagrams help us understand that mixed numbers and improper fractions are different forms of the same numbers, as Example 4 illustrates.

EXAMPLE 4

Draw diagrams to show that $2\frac{1}{3} = \frac{7}{3}$.

Solution First, represent the mixed number and the improper fraction in diagrams.

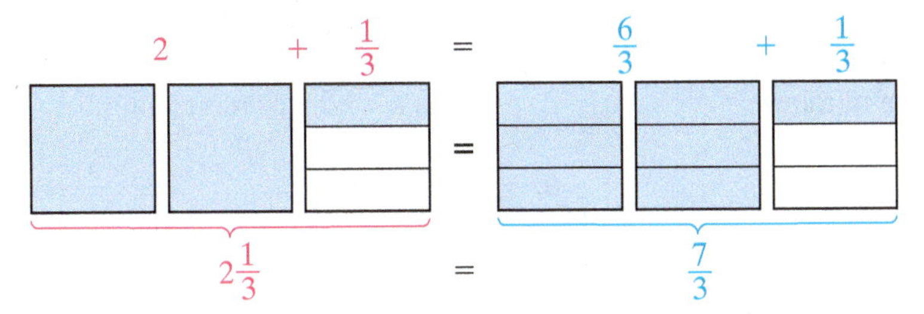

Both diagrams represent $2 + \frac{1}{3}$, so the numbers $2\frac{1}{3}$ and $\frac{7}{3}$ must be equal.

PRACTICE 4

By means of diagrams, explain why $1\frac{2}{3} = \frac{5}{3}$.

In Example 4 each unit (or square) corresponds to 1 whole, which is also three-thirds. That is why the total number of *thirds* in $2\frac{1}{3}$ is $(2 \times 3) + 1$, or 7. The number of *wholes* in $\frac{7}{3}$ is 2 wholes, with $\frac{1}{3}$ of a whole left over. We can generalize these observations into two rules.

To Change a Mixed Number to an Improper Fraction

- Multiply the denominator of the fraction by the whole-number part of the mixed number.
- Add the numerator of the fraction to this product.
- Write this sum over the denominator to form the improper fraction.

EXAMPLE 5

Write each of the following mixed numbers as an improper fraction.

a. $3\frac{2}{9}$ **b.** $12\frac{1}{4}$

Solution

a. $3\frac{2}{9} = \frac{(9 \times 3) + 2}{9}$ Multiply the denominator 9 by the whole number 3, adding the numerator 2. Place over the denominator.

$= \frac{27 + 2}{9} = \frac{29}{9}$ Simplify the numerator.

b. $12\frac{1}{4} = \frac{(4 \times 12) + 1}{4}$

$= \frac{48 + 1}{4} = \frac{49}{4}$

PRACTICE 5

Express each mixed number as an improper fraction.

a. $5\frac{1}{3}$ **b.** $20\frac{2}{5}$

To Change an Improper Fraction to a Mixed Number

- Divide the numerator by the denominator.
- If there is a remainder, write it over the denominator.

EXAMPLE 6

Write each improper fraction as a mixed or whole number.

a. $\frac{11}{2}$ **b.** $\frac{20}{20}$ **c.** $\frac{42}{5}$

Solution

a. $\frac{11}{2} = 2\overline{)11}$ with quotient 5 R1 — Divide the numerator by the denominator.

$\frac{11}{2} = 5\frac{1}{2}$ — Write the remainder over the denominator.

In other words, 5 R1 means that in $\frac{11}{2}$ there are 5 wholes with $\frac{1}{2}$ of a whole left over.

b. $\frac{20}{20} = 1$

c. $\frac{42}{5} = 8\frac{2}{5}$

PRACTICE 6

Express as a whole or mixed number.

a. $\frac{4}{2}$ **b.** $\frac{50}{9}$ **c.** $\frac{8}{3}$

Changing an improper fraction to a mixed number is important when we are dividing whole numbers: It allows us to express any remainder as a fraction. Previously, we would have said that the problems $2\overline{)7}$ and $4\overline{)13}$ both have the answer 3 R1. But by interpreting these problems as improper fractions, we see that their answers are different.

$$\frac{7}{2} = 3\frac{1}{2} \quad \text{but} \quad \frac{13}{4} = 3\frac{1}{4}$$

When a number is expressed as a mixed number, we know its size more readily than when it is expressed as an improper fraction. For instance, consider the mixed number $11\frac{7}{8}$. We immediately see that it is larger than 11 and smaller than 12 (that is, between 11 and 12). We could not reach this conclusion so easily if we were to examine only $\frac{95}{8}$, its improper form. However, there are situations—when we multiply or divide fractions—in which the use of improper fractions is preferable.

Equivalent Fractions

Some fractions that at first glance appear to be different from one another are really the same.

For instance, suppose that we cut a pizza into 8 equal slices, and then eat 4 of the slices. The shaded portion of the diagram at the right represents the amount eaten. Can you explain why in this diagram the fractions $\frac{4}{8}$ and $\frac{1}{2}$ describe the same part of the whole pizza? We say that these fractions are **equivalent**.

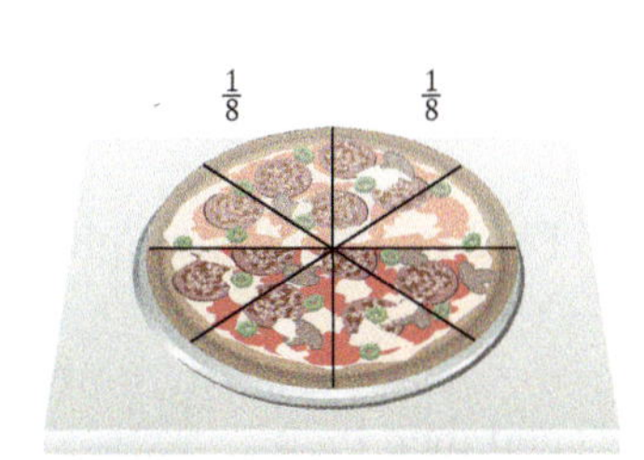

Any fraction has infinitely many equivalent fractions. To see why, let's consider the fraction $\frac{1}{3}$. We can draw different diagrams representing one-third of a whole.

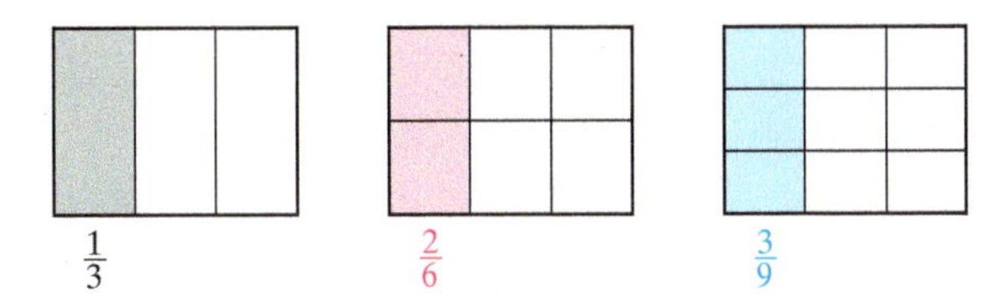

All the shaded portions of the diagrams are identical, so $\frac{1}{3} = \frac{2}{6} = \frac{3}{9}$.

A faster way to generate fractions equivalent to $\frac{1}{3}$ is to multiply both its numerator and denominator by the *same* whole number. Any whole number except 0 will do.

$$\frac{1}{3} = \frac{1 \cdot 2}{3 \cdot 2} = \frac{2}{6}$$

$$\frac{1}{3} = \frac{1 \cdot 3}{3 \cdot 3} = \frac{3}{9}$$

$$\frac{1}{3} = \frac{1 \cdot 4}{3 \cdot 4} = \frac{4}{12}$$

$$\frac{1}{3} = \frac{1 \cdot 5}{3 \cdot 5} = \frac{5}{15}$$

So $\frac{1}{3} = \frac{2}{6} = \frac{3}{9} = \frac{4}{12} = \frac{5}{15} = \dots$.

Can you explain how you would generate fractions equivalent to $\frac{3}{5}$?

To Find an Equivalent Fraction

Multiply the numerator and denominator of $\frac{a}{b}$ by the same whole number n,

$$\frac{a}{b} = \frac{a \cdot n}{b \cdot n},$$

where both b and n are nonzero.

An important property of equivalent fractions is that their **cross products** are always equal.

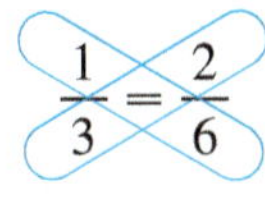

In this case, $1 \cdot 6 = 3 \cdot 2 = 6$

EXAMPLE 7

Find two fractions equivalent to $\frac{1}{7}$.

Solution Let's multiply the numerator and denominator by 2 and then by 6.

$$\frac{1}{7} = \frac{1 \cdot 2}{7 \cdot 2} = \frac{2}{14} \quad \text{and} \quad \frac{1}{7} = \frac{1 \cdot 6}{7 \cdot 6} = \frac{6}{42}$$

We use cross products to check.

$$\frac{1}{7} \stackrel{?}{=} \frac{2}{14}$$
$$1 \cdot 14 \stackrel{?}{=} 7 \cdot 2$$
$$14 \stackrel{\checkmark}{=} 14$$

So $\frac{1}{7}$ and $\frac{2}{14}$ are equivalent.

$$\frac{1}{7} \stackrel{?}{=} \frac{6}{42}$$
$$1 \cdot 42 \stackrel{?}{=} 7 \cdot 6$$
$$42 \stackrel{\checkmark}{=} 42$$

So $\frac{1}{7}$ and $\frac{6}{42}$ are equivalent.

PRACTICE 7

Identify three fractions equivalent to $\frac{2}{5}$.

EXAMPLE 8

Write $\frac{3}{7}$ as an equivalent fraction whose denominator is 35.

Solution $\frac{3}{7} = \frac{3 \cdot 5}{7 \cdot 5} = \frac{15}{35}$ Multiply the numerator and denominator by 5.

Therefore, $\frac{15}{35}$ is equivalent to $\frac{3}{7}$. To check, we find the cross products: Both $3 \cdot 35$ and $7 \cdot 15$ equal 105.

PRACTICE 8

Express $\frac{5}{8}$ as a fraction whose denominator is 72.

Writing a Fraction in Simplest Form

In the preceding section, we showed that $\frac{4}{8}$ and $\frac{1}{2}$ are equivalent fractions. Note that we could have written $\frac{4}{8}$ in its simplest form by dividing its numerator and denominator by 4.

$$\frac{4}{8} = \frac{4 \div 4}{8 \div 4} = \frac{1}{2}$$ 1 and 2 have no common factor except 1.

A fraction is said to be in **simplest form** (or **reduced to lowest terms**) when the only common factor of its numerator and its denominator is 1.

To Simplify (Reduce) a Fraction

Divide the numerator and denominator of $\frac{a}{b}$ by the same whole number n,

$$\frac{a}{b} = \frac{a \div n}{b \div n},$$

where both b and n are nonzero.

EXAMPLE 9

Express $\frac{3}{15}$ in simplest form.

Solution To reduce this fraction, we can divide both its numerator and its denominator by 3.

$$\frac{3}{15} = \frac{3 \div 3}{15 \div 3} = \frac{1}{5}$$

To be sure that we have not made an error, let's check whether the cross products are equal: $3 \cdot 5 = 15$ and $1 \cdot 15 = 15$.

PRACTICE 9

Reduce $\frac{14}{21}$ to lowest terms.

In Example 9, we simplified the fraction $\frac{3}{15}$. How would you simplify the mixed number $2\frac{3}{15}$?

To reduce a fraction to lowest terms, we divide the numerator and denominator by all the factors that they have in common. To find these common factors, it is often helpful to express both the numerator and denominator as the product of prime factors. We can then divide out or *cancel* all common factors.

EXAMPLE 10

Write $\frac{42}{28}$ in lowest terms.

Solution $\frac{42}{28} = \frac{2 \cdot 3 \cdot 7}{2 \cdot 2 \cdot 7}$ Express the numerator and denominator as the product of primes.

$= \frac{\overset{1}{\cancel{2}} \cdot 3 \cdot \overset{1}{\cancel{7}}}{\underset{1}{\cancel{2}} \cdot 2 \cdot \underset{1}{\cancel{7}}}$ Divide out the common factors, noting that 1 remains.

$= \frac{3}{2}$ Multiply the remaining factors.

PRACTICE 10

Simplify $\frac{42}{18}$.

EXAMPLE 11

Suppose that a couple's annual income is \$75,000. If they pay \$9,000 for rent and \$3,000 for food per year, rent and food account for what fraction of their income? Simplify the answer.

Solution First, we must find the total part of the income that is paid for rent and food per year.

$$\underset{\text{Rent}}{\$9{,}000} + \underset{\text{Food}}{\$3{,}000} = \underset{\text{Total part}}{\$12{,}000}$$

The total part is \$12,000 and the whole is \$75,000, so the fraction is $\frac{12{,}000}{75{,}000}$. We can simplify this fraction in the following way:

$$\frac{12{,}000}{75{,}000} = \frac{12{,}0\not{0}\not{0}}{75{,}0\not{0}\not{0}} = \frac{12}{75}$$

Note that canceling a 0 is the same as dividing by 10.

$$= \frac{3 \cdot 4}{3 \cdot 25} = \frac{\overset{1}{\not{3}} \cdot 4}{\underset{1}{\not{3}} \cdot 25} = \frac{4}{25}$$

Therefore, $\frac{4}{25}$ of the couple's income goes for rent and food.

PRACTICE 11

An acre is a unit of area approximately equal to 4,800 square yards. A developer is selling parcels of land of 50 yards by 30 yards. What fraction of an acre is each parcel?

Comparing Fractions

Some situations require us to *compare* fractions, that is, to rank them in order of size.

For instance, suppose that $\frac{5}{8}$ of one airline's flights arrive on time, in contrast to $\frac{3}{5}$ of another airline's flights. To decide which airline has a better record for on time arrivals, we need to compare the fractions.

Or to take another example, suppose that the drinking water in your home, according to a lab report, has 2 parts per million (ppm) of lead. Is the water safe to drink if the federal limit on lead in drinking water is 15 parts per billion (ppb)? Again, we need to compare fractions.

One way to handle such problems is to draw diagrams corresponding to the fractions in question. The larger fraction corresponds to the larger shaded region.

For instance, the diagrams to the right show that $\frac{3}{4}$ is greater than $\frac{1}{4}$. The symbol $>$ stands for "greater than."

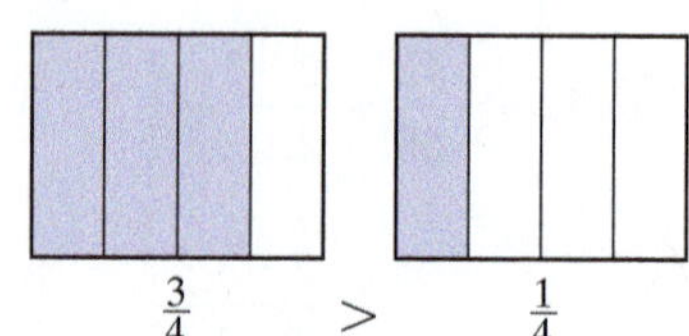

$\frac{3}{4} > \frac{1}{4}$

Both $\frac{3}{4}$ and $\frac{1}{4}$ have the same denominator, so we can rank them simply by comparing their numerators.

$$\frac{3}{4} > \frac{1}{4} \quad \text{because} \quad 3 > 1$$

For **like fractions**, the fraction with the larger numerator is the larger fraction.

Definitions

Like fractions are fractions with the same denominator.

Unlike fractions are fractions with different denominators.

To Compare Fractions

- If the fractions are like, compare their numerators.
- If the fractions are unlike, write them as equivalent fractions with the same denominator and then compare their numerators.

EXAMPLE 12

Compare $\frac{7}{15}$ and $\frac{4}{9}$.

Solution These fractions are unlike because they have different denominators. Therefore, we need to express them as equivalent fractions having the same denominator. But what should that denominator be?

One common denominator that we can use is the *product of the denominators:* $15 \cdot 9 = 135$.

$$\frac{7}{15} = \frac{7 \cdot 9}{15 \cdot 9} = \frac{63}{135}$$ $135 = 15 \cdot 9$, so the new numerator is $7 \cdot 9$ or 63.

$$\frac{4}{9} = \frac{4 \cdot 15}{9 \cdot 15} = \frac{60}{135}$$ $135 = 9 \cdot 15$, so the new numerator is $4 \cdot 15$ or 60.

Next, we compare the numerators of the like fractions that we just found.

Because $63 > 60$, $\frac{63}{135} > \frac{60}{135}$. Therefore, $\frac{7}{15} > \frac{4}{9}$.

Another common denominator that we can use is the least common multiple of the denominators.

$$15 = 3 \times 5 \qquad 9 = 3 \times 3 = 3^2$$

The LCM is $3^2 \times 5 = 9 \times 5 = 45$. We then compute the equivalent fractions.

$$\frac{7}{15} = \frac{7 \cdot 3}{15 \cdot 3} = \frac{21}{45}$$ Multiply the numerator and denominator by 3.

$$\frac{4}{9} = \frac{4 \cdot 5}{9 \cdot 5} = \frac{20}{45}$$ Multiply the numerator and denominator by 5.

Because $\frac{21}{45} > \frac{20}{45}$, we know that $\frac{7}{15} > \frac{4}{9}$.

PRACTICE 12

Which is larger, $\frac{13}{24}$ or $\frac{11}{16}$?

Note that in Example 12 we computed the LCM of the two denominators. This type of computation is used frequently in working with fractions.

Definition

For two or more fractions, their **least common denominator** (LCD) is the least common multiple of their denominators.

In Example 13, pay particular attention to how we use the LCD.

EXAMPLE 13

Order from smallest to largest: $\frac{3}{4}, \frac{7}{10}$, and $\frac{29}{40}$

Solution Because these fractions are unlike, we need to find equivalent fractions with a common denominator. Let's use their LCD as that denominator.

$$4 = 2 \times 2 = 2^2$$
$$10 = 2 \times 5$$
$$40 = 2 \times 2 \times 2 \times 5 = 2^3 \times 5$$

The LCD $= 2^3 \times 5 = 8 \times 5 = 40$. Check: 4 and 10 are both factors of 40.

We write each fraction with a denominator of 40.

$$\frac{3}{4} = \frac{3 \cdot 10}{4 \cdot 10} = \frac{30}{40} \qquad \frac{7}{10} = \frac{7 \cdot 4}{10 \cdot 4} = \frac{28}{40} \qquad \frac{29}{40} = \frac{29}{40}$$

Then we order the fractions from smallest to largest. (The symbol $<$ stands for "less than.")

$$\frac{28}{40} < \frac{29}{40} < \frac{30}{40} \quad \text{or} \quad \frac{7}{10} < \frac{29}{40} < \frac{3}{4}$$

PRACTICE 13

Arrange $\frac{9}{10}, \frac{23}{30}$, and $\frac{8}{15}$ from smallest to largest.

EXAMPLE 14

About $\frac{7}{10}$ of Earth's surface is covered by water and $\frac{1}{20}$ is covered by desert. Does water or desert cover more of Earth?

Solution We need to compare $\frac{7}{10}$ with $\frac{1}{20}$. The LCD is 20.

$$\frac{7}{10} = \frac{14}{20}$$
$$\frac{1}{20} = \frac{1}{20}$$

Since $\frac{14}{20} > \frac{1}{20}, \frac{7}{10} > \frac{1}{20}$. Therefore, water covers more of Earth than desert does.

PRACTICE 14

In 2005, about $\frac{1}{22}$ of the commercial radio stations in the United States had a top-40 format, and $\frac{2}{11}$ had a country format. In that year, were there more top-40 stations or country stations? (*Source: The M Street Radio Directory*)

2.2 Exercises

FOR EXTRA HELP       

PRACTICE WATCH DOWNLOAD READ REVIEW

Mathematically Speaking

Fill in each blank with the most appropriate term or phrase from the given list.

improper fraction	proper fraction	mixed
greatest common factor	simplify	composite
convert	least common denominator	
equivalent	like fractions	

1. A fraction whose numerator is smaller than its denominator is called a(n) _______.

2. The improper fraction $\frac{5}{2}$ can be expressed as a(n) _______ number.

3. The fractions $\frac{6}{8}$ and $\frac{3}{4}$ are _______.

4. Divide the numerator and denominator of a fraction by the same whole number in order to _______ it.

5. Fractions with the same denominator are said to be _______.

6. The _______ of two or more fractions is the least common multiple of their denominators.

Identify a fraction or mixed number represented by the shaded portion of each figure.

7. 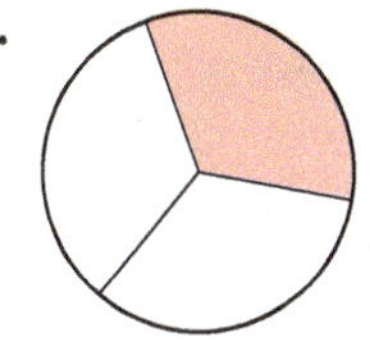

8. 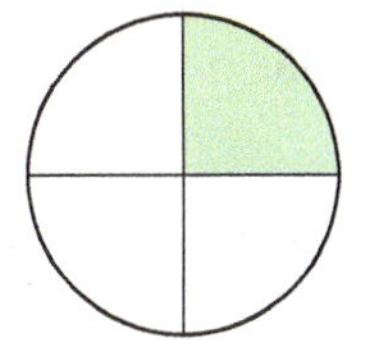

9. 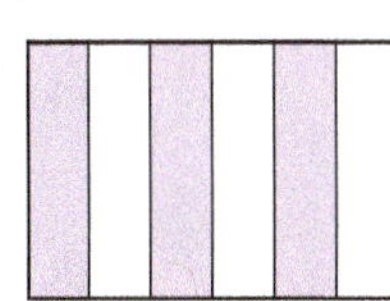

10. 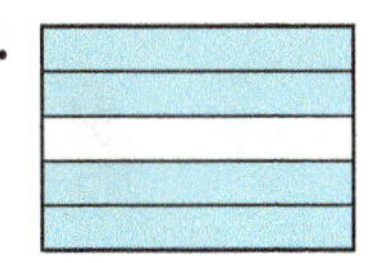

11. 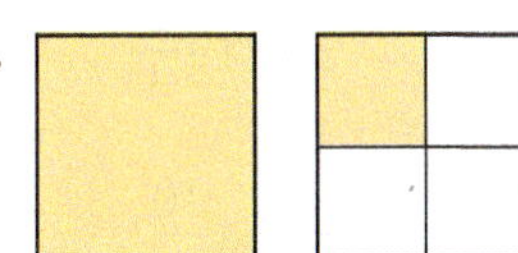

12. 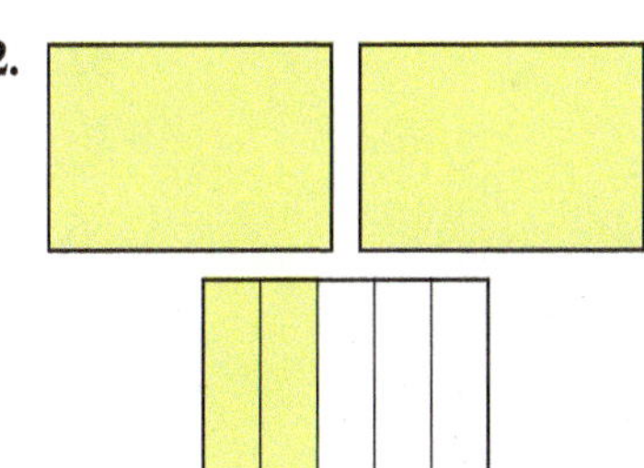

13. 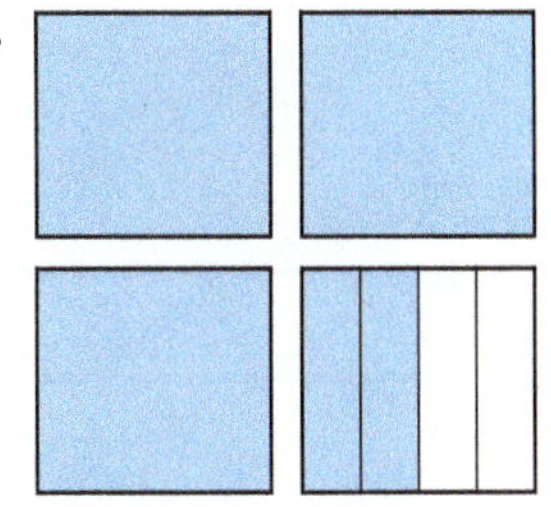

14. 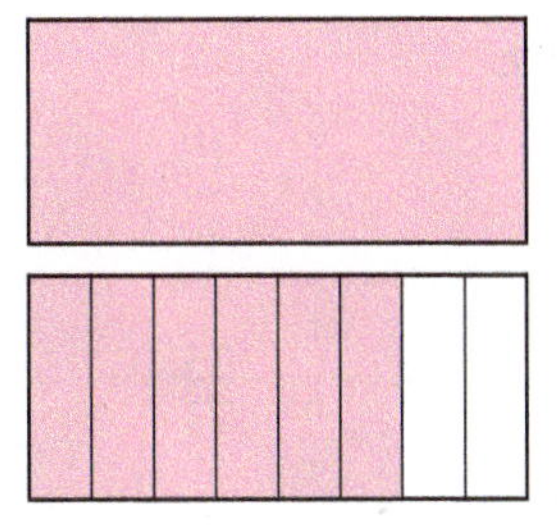

Draw a diagram to represent each fraction or mixed number.

15. $\frac{5}{7}$

16. $\frac{6}{11}$

17. $\frac{2}{9}$

18. $\frac{4}{10}$

19. $\frac{6}{6}$

20. $\frac{11}{11}$

21. $\frac{6}{5}$

22. $\frac{8}{3}$

23. $2\frac{1}{2}$

24. $4\frac{1}{5}$

25. $2\frac{1}{3}$

26. $3\frac{4}{9}$

Indicate whether each number is a proper fraction, an improper fraction, or a mixed number.

27. $\frac{2}{5}$

28. $\frac{7}{12}$

29. $\frac{10}{9}$

30. $\frac{11}{10}$

31. $16\frac{2}{3}$

32. $12\frac{1}{2}$

33. $\frac{5}{5}$

34. $\frac{4}{4}$

35. $\frac{4}{9}$

36. $\frac{5}{6}$

37. $66\frac{2}{3}$

38. $10\frac{3}{4}$

Write each number as an improper fraction.

39. $2\frac{3}{5}$

40. $1\frac{1}{3}$

41. $6\frac{1}{9}$

42. $10\frac{2}{3}$

43. $11\frac{2}{5}$

44. $12\frac{3}{4}$

45. 5

46. 8

47. $7\frac{3}{8}$

48. $6\frac{5}{6}$

49. $9\frac{7}{9}$

50. $10\frac{1}{2}$

51. $13\frac{1}{2}$

52. $20\frac{1}{8}$

53. $19\frac{3}{5}$

54. $11\frac{5}{7}$

55. 14

56. 10

57. $4\frac{10}{11}$

58. $2\frac{7}{13}$

59. $8\frac{3}{14}$

60. $4\frac{1}{6}$

61. $8\frac{2}{25}$

62. $14\frac{1}{10}$

Express each fraction as a mixed or whole number.

63. $\frac{4}{3}$

64. $\frac{6}{5}$

65. $\frac{10}{9}$

66. $\frac{12}{5}$

67. $\frac{9}{3}$

68. $\frac{12}{12}$

69. $\frac{15}{15}$

70. $\frac{62}{3}$

71. $\frac{99}{5}$

72. $\frac{31}{2}$

73. $\frac{82}{9}$

74. $\frac{100}{100}$

75. $\frac{45}{45}$

76. $\frac{40}{3}$

77. $\frac{74}{9}$

78. $\frac{41}{8}$

79. $\frac{27}{2}$ 80. $\frac{58}{11}$ 81. $\frac{100}{9}$ 82. $\frac{19}{1}$

83. $\frac{27}{1}$ 84. $\frac{72}{9}$ 85. $\frac{56}{7}$ 86. $\frac{38}{3}$

Find two fractions equivalent to each fraction.

87. $\frac{1}{8}$ 88. $\frac{3}{10}$ 89. $\frac{2}{11}$ 90. $\frac{1}{10}$

91. $\frac{3}{4}$ 92. $\frac{5}{6}$ 93. $\frac{1}{9}$ 94. $\frac{3}{5}$

Write an equivalent fraction with the given denominator.

95. $\frac{3}{4} = \frac{}{12}$ 96. $\frac{2}{9} = \frac{}{18}$ 97. $\frac{5}{8} = \frac{}{24}$ 98. $\frac{7}{10} = \frac{}{20}$

99. $4 = \frac{}{10}$ 100. $5 = \frac{}{15}$ 101. $\frac{3}{5} = \frac{}{60}$ 102. $\frac{4}{9} = \frac{}{63}$

103. $\frac{5}{8} = \frac{}{64}$ 104. $\frac{3}{10} = \frac{}{40}$ 105. $3 = \frac{}{18}$ 106. $2 = \frac{}{21}$

107. $\frac{4}{9} = \frac{}{81}$ 108. $\frac{7}{8} = \frac{}{24}$ 109. $\frac{6}{7} = \frac{}{49}$ 110. $\frac{5}{6} = \frac{}{48}$

111. $\frac{2}{17} = \frac{}{51}$ 112. $\frac{1}{3} = \frac{}{90}$ 113. $\frac{7}{12} = \frac{}{84}$ 114. $\frac{1}{4} = \frac{}{100}$

115. $\frac{2}{3} = \frac{}{48}$ 116. $\frac{7}{8} = \frac{}{56}$ 117. $\frac{3}{10} = \frac{}{100}$ 118. $\frac{5}{6} = \frac{}{144}$

Simplify, if possible.

119. $\frac{6}{9}$ 120. $\frac{9}{12}$ 121. $\frac{10}{10}$ 122. $\frac{21}{21}$

123. $\frac{5}{15}$ 124. $\frac{4}{24}$ 125. $\frac{9}{20}$ 126. $\frac{25}{49}$

127. $\frac{25}{100}$ 128. $\frac{75}{100}$ 129. $\frac{125}{1{,}000}$ 130. $\frac{875}{1{,}000}$

131. $\frac{20}{16}$ 132. $\frac{15}{9}$ 133. $\frac{66}{32}$ 134. $\frac{30}{18}$

135. $\frac{18}{32}$ 136. $\frac{36}{45}$ 137. $\frac{7}{24}$ 138. $\frac{19}{51}$

139. $\frac{27}{9}$ 140. $\frac{36}{144}$ 141. $\frac{12}{84}$ 142. $\frac{21}{36}$

143. $3\frac{38}{57}$ 144. $11\frac{51}{102}$ 145. $2\frac{100}{100}$ 146. $1\frac{144}{144}$

Between each pair of numbers, insert the appropriate sign: <, =, or >.

147. $\frac{7}{20}\quad\frac{11}{20}$

148. $\frac{5}{10}\quad\frac{3}{10}$

149. $\frac{1}{8}\quad\frac{1}{9}$

150. $\frac{5}{6}\quad\frac{7}{8}$

151. $\frac{2}{3}\quad\frac{6}{9}$

152. $\frac{9}{12}\quad\frac{3}{4}$

153. $2\frac{1}{3}\quad 2\frac{9}{15}$

154. $2\frac{3}{7}\quad 1\frac{1}{2}$

Arrange in increasing order.

155. $\frac{1}{2}, \frac{1}{3}, \frac{1}{4}$

156. $\frac{3}{2}, \frac{3}{3}, \frac{3}{4}$

157. $\frac{2}{3}, \frac{7}{12}, \frac{5}{6}$

158. $\frac{3}{4}, \frac{5}{6}, \frac{7}{8}$

159. $\frac{3}{5}, \frac{2}{3}, \frac{8}{9}$

160. $\frac{5}{8}, \frac{1}{2}, \frac{4}{11}$

Mixed Practice

Solve.

161. Choose the number whose value is between the other two: $\frac{7}{10}, \frac{8}{9}, \frac{5}{6}$.

162. Express $\frac{32}{6}$ as a mixed number.

163. Find two fractions equivalent to $\frac{2}{9}$.

164. Draw a diagram to represent $\frac{9}{10}$.

165. Write an equivalent fraction for $\frac{4}{5}$ with denominator 15.

166. Write $2\frac{3}{8}$ as an improper fraction.

Applications

Solve. Write your answer in simplest form.

167. During the last 5 days, a student spent 11 hours studying mathematics at home. On the average, how much time is this per day?

168. A recipe for pasta with garlic and oil calls for 6 garlic cloves, peeled and chopped. If the recipe serves 4, how many garlic cloves on the average are in each serving?

169. As of 2005, the Nobel Prize was awarded to 33 women and 725 men. (*Source:* nobelprize.org)

a. What fraction of the Nobel prize winners were women?

b. What fraction were men?

170. In 2010, it is projected that there will be 3,256,000 teachers in public elementary and secondary schools and 424,000 teachers in private elementary and secondary schools. (*Source:* Natonal Center for Education Statistics, *Digest of Education Statistics 2005*)

a. What fraction of these teachers will be in public schools?

b. What fraction will be in private schools?

171. Of the 206 bones in the human skeleton, 106 are in the hands and feet. What fraction of these bones are not in the hands and feet? (*Source:* Henry Gray, *Anatomy of the Human Body*)

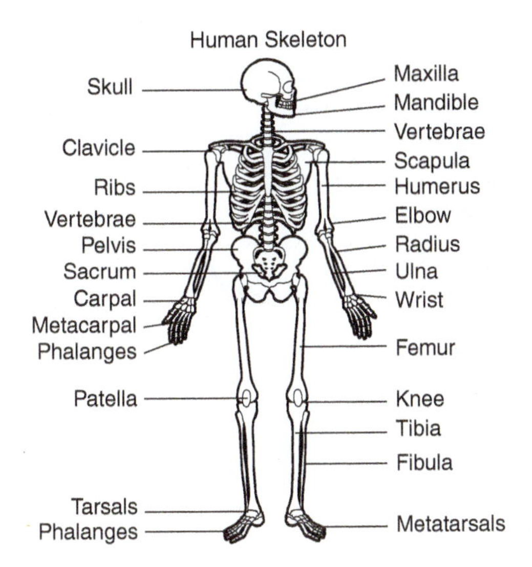

172. Perhaps because of an ankle injury, Rasheed Wallace, the Detroit Pistons power forward, made just 3 of 10 shots in game 1 of the Eastern Conference finals. Of the shots that Wallace took, what fraction did he not make? (*Source:* Yahoo! Sports 2006)

173. The gutter on your roof overflows whenever more than $\frac{1}{4}$ inch of rain falls. Yesterday, $\frac{23}{100}$ inch of rain fell. Did the gutter overflow? Explain.

174. In a course on probability and statistics, a student learns that when rolling a pair of dice, the probability of getting a 5 is $\frac{1}{9}$, and the probability of getting a 6 is $\frac{5}{36}$. Does getting a 5 or a 6 have a greater probability? Explain.

175. According to projections, $\frac{2}{5}$ of the total energy consumption in the United States in the year 2020 will come from petroleum products, $\frac{3}{40}$ will come from nuclear power, and $\frac{7}{30}$ will come from natural gas. (*Source: Annual Energy Outlook 2006*)

a. The greatest consumption of energy will come from which energy resource?

b. Will more nuclear power or natural gas be consumed in 2020?

176. When fog hit the New York City area, visibility was reduced to one-sixteenth mile at Kennedy Airport, one-eighth mile at LaGuardia Airport, and one-half mile at Newark Airport.

a. Which of the three airports had the best visibility?

b. Which of the three airports had the worst visibility?

177. In a recent year, the weights (in pounds) of the six centers that played for the Sabres, the hockey team from Buffalo, were 178, 186, 180, 217, 194, and 187. What was the average weight of these centers? (*Source:* http://sabers.com).

178. The following chart gives the age of the first six American presidents at the time of their inauguration.

President	Washington	J. Adams	Jefferson	Madison	Monroe	J. Q. Adams
Age (in years)	57	61	57	57	58	57

What was their average age at inauguration? (*Source: Significant American Presidents of the United States*)

• *Check your answers on page A-3.*

MINDSTRETCHERS

Mathematical Reasoning

1. Identify the fraction that the shaded portion of the figure to the right represents.

Groupwork

2. Working with a partner, determine how many fractions there are between the numbers 1 and 2.

Critical Thinking

3. Consider the three equivalent fractions shown. Note that the numerators and denominators are made up of the digits 1, 2, 3, 4, 5, 6, 7, 8, and 9—each appearing once.

$$\frac{3}{6} = \frac{7}{14} = \frac{29}{58}$$

a. Verify that these fractions are equivalent by making sure that their cross products are equal.

b. Complete the following fractions to form another trio of equivalent fractions that use the same nine digits only once.

$$\frac{2}{4} = \quad =$$

2.3 Adding and Subtracting Fractions

OBJECTIVES

- To add and subtract fractions and mixed numbers
- To estimate sums and differences involving mixed numbers
- To solve word problems involving the addition or subtraction of fractions or mixed numbers

In Section 2.2 we examined what fractions mean, how they are written, and how they are compared. In the rest of this chapter, we discuss computations involving fractions, beginning with sums and differences.

Adding and Subtracting Like Fractions

Let's first discuss how to add and subtract like fractions. Suppose that an employee spends $\frac{1}{7}$ of his weekly salary for food and $\frac{2}{7}$ for rent. What part of his salary does he spend for food and rent combined? A diagram can help us understand what is involved. First we shade one-seventh of the diagram, then another two-sevenths. We see in the diagram that the total shaded area is three-sevenths, $\frac{1}{7} + \frac{2}{7} = \frac{3}{7}$. Note that we added the original numerators to get the numerator of the answer but that *the denominator stayed the same*.

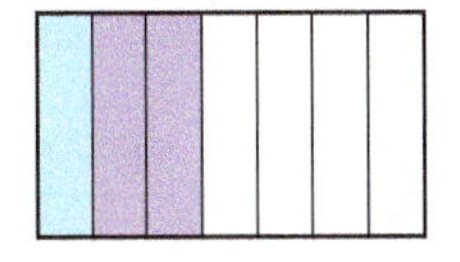

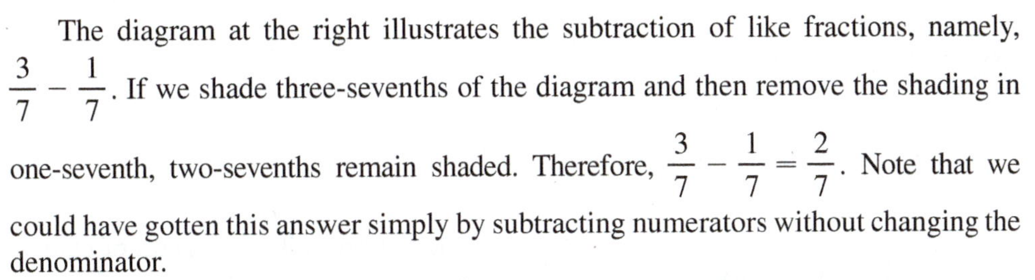

The diagram at the right illustrates the subtraction of like fractions, namely, $\frac{3}{7} - \frac{1}{7}$. If we shade three-sevenths of the diagram and then remove the shading in one-seventh, two-sevenths remain shaded. Therefore, $\frac{3}{7} - \frac{1}{7} = \frac{2}{7}$. Note that we could have gotten this answer simply by subtracting numerators without changing the denominator.

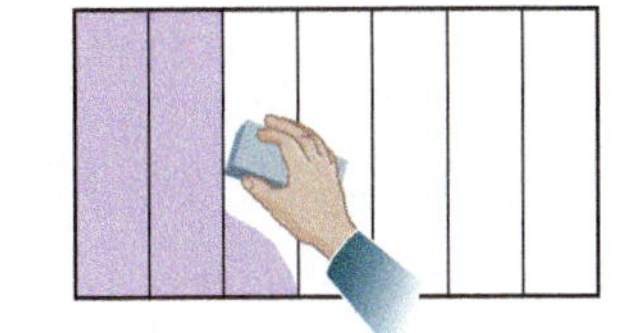

The following rule summarizes how to add or subtract fractions, *provided that they have the same denominator*.

To Add (or Subtract) Like Fractions

- Add (or subtract) the numerators.
- Use the given denominator.
- Write the answer in simplest form.

EXAMPLE 1

Add: $\frac{7}{12} + \frac{2}{12}$

Solution Applying the rule, we get $\frac{7}{12} + \frac{2}{12} = \frac{7+2}{12} = \frac{9}{12}$

$= \frac{3 \cdot 3}{4 \cdot 3} = \frac{3 \cdot \overset{1}{\cancel{3}}}{4 \cdot \underset{1}{\cancel{3}}} = \frac{3}{4}$.

Add the numerators.

Keep the same denominator.

Simplest form

PRACTICE 1

Find the sum of $\frac{7}{15}$ and $\frac{3}{15}$.

Tip Be careful *not* to add the denominators when adding fractions.

EXAMPLE 2

Find the sum of $\frac{12}{16}$, $\frac{3}{16}$, and $\frac{9}{16}$.

Solution

Answer as a mixed number

$$\frac{12}{16} + \frac{3}{16} + \frac{9}{16} = \frac{24}{16} = \frac{3}{2}, \text{ or } 1\frac{1}{2}$$

So the sum of $\frac{12}{16}$, $\frac{3}{16}$, and $\frac{9}{16}$ is $1\frac{1}{2}$.

PRACTICE 2

Add: $\frac{13}{40}$, $\frac{11}{40}$ and $\frac{23}{40}$

EXAMPLE 3

Find the difference between $\frac{11}{7}$ and $\frac{3}{7}$.

Solution

Subtract the numerators.

$$\frac{11}{7} - \frac{3}{7} = \frac{11-3}{7} = \frac{8}{7}, \text{ or } 1\frac{1}{7}$$

Keep the same denominator.

PRACTICE 3

Subtract: $\frac{19}{20} - \frac{11}{20}$

EXAMPLE 4

In the following diagram,

a. how far is it from the college to the library via city hall?

b. which route from the college to the library is shorter—via city hall or via the hospital? By how much?

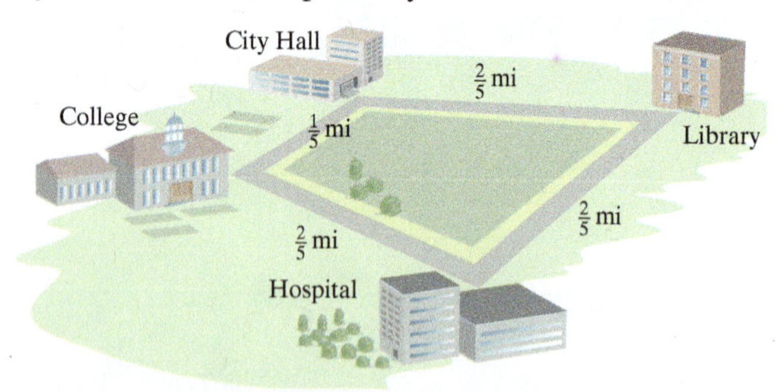

Solution **a.** Examining the diagram, we see that

- the distance from the college to city hall is $\frac{1}{5}$ mile, and
- the distance from city hall to the library is $\frac{2}{5}$ mile.

To find the distance from the college to the library via city hall, we add.

$$\frac{1}{5} + \frac{2}{5} = \frac{3}{5}$$

So this distance is $\frac{3}{5}$ mile.

PRACTICE 4

A doctor prescribed $\frac{9}{20}$ gram of pain medication for a patient to take every 4 hours.

a. If the dosage were increased by $\frac{3}{20}$ gram, what would the new dosage be?

b. If the original dosage were decreased by $\frac{1}{20}$ gram, find the new dosage.

b. To find the distance from the college to the library via the hospital, we again add.

$$\frac{2}{5} + \frac{2}{5} = \frac{4}{5}$$

So this distance is $\frac{4}{5}$ mile. Since $\frac{3}{5} < \frac{4}{5}$, the route from the college to the library via city hall is shorter than the route via the hospital. Now we find the difference.

$$\frac{4}{5} - \frac{3}{5} = \frac{1}{5}$$

Therefore, the route via city hall is $\frac{1}{5}$ mile shorter than the route via the hospital.

Adding and Subtracting Unlike Fractions

Adding (or subtracting) **unlike fractions** is more complicated than adding (or subtracting) like fractions. An extra step is required: changing the unlike fractions to equivalent like fractions. For instance, suppose that we want to add $\frac{1}{10}$ and $\frac{2}{15}$. Even though we can use any common denominator for these fractions, let's use their *least* common denominator to find equivalent fractions.

$$\begin{aligned} 10 &= 2 \cdot 5 \\ 15 &= 3 \cdot 5 \\ \text{LCD} &= 2 \cdot 3 \cdot 5 = 30 \end{aligned}$$

Let's rewrite the fractions vertically as equivalent fractions with the denominator 30.

$$\begin{array}{rcccr} \frac{1}{10} & = & \frac{1 \cdot 3}{10 \cdot 3} & = & \frac{3}{30} \\ +\frac{2}{15} & = & \frac{2 \cdot 2}{15 \cdot 2} & = & +\frac{4}{30} \\ \hline \end{array}$$

Now we add the equivalent like fractions.

$$\begin{array}{r} \frac{3}{30} \\ +\frac{4}{30} \\ \hline \frac{7}{30} \end{array}$$

So $\frac{1}{10} + \frac{2}{15} = \frac{7}{30}$.

We can also add and subtract unlike fractions horizontally.

$$\frac{1}{10} + \frac{2}{15} = \frac{3}{30} + \frac{4}{30} = \frac{3+4}{30} = \frac{7}{30}$$

To Add (or Subtract) Unlike Fractions

- Write the fractions as equivalent fractions with the same denominator, usually the LCD.
- Add (or subtract) the numerators, keeping the same denominator.
- Write the answer in simplest form.

EXAMPLE 5

Add: $\frac{5}{12} + \frac{5}{16}$

Solution First, we find the LCD, which is 48. After finding equivalent fractions, we add the numerators, keeping the same denominator.

$$\begin{array}{rcrcr} \frac{5}{12} & = & \frac{5 \cdot 4}{12 \cdot 4} & = & \frac{20}{48} \\ +\frac{5}{16} & = & \frac{5 \cdot 3}{16 \cdot 3} & = & +\frac{15}{48} \\ \hline & & & & \frac{35}{48} \end{array}$$

← Already in lowest terms

PRACTICE 5

Add: $\frac{11}{12} + \frac{3}{4}$

EXAMPLE 6

Subtract $\frac{1}{12}$ from $\frac{1}{3}$.

Solution Because 3 is a factor of 12, the LCD is 12. Again, let's set up the problem vertically.

$$\begin{array}{rcr} \frac{1}{3} & = & \frac{4}{12} \\ -\frac{1}{12} & = & -\frac{1}{12} \\ \hline & & \frac{3}{12} = \frac{1}{4} \end{array}$$

Subtract the numerators, keeping the same denominator.

Reduce $\frac{3}{12}$ to lowest terms.

PRACTICE 6

Calculate: $\frac{4}{5} - \frac{1}{2}$

EXAMPLE 7

Combine: $\frac{1}{3} + \frac{1}{6} - \frac{3}{10}$

Solution First, we find the LCD of all three fractions. The LCD is 30.

$$\frac{1}{3} = \frac{10}{30}, \quad \frac{1}{6} = \frac{5}{30}, \quad \text{and} \quad \frac{3}{10} = \frac{9}{30}.$$

So $\frac{1}{3} + \frac{1}{6} - \frac{3}{10} = \frac{10}{30} + \frac{5}{30} - \frac{9}{30} = \frac{10 + 5 - 9}{30} = \frac{6}{30} = \frac{1}{5}$.

PRACTICE 7

Combine: $\frac{1}{3} - \frac{2}{9} + \frac{7}{8}$

EXAMPLE 8

Find the perimeter of the piece of stained glass.

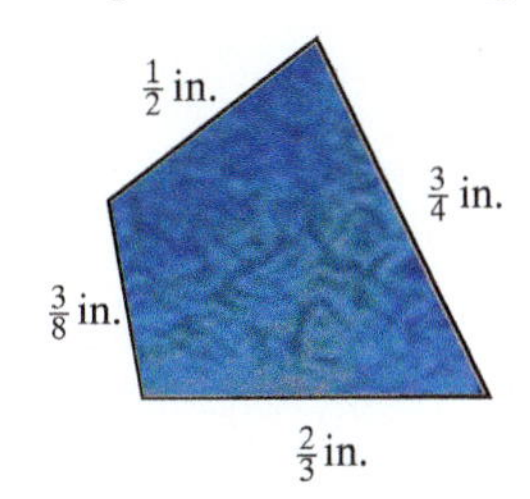

Solution Recall that the perimeter of a figure is the sum of the lengths of its sides.

$$\text{Perimeter} = \frac{3}{8} + \frac{1}{2} + \frac{3}{4} + \frac{2}{3}$$

$$\frac{3}{8} = \frac{9}{24} \leftarrow \text{LCD}$$

$$\frac{1}{2} = \frac{12}{24}$$

$$\frac{3}{4} = \frac{18}{24}$$

$$+\frac{2}{3} = +\frac{16}{24}$$

$$\frac{55}{24}, \text{ or } 2\frac{7}{24}$$

The perimeter of the piece of stained glass is $2\frac{7}{24}$ inches.

PRACTICE 8

What is the perimeter of the triangular park?

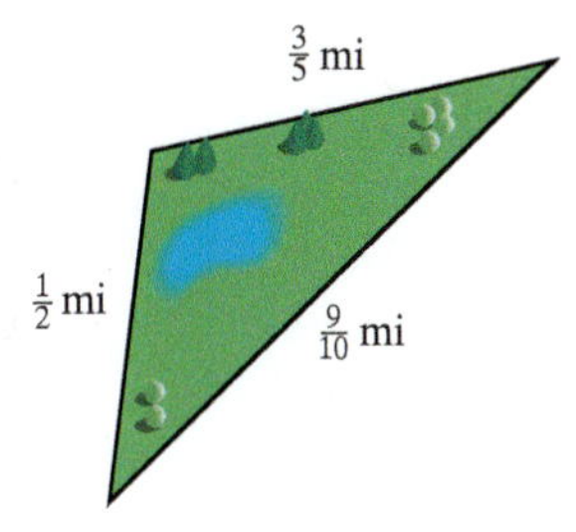

Adding Mixed Numbers

Now let's consider how to add **mixed numbers**, starting with those that have the same denominator.

Suppose, for instance, that we want to add $1\frac{1}{5}$ and $2\frac{1}{5}$. Let's draw a diagram to represent this sum.

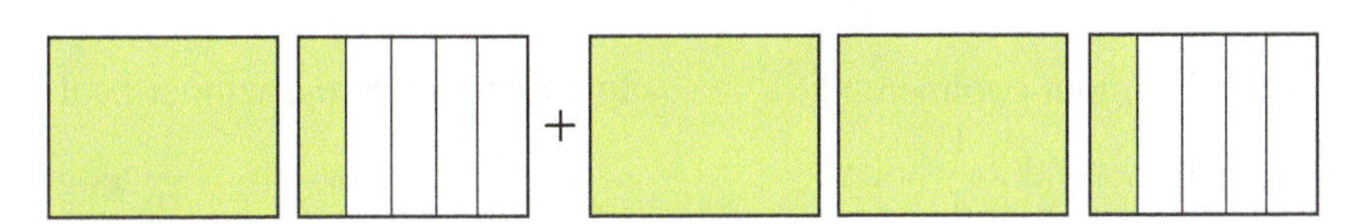

We can rearrange the elements of the diagram by combining the whole numbers and the fractions separately.

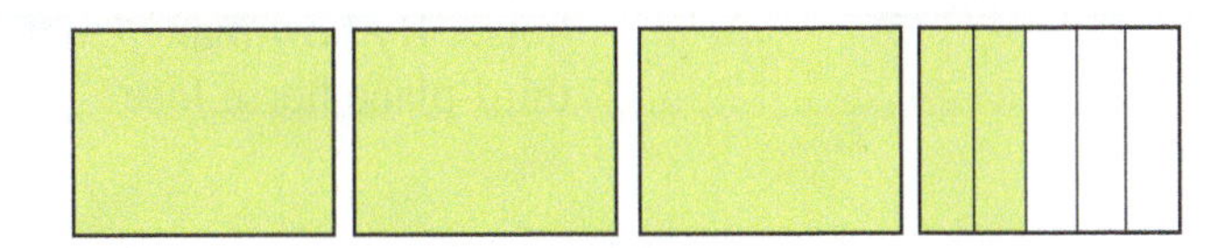

This diagram shows that the sum is $3\frac{2}{5}$.

Note that we can also write and solve this problem vertically.

$$\begin{array}{r} 1\frac{1}{5} \\ +\ 2\frac{1}{5} \\ \hline 3\frac{2}{5} \end{array}$$

← Sum of the fractions

↑ Sum of the whole numbers

EXAMPLE 9

Add: $8\frac{5}{9} + 10\frac{1}{9}$

Solution

$$\begin{array}{r} 8\frac{5}{9} \\ +\ 10\frac{1}{9} \\ \hline 18\frac{6}{9} \end{array} = 18\frac{2}{3}$$

PRACTICE 9

Add: $25\frac{3}{10} + 9\frac{1}{10}$

EXAMPLE 10

Find the sum of $3\frac{3}{5}$, $2\frac{4}{5}$, and 6.

Solution Add the fractions and then add the whole numbers.

$$\begin{array}{r} 3\frac{3}{5} \\ 2\frac{4}{5} \\ +\ 6 \\ \hline 11\frac{7}{5} \end{array} = 12\frac{2}{5}$$

Since $\frac{7}{5} = 1\frac{2}{5}$, we get $11\frac{7}{5} = 11 + 1\frac{2}{5} = 12\frac{2}{5}$.

So the sum is $12\frac{2}{5}$.

PRACTICE 10

Find the sum of $2\frac{5}{16}$, $1\frac{3}{16}$, and 4.

EXAMPLE 11

Two movies are shown back-to-back on TV without commercial interruption. The first runs $1\frac{3}{4}$ hours, and the second $2\frac{1}{4}$ hours. How long will it take to watch both movies?

Solution

We need to add $1\frac{3}{4}$ and $2\frac{1}{4}$.

$$\begin{array}{r} 1\frac{3}{4} \\ +2\frac{1}{4} \\ \hline 3\frac{4}{4} \end{array} = 3 + 1 = 4$$

Therefore, it will take 4 hours to watch the two movies.

PRACTICE 11

In a horse race, the winner beat the second-place horse by $1\frac{1}{2}$ lengths, and the second-place horse finished $2\frac{1}{2}$ lengths ahead of the third-place horse. By how many lengths did the third-place horse lose?

We have previously shown that when we add fractions with different denominators, we must first change the unlike fractions to equivalent like fractions. The same applies to adding mixed numbers that have different denominators.

To Add Mixed Numbers

- Write the fractions as equivalent fractions with the same denominator, usually the LCD.
- Add the fractions.
- Add the whole numbers.
- Write the answer in simplest form.

EXAMPLE 12

Find the sum of $3\frac{1}{5}$ and $7\frac{2}{3}$.

Solution The LCD is 15. Add the fractions and then add the whole numbers.

$$\begin{array}{rcr} 3\frac{1}{5} & = & 3\frac{3}{15} \\ +\ 7\frac{2}{3} & = & +\ 7\frac{10}{15} \\ \hline & & 10\frac{13}{15} \end{array}$$

The sum of $3\frac{1}{5}$ and $7\frac{2}{3}$ is $10\frac{13}{15}$.

PRACTICE 12

Add $4\frac{1}{8}$ to $3\frac{1}{2}$.

EXAMPLE 13

Find the sum of $1\frac{2}{3}$, $8\frac{1}{4}$, and $3\frac{4}{5}$.

Solution Set up the problem vertically and use the LCD, which is 60. Add the fractions and then add the whole numbers.

$$\begin{array}{rcr} 1\frac{2}{3} & = & 1\frac{40}{60} \\ 8\frac{1}{4} & = & 8\frac{15}{60} \\ +\ 3\frac{4}{5} & = & +\ 3\frac{48}{60} \\ \hline \end{array}$$

$$12\frac{103}{60} = 12 + 1\frac{43}{60} = 13\frac{43}{60}$$

PRACTICE 13

What is the sum of $5\frac{5}{8}$, $3\frac{1}{6}$, and $2\frac{5}{12}$?

Subtracting Mixed Numbers

Now let's discuss how to subtract mixed numbers, beginning with those that have the same denominator.

For instance, suppose that we want to subtract $2\frac{1}{5}$ from $3\frac{2}{5}$. We draw a diagram to represent $3\frac{2}{5}$.

 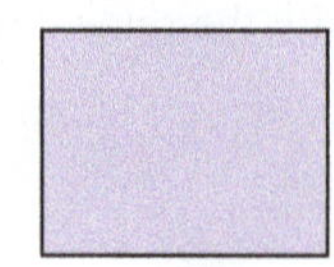 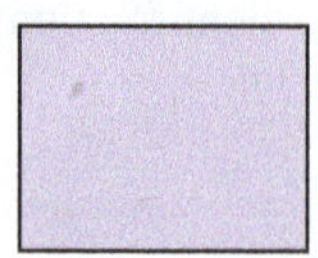 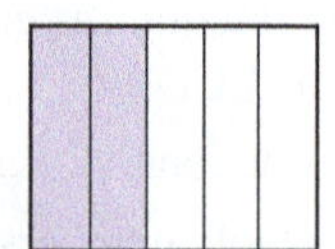

If we remove the shading from $2\frac{1}{5}$, then $1\frac{1}{5}$ remains shaded.

 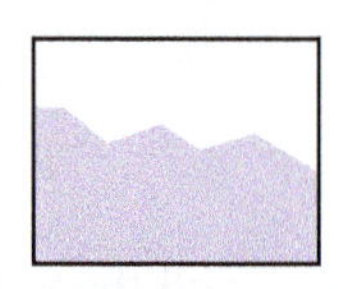  

So the difference is $1\frac{1}{5}$.

We can also write and solve this problem vertically.

$$\begin{array}{r} 3\frac{2}{5} \\ -\,2\frac{1}{5} \\ \hline 1\frac{1}{5} \end{array}$$

$\frac{1}{5}$ ← Difference of the fractions

1 ← Difference of the whole numbers

EXAMPLE 14

Subtract: $4\frac{5}{6} - 2\frac{1}{6}$

Solution We set up the problem vertically. Subtract the fractions and then subtract the whole numbers.

$$\begin{array}{r} 4\frac{5}{6} \\ -\,2\frac{1}{6} \\ \hline 2\frac{4}{6} \end{array} = 2\frac{2}{3}$$

Therefore, the difference is $2\frac{2}{3}$.

PRACTICE 14

Subtract $5\frac{3}{10}$ from $9\frac{7}{10}$.

EXAMPLE 15

A construction job was scheduled to last $5\frac{3}{4}$ days, but was finished in $4\frac{1}{4}$ days. How many days ahead of schedule was the job?

Solution

This question asks us to subtract $4\frac{1}{4}$ from $5\frac{3}{4}$.

$$\begin{array}{r} 5\frac{3}{4} \\ -4\frac{1}{4} \\ \hline 1\frac{2}{4} = 1\frac{1}{2} \end{array}$$

So the job was $1\frac{1}{2}$ days ahead of schedule.

PRACTICE 15

A photograph is displayed in a frame. What is the difference between the height of the frame and the height of the photo?

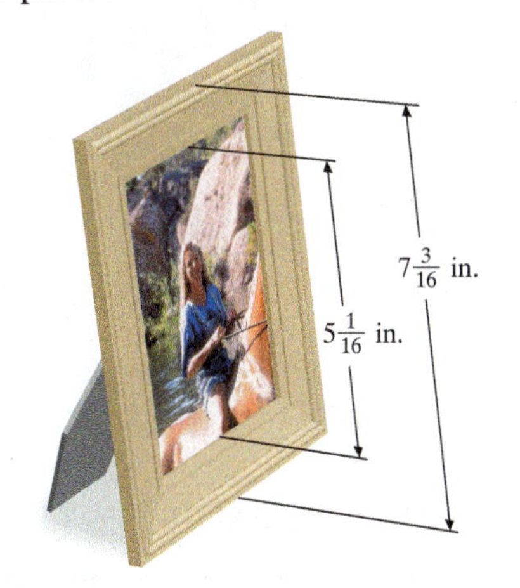

Subtracting mixed numbers that have different denominators is similar to adding mixed numbers.

EXAMPLE 16

Subtract $2\frac{7}{100}$ from $5\frac{9}{10}$.

Solution As usual, we use the LCD (which is 100) to find equivalent fractions. Then we subtract the equivalent mixed numbers with the same denominator. Again, let's set up the problem vertically. Subtract the fractions and then subtract the whole numbers.

$$\begin{array}{rcr} 5\frac{9}{10} & = & 5\frac{90}{100} \\ -\,2\frac{7}{100} & = & -\,2\frac{7}{100} \\ \hline & & 3\frac{83}{100} \end{array}$$

The answer is $3\frac{83}{100}$.

PRACTICE 16

Calculate: $8\frac{2}{3} - 4\frac{1}{12}$

EXAMPLE 17

Find the length of the flower bed.

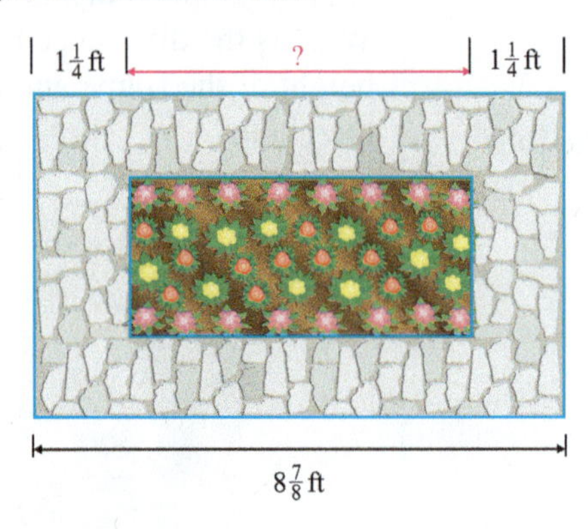

Solution The total length of the rectangular area is $8\frac{7}{8}$ feet.

To find the length of the flower bed, we need to add $1\frac{1}{4}$ feet and $1\frac{1}{4}$ feet and then subtract this sum from $8\frac{7}{8}$ feet.

$$\begin{array}{r} 1\frac{1}{4} \\ +\ 1\frac{1}{4} \\ \hline 2\frac{2}{4} = 2\frac{1}{2} \end{array} \qquad \begin{array}{rcr} 8\frac{7}{8} & = & 8\frac{7}{8} \\ -\ 2\frac{1}{2} & = & -\ 2\frac{4}{8} \\ \hline & & 6\frac{3}{8} \end{array}$$

So the length of the flower bed is $6\frac{3}{8}$ feet. We can check this answer by adding $1\frac{1}{4}$, $6\frac{3}{8}$, and $1\frac{1}{4}$, getting $8\frac{7}{8}$.

PRACTICE 17

The figure below is called a **trapezoid**. Suppose that this trapezoid's perimeter is $20\frac{1}{2}$ miles. How long is the left side?

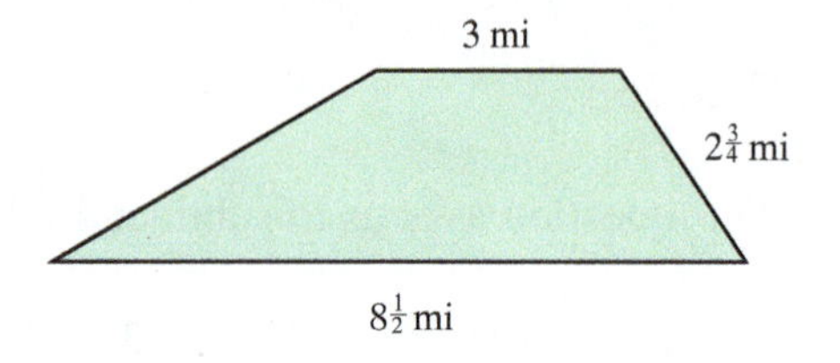

Recall from our discussion of subtracting whole numbers that, in problems in which a digit in the subtrahend is larger than the corresponding digit in the minuend, we need to borrow.

$$\begin{array}{r} \scriptstyle 2\ 1\ \ \ \\ \not{3}\ 2\ 9 \\ -\ 8\ 7 \\ \hline 2\ 4\ 2 \end{array}$$

A similar situation can arise when we are subtracting mixed numbers. If the fraction on the bottom is larger than the fraction on top, we *rename* (or *borrow from*) the whole number on top.

EXAMPLE 18

Subtract: $6 - 1\frac{1}{3}$

Solution Let's rewrite the problem vertically.

$$\begin{array}{r} 6 \\ -1\frac{1}{3} \\ \hline \end{array}$$ There is no fraction on top from which to subtract $\frac{1}{3}$.

$$\begin{array}{r} 5\frac{3}{3} \\ -1\frac{1}{3} \\ \hline \end{array}$$ Rename 6 as 5 + 1, or $5 + \frac{3}{3}$, or $5\frac{3}{3}$.

$$\begin{array}{r} 5\frac{3}{3} \\ -1\frac{1}{3} \\ \hline 4\frac{2}{3} \end{array}$$ Now subtract.

So $6 - 1\frac{1}{3} = 4\frac{2}{3}$.

As in any subtraction problem, we can check our answer by addition.

$$4\frac{2}{3} + 1\frac{1}{3} = 5\frac{3}{3} = 6$$

PRACTICE 18

Subtract: $9 - 7\frac{5}{7}$

In Example 18, the answer is $4\frac{2}{3}$. Would we get the same answer if we compute $6\frac{1}{3} - 1$?

We have already discussed subtracting mixed numbers without renaming (borrowing) as well as subtracting a mixed number from a whole number. Now let's consider the general rule for subtracting mixed numbers.

To Subtract Mixed Numbers

- Write the fractions as equivalent fractions with the same denominator, usually the LCD.
- Rename (or borrow from) the whole number on top if the fraction on the bottom is larger than the fraction on top.
- Subtract the fractions.
- Subtract the whole numbers.
- Write the answer in simplest form.

EXAMPLE 19

Compute: $13\frac{2}{9} - 7\frac{8}{9}$

Solution First, we write the problem vertically.

$$\begin{array}{r} 13\frac{2}{9} \\ -7\frac{8}{9} \\ \hline \end{array}$$

Because $\frac{8}{9}$ is larger than $\frac{2}{9}$, we need to rename $13\frac{2}{9}$.

$$13\frac{2}{9} = 12 + 1 + \frac{2}{9} = 12 + \frac{9}{9} + \frac{2}{9} = 12\frac{11}{9}$$

$$\begin{array}{r} 12\frac{11}{9} \\ -7\frac{8}{9} \\ \hline \end{array}$$

Finally, we subtract and then write the answer in simplest form.

$$\begin{array}{rcr} 13\frac{2}{9} & = & 12\frac{11}{9} \\ -7\frac{8}{9} & = & -7\frac{8}{9} \\ \hline & & 5\frac{3}{9}, \text{ or } 5\frac{1}{3} \end{array}$$

PRACTICE 19

Find the difference between $15\frac{1}{12}$ and $9\frac{11}{12}$.

EXAMPLE 20

Find the difference between $10\frac{1}{4}$ and $1\frac{5}{12}$.

Solution First, we write the equivalent fractions, using the LCD.

$$\begin{array}{rcr} 10\frac{1}{4} & = & 10\frac{3}{12} \\ -1\frac{5}{12} & = & -1\frac{5}{12} \\ \hline \end{array}$$

Then, we subtract.

$$\begin{array}{rcr} 10\frac{3}{12} & = & 9\frac{15}{12} \\ -1\frac{5}{12} & = & -1\frac{5}{12} \\ \hline & & 8\frac{10}{12} = 8\frac{5}{6} \end{array}$$

We rename $10\frac{3}{12}$: $10\frac{3}{12} = 9 + \frac{12}{12} + \frac{3}{12} = 9\frac{15}{12}$

PRACTICE 20

Find the difference between $16\frac{3}{5}$ and $3\frac{1}{10}$.

EXAMPLE 21

In Oregon's Columbia River Gorge, a hiker walks along the Eagle Creek Trail, headed for Punchbowl Falls $2\frac{1}{10}$ miles away. After reaching Metlako Falls, does he have more or less than $\frac{1}{2}$ mile left to go? (*Source:* USDA Forest Service)

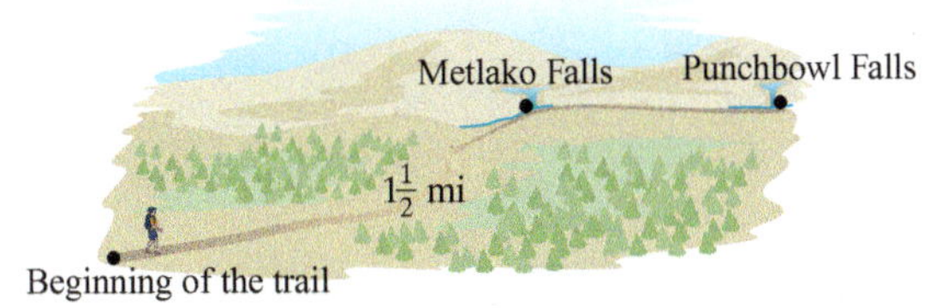

Eagle Creek Trail

Solution First, we must find the difference between the length of the trail from its begining to Punchbowl Falls, namely, $2\frac{1}{10}$ miles, and the distance already hiked, $1\frac{1}{2}$ miles.

$$\begin{array}{rcrcr} 2\frac{1}{10} & = & 2\frac{1}{10} & = & 1\frac{11}{10} \\ -1\frac{1}{2} & = & -1\frac{5}{10} & = & -1\frac{5}{10} \\ \hline & & & & \frac{6}{10} = \frac{3}{5} \end{array}$$

So the distance left to hike is $\frac{3}{5}$ mile. Finally, we compare $\frac{3}{5}$ mile and $\frac{1}{2}$ mile.

$$\frac{3}{5} = \frac{6}{10} \qquad \frac{1}{2} = \frac{5}{10}$$

Because $6 > 5$, $\frac{6}{10} > \frac{5}{10}$. Therefore, $\frac{3}{5} > \frac{1}{2}$, and the hiker has more than $\frac{1}{2}$ mile left to go from Metlako Falls to Punchbowl Falls.

PRACTICE 21

A homeowner purchased a roll of wallpaper that unrolls to $30\frac{1}{2}$ yards long and used $26\frac{7}{8}$ yards from the roll to paper a room. Is there enough paper left on the roll for a job that requires 4 yards of paper?

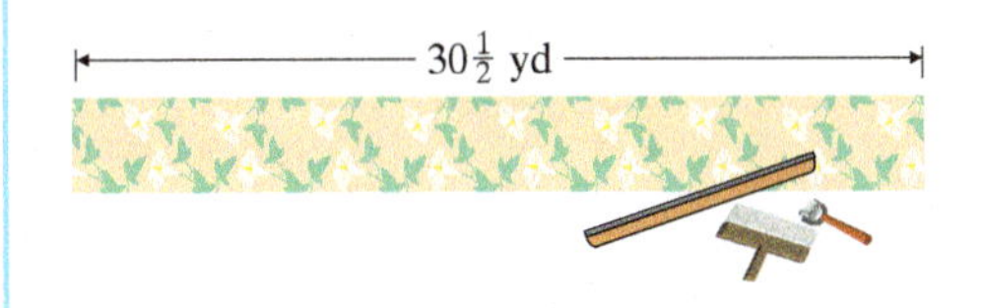

Another Method of Adding and Subtracting Mixed Numbers

Recall that any mixed number can be rewritten as an improper fraction. So when adding or subtracting mixed numbers, we can first express them as improper fractions. In a subtraction problem, this method has an advantage over the method previously discussed; namely, we never have to borrow. However, expressing mixed numbers as improper fractions may have the disadvantage of involving unnecessarily large numbers, as the following examples show.

EXAMPLE 22

Add: $14\frac{1}{6} + 8\frac{2}{3}$

Solution We begin by writing each mixed number as an improper fraction.

$14\frac{1}{6} + 8\frac{2}{3} = \frac{85}{6} + \frac{26}{3}$ Express $14\frac{1}{6}$ and $8\frac{2}{3}$ as improper fractions.

$= \frac{85}{6} + \frac{52}{6}$ Write the fractions to be added as equivalent fractions.

$= \frac{137}{6}$ Add the like fractions.

$= 22\frac{5}{6}$ Express the improper fraction as a mixed number.

PRACTICE 22

Find the sum: $7\frac{4}{5} + 2\frac{3}{4}$

Can you show that this method gives the same sum that we would have gotten if we had not expressed the mixed numbers as improper fractions? Explain.

EXAMPLE 23

Find the difference: $8\frac{5}{6} - 4\frac{9}{10}$

Solution $8\frac{5}{6} - 4\frac{9}{10} = \frac{53}{6} - \frac{49}{10}$ Write as improper fractions.

$= \frac{265}{30} - \frac{147}{30}$ Write as equivalent fractions.

$= \frac{118}{30}$ Subtract the like fractions.

$= 3\frac{28}{30}$ Express as a mixed number.

$= 3\frac{14}{15}$ Simplify.

Check that this answer is the same as we would have gotten without changing the mixed numbers to improper fractions.

PRACTICE 23

Subtract: $13\frac{1}{4} - 11\frac{7}{8}$

Estimating Sums and Differences of Mixed Numbers

When adding or subtracting mixed numbers, we can check by *estimating,* determining whether our estimate and our answer are close. Note that when we round mixed numbers, we round to the nearest whole number.

Checking a Sum by Estimating

$$\begin{array}{rcr} 1\frac{1}{5} & \rightarrow & 1 \\ +2\frac{3}{5} & \rightarrow & +3 \\ \hline 3\frac{4}{5} & & 4 \end{array}$$

Because $\frac{1}{5} < \frac{1}{2}$, round *down* to the whole number 1.

Because $\frac{3}{5} > \frac{1}{2}$, round *up* to the whole number 3.

Our answer, $3\frac{4}{5}$, is close to 4, the sum of the rounded addends (1 and 3).

Checking a Difference by Estimating

$$\begin{array}{rcr} 3\frac{2}{5} & \rightarrow & 3 \\ -1\frac{1}{5} & \rightarrow & -1 \\ \hline 2\frac{1}{5} & & 2 \end{array}$$

Because $\frac{2}{5} < \frac{1}{2}$, round *down* to 3.

Round *down* to 1.

Our answer, $2\frac{1}{5}$, is close to 2, the difference of the rounded numbers (3 and 1).

EXAMPLE 24

Combine and check: $5\frac{1}{3} - \left(2\frac{4}{5} + 1\frac{1}{10}\right)$

Solution Following the order of operations rule, we begin by adding the two mixed numbers in parentheses.

$$\begin{array}{rcr} 2\frac{4}{5} & = & 2\frac{8}{10} \\ +1\frac{1}{10} & = & +1\frac{1}{10} \\ \hline & & 3\frac{9}{10} \end{array}$$

Next we subtract this sum from $5\frac{1}{3}$.

$$\begin{array}{rcrcr} 5\frac{1}{3} & = & 5\frac{10}{30} & = & 4\frac{40}{30} \\ -3\frac{9}{10} & = & -3\frac{27}{30} & = & -3\frac{27}{30} \\ \hline & & & & 1\frac{13}{30} \end{array}$$

So $5\frac{1}{3} - \left(2\frac{4}{5} + 1\frac{1}{10}\right) = 1\frac{13}{30}$.

Now let's check this answer by estimating:

$$5\frac{1}{3} - \left(2\frac{4}{5} + 1\frac{1}{10}\right)$$

$$\downarrow \quad \downarrow \quad \downarrow$$

$$5 - (3 + 1) = 5 - 4 = 1$$

The estimate, 1, is sufficiently close to $1\frac{13}{30}$ to confirm our answer.

PRACTICE 24

Calculate and check:

$8\frac{1}{4} - \left(3\frac{2}{5} - 1\frac{9}{10}\right)$

2.3 Exercises

FOR EXTRA HELP

Mathematically Speaking

Fill in each blank with the most appropriate term or phrase from the given list.

denominators	borrow	equivalent
subtract	numerators	improper

1. To add like fractions, add the _______.

2. To subtract unlike fractions, rewrite them as _______ fractions with the same denominator.

3. When subtracting $2\frac{4}{5}$ from $7\frac{1}{5}$, _______ from the 7 on the top.

4. Fractions with equal numerators and _______ are equivalent to 1.

Add and simplify.

5. $\frac{5}{8}+\frac{5}{8}$

6. $\frac{7}{10}+\frac{9}{10}$

7. $\frac{11}{12}+\frac{7}{12}$

8. $\frac{71}{100}+\frac{79}{100}$

9. $\frac{1}{5}+\frac{1}{5}+\frac{2}{5}$

10. $\frac{1}{7}+\frac{3}{7}+\frac{2}{7}$

11. $\frac{3}{20}+\frac{1}{20}+\frac{8}{20}$

12. $\frac{1}{10}+\frac{3}{10}+\frac{1}{10}$

13. $\frac{2}{3}+\frac{1}{2}$

14. $\frac{1}{4}+\frac{2}{5}$

15. $\frac{1}{2}+\frac{3}{8}$

16. $\frac{1}{6}+\frac{2}{3}$

17. $\frac{7}{10}+\frac{7}{100}$

18. $\frac{5}{6}+\frac{1}{12}$

19. $\frac{4}{5}+\frac{1}{8}$

20. $\frac{3}{4}+\frac{3}{7}$

21. $\frac{4}{9}+\frac{5}{6}$

22. $\frac{9}{10}+\frac{4}{5}$

23. $\frac{87}{100}+\frac{3}{10}$

24. $\frac{7}{20}+\frac{3}{4}$

25. $\frac{1}{3}+\frac{1}{4}+\frac{1}{6}$

26. $\frac{1}{5}+\frac{1}{6}+\frac{1}{3}$

27. $\frac{3}{8}+\frac{1}{10}+\frac{3}{16}$

28. $\frac{3}{10}+\frac{1}{3}+\frac{1}{9}$

29. $\frac{2}{9}+\frac{5}{8}+\frac{1}{4}$

30. $\frac{1}{2}+\frac{1}{3}+\frac{1}{4}$

31. $\frac{7}{8}+\frac{1}{5}+\frac{1}{4}$

32. $\frac{1}{10}+\frac{2}{5}+\frac{5}{6}$

Add and simplify. Then check by estimating.

33. $1+2\frac{1}{3}$

34. $4\frac{1}{5}+2$

35. $8\frac{1}{10}+7\frac{3}{10}$

36. $6\frac{1}{12}+4\frac{1}{12}$

37. $7\frac{3}{10}+6\frac{9}{10}$

38. $8\frac{2}{3}+6\frac{2}{3}$

39. $5\frac{1}{6}+9\frac{5}{6}$

40. $2\frac{3}{10}+7\frac{9}{10}$

41. $5\frac{1}{4}+5\frac{1}{6}$

42. $17\frac{3}{8}+20\frac{1}{5}$

43. $3\frac{1}{3}+\frac{2}{5}$

44. $4\frac{7}{10}+\frac{7}{20}$

45. $8\frac{1}{5} + 5\frac{2}{3}$

46. $4\frac{1}{9} + 20\frac{7}{10}$

47. $\frac{2}{3} + 6\frac{1}{8}$

48. $\frac{1}{6} + 3\frac{2}{5}$

49. $9\frac{2}{3} + 10\frac{7}{12}$

50. $20\frac{3}{5} + 4\frac{1}{2}$

51. $6\frac{1}{10} + 3\frac{93}{100}$

52. $4\frac{8}{9} + 5\frac{1}{3}$

53. $4\frac{1}{2} + 6\frac{7}{8}$

54. $10\frac{5}{6} + 8\frac{1}{4}$

55. $30\frac{21}{100} + 5\frac{17}{20}$

56. $8\frac{3}{10} + 2\frac{321}{1{,}000}$

57. $80\frac{1}{3} + \frac{3}{4} + 10\frac{1}{2}$

58. $\frac{1}{3} + 25\frac{7}{24} + 100\frac{1}{2}$

59. $2\frac{1}{3} + 2 + 2\frac{1}{6}$

60. $4\frac{1}{8} + 4\frac{3}{16} + \frac{5}{4}$

61. $6\frac{7}{8} + 2\frac{3}{4} + 1\frac{1}{5}$

62. $1\frac{2}{3} + 5\frac{5}{6} + 3\frac{1}{4}$

63. $2\frac{1}{2} + 5\frac{1}{4} + 3\frac{5}{8}$

64. $4\frac{2}{3} + 2\frac{11}{36} + 1\frac{1}{2}$

Subtract and simplify.

65. $\frac{4}{5} - \frac{3}{5}$

66. $\frac{7}{9} - \frac{5}{9}$

67. $\frac{7}{10} - \frac{3}{10}$

68. $\frac{11}{12} - \frac{5}{12}$

69. $\frac{23}{100} - \frac{7}{100}$

70. $\frac{3}{2} - \frac{1}{2}$

71. $\frac{3}{4} - \frac{1}{4}$

72. $\frac{7}{9} - \frac{4}{9}$

73. $\frac{12}{5} - \frac{2}{5}$

74. $\frac{1}{8} - \frac{1}{8}$

75. $\frac{3}{4} - \frac{2}{3}$

76. $\frac{2}{5} - \frac{1}{6}$

77. $\frac{4}{9} - \frac{1}{6}$

78. $\frac{9}{10} - \frac{3}{100}$

79. $\frac{4}{5} - \frac{3}{4}$

80. $\frac{5}{6} - \frac{1}{8}$

81. $\frac{4}{7} - \frac{1}{2}$

82. $\frac{2}{5} - \frac{2}{9}$

83. $\frac{4}{9} - \frac{3}{8}$

84. $\frac{11}{12} - \frac{1}{3}$

85. $\frac{6}{8} - \frac{1}{2}$

86. $\frac{5}{6} - \frac{2}{3}$

Subtract and simplify. Then check either by adding or by estimating.

87. $5\frac{3}{7} - 1\frac{1}{7}$

88. $6\frac{2}{3} - 1\frac{1}{3}$

89. $3\frac{7}{8} - 2\frac{1}{8}$

90. $10\frac{5}{6} - 2\frac{5}{6}$

91. $20\frac{1}{2} - \frac{1}{2}$

92. $7\frac{3}{4} - \frac{1}{4}$

93. $8\frac{1}{10} - 4$

94. $2\frac{1}{3} - 2$

95. $6 - 2\frac{2}{3}$

96. $4 - 1\frac{1}{5}$

97. $8 - 4\frac{7}{10}$

98. $2 - 1\frac{1}{2}$

99. $10 - 3\frac{2}{3}$

100. $5 - 4\frac{9}{10}$

101. $6 - \frac{1}{2}$

102. $9 - \frac{3}{4}$

103. $7\frac{1}{4} - 2\frac{3}{4}$

104. $5\frac{1}{10} - 2\frac{3}{10}$

105. $6\frac{1}{8} - 2\frac{7}{8}$

106. $3\frac{1}{5} - 1\frac{4}{5}$

107. $12\frac{2}{5} - \frac{3}{5}$

108. $3\frac{7}{10} - \frac{9}{10}$

109. $8\frac{1}{3} - 1\frac{2}{3}$

110. $2\frac{1}{5} - \frac{4}{5}$

111. $13\frac{1}{2} - 5\frac{2}{3}$

112. $7\frac{1}{10} - 2\frac{1}{7}$

113. $9\frac{3}{8} - 5\frac{5}{6}$

114. $2\frac{1}{10} - 1\frac{27}{100}$

115. $20\frac{2}{9} - 4\frac{5}{6}$

116. $9\frac{13}{100} - 6\frac{7}{10}$

117. $3\frac{4}{5} - \frac{5}{6}$

118. $1\frac{2}{8} - \frac{2}{6}$

119. $1\frac{3}{4} - 1\frac{1}{2}$

120. $2\frac{1}{2} - 1\frac{3}{4}$

121. $10\frac{1}{12} - 4\frac{2}{3}$

122. $7\frac{1}{4} - 1\frac{5}{16}$

123. $22\frac{7}{8} - 8\frac{9}{10}$

124. $9\frac{1}{10} - 3\frac{1}{2}$

125. $3\frac{1}{8} - 2\frac{3}{4}$

126. $3\frac{1}{4} - 2\frac{5}{16}$

Combine and simplify.

127. $\frac{5}{8} + \frac{9}{10} - \frac{1}{4}$

128. $\frac{2}{3} - \frac{1}{5} + \frac{1}{2}$

129. $12\frac{1}{6} + 5\frac{9}{10} - 1\frac{3}{10}$

130. $7\frac{1}{3} - 2\frac{4}{5} - 1\frac{1}{3}$

131. $15\frac{1}{2} - 3\frac{4}{5} - 6\frac{1}{2}$

132. $4\frac{1}{10} + 2\frac{9}{10} - 3\frac{3}{4}$

133. $20\frac{1}{10} - \left(\frac{1}{20} + 1\frac{1}{2}\right)$

134. $19\frac{1}{6} - \left(8\frac{9}{10} - \frac{1}{5}\right)$

Mixed Practice

Perform the indicated operations and simplify.

135. Subtract $1\frac{7}{8}$ from 6.

136. Add: $6\frac{1}{10} + 3\frac{7}{15}$

137. Calculate: $12\frac{2}{3} - \left(8\frac{5}{6} - 4\frac{1}{2}\right)$

138. Find the sum of $\frac{3}{8}$, $\frac{1}{2}$, and $\frac{1}{3}$.

139. Find the difference between $4\frac{3}{5}$ and $1\frac{2}{3}$.

140. Subtract: $\frac{9}{10} - \frac{1}{4}$

Applications

Solve. Write the answer in simplest form.

141. A $\frac{7}{8}$-inch nail was hammered through a $\frac{3}{4}$-inch door. How far did it extend from the door?

142. A building occupies $\frac{1}{4}$ acre on a $\frac{7}{8}$-acre plot of land. What is the area of the land not occupied by the building?

143. The Kentucky Derby, Belmont Stakes, and the Preakness Stakes are three prestigious horse races that comprise the Triple Crown. (*Source:* http://infoplease.com)

a. Horses run $1\frac{3}{16}$ mile in the Preakness Stakes. If the Preakness Stakes is $\frac{5}{16}$ mile shorter than the Belmont Stakes, how far do horses run in the Belmont Stakes?

b. Horses run $1\frac{1}{4}$ miles in the Kentucky Derby. How much farther do horses run in the Belmont Stakes than in the Kentucky Derby?

144. In the year 2010, the total amount of electricity consumed anywhere in the world is projected to be approximately 16 billion kilowatt-hours. Of this amount, $\frac{1}{4}$ is expected to be consumed in the United States, $\frac{1}{9}$ in China, and $\frac{1}{20}$ in Japan. (*Source: International Energy Outlook 2004*)

a. According to these projections, the combined electrical consumption of China and Japan will be what fraction of the world consumption?

b. As a fraction of world consumption, how much greater is the U.S. consumption than the combined consumption in China and Japan?

145. The first game of a baseball doubleheader lasted $2\frac{1}{4}$ hours. The second game began after a $\frac{1}{4}$-hour break and lasted $2\frac{1}{2}$ hours. How long did the doubleheader take to play?

146. Three student candidates competed in a student government election. The winner got $\frac{5}{8}$ of the votes, and the second-place candidate got $\frac{1}{4}$ of the votes. If the rest of the votes went to the third candidate, what fraction of the votes did that student get?

147. In the hallway pictured, how much greater is the length than the width?

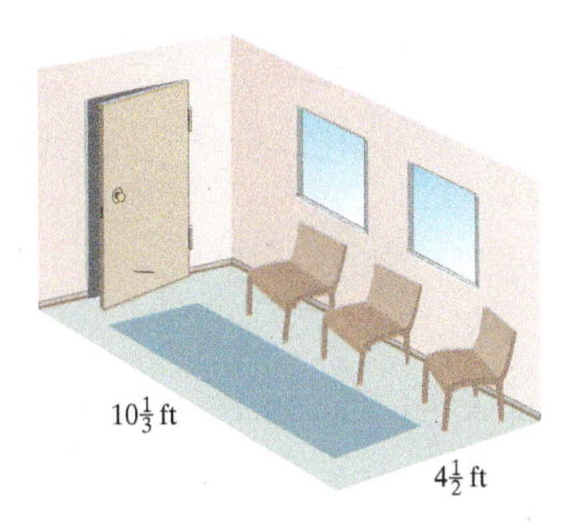

148. Find the perimeter of the figure shown.

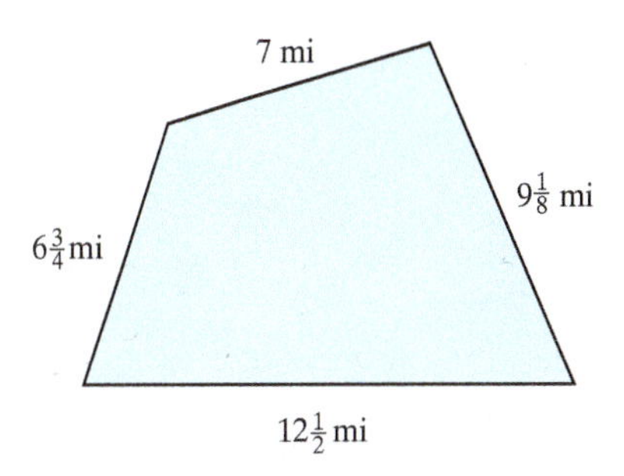

149. In testing a new drug, doctors found that $\frac{1}{2}$ of the patients given the drug improved, $\frac{2}{5}$ showed no change in their condition, and the remainder got worse. What fraction got worse?

150. According to a growth chart for young girls, their average weight is $38\frac{3}{4}$ pounds at age 4, $47\frac{1}{2}$ pounds at age 6, and $60\frac{3}{4}$ pounds at age 8. On the average, do girls gain more weight from age 4 to age 6 or from age 6 to age 8? (*Source:* http://www.babybag.com)

151. Suppose that four packages are placed on a scale, as shown. If the scale balances, how heavy is the small package on the right?

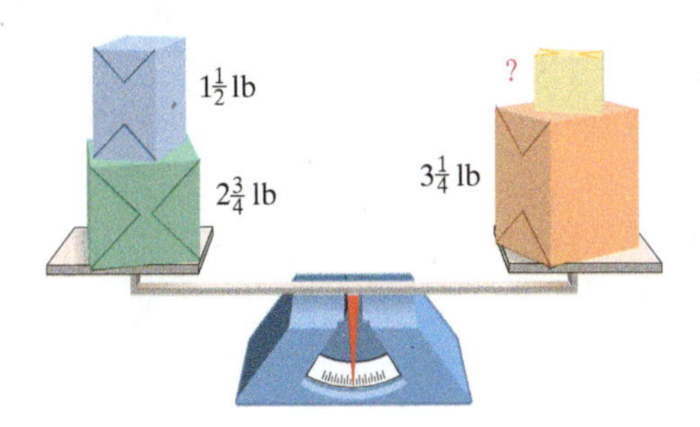

152. If the scale pictured balances, how heavy is the small package on the left?

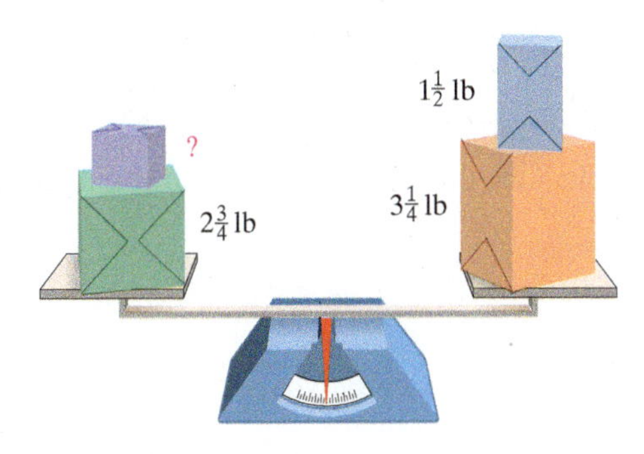

• *Check your answers on page A-4.*

MINDSTRETCHERS

Groupwork

1. Working with a partner, complete the following magic square in which each row, column, and diagonal adds up to the same number.

$1\frac{1}{4}$		
	1	
$\frac{11}{12}$		$\frac{3}{4}$

Mathematical Reasoning

2. A fraction with 1 as the numerator is called a **unit fraction**. For example, $\frac{1}{7}$ is a unit fraction. Write $\frac{3}{7}$ as the sum of three unit fractions, using no unit fraction more than once.

$$\frac{3}{7} = \frac{1}{\square} + \frac{1}{\square} + \frac{1}{\square}$$

Writing

3. Consider the following two ways of subtracting $2\frac{4}{5}$ from $4\frac{1}{5}$.

Method 1

$$4\frac{1}{5} = 3 + \frac{5}{5} + \frac{1}{5} = 3\frac{6}{5}$$
$$-2\frac{4}{5} \qquad = -2\frac{4}{5}$$
$$1\frac{2}{5}$$

Method 2

$$4\frac{1}{5} \rightarrow 4 + \frac{1}{5} + \frac{1}{5} = 4\frac{2}{5}$$
$$-2\frac{4}{5} \rightarrow 2 + \frac{4}{5} + \frac{1}{5} = -3$$
$$1\frac{2}{5}$$

a. Explain the difference between the two methods.

b. Explain which method you prefer.

c. Explain why you prefer that method.

2.4 Multiplying and Dividing Fractions

OBJECTIVES

- To multiply and divide fractions and mixed numbers
- To estimate products and quotients involving mixed numbers
- To solve word problems involving the multiplication or division of fractions or mixed numbers

This section begins with a discussion of multiplying fractions. We then move on to multiplying mixed numbers and conclude with dividing fractions and mixed numbers.

Multiplying Fractions

Many situations require us to multiply fractions. For instance, suppose that a mixture in a chemistry class calls for $\frac{4}{5}$ gram of sodium chloride. If we make only $\frac{2}{3}$ of that mixture, we need

$$\frac{2}{3} \text{ of } \frac{4}{5}$$
$$\downarrow$$
$$\frac{2}{3} \times \frac{4}{5}$$

that is, $\frac{2}{3} \times \frac{4}{5}$ gram of sodium chloride.

To illustrate how to find this product, we diagram these two fractions.

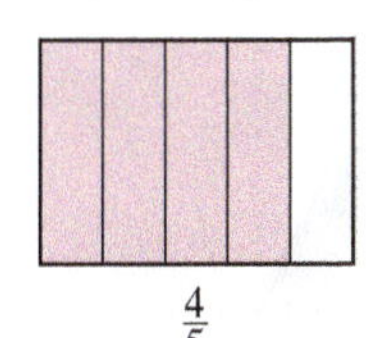
$\frac{4}{5}$

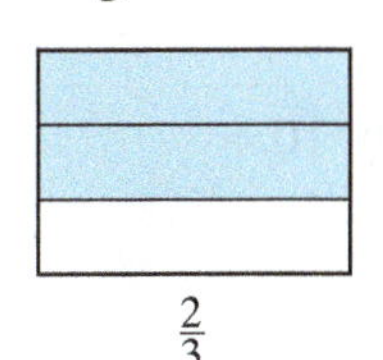
$\frac{2}{3}$

In the following diagram, we are taking $\frac{2}{3}$ of the $\frac{4}{5}$.

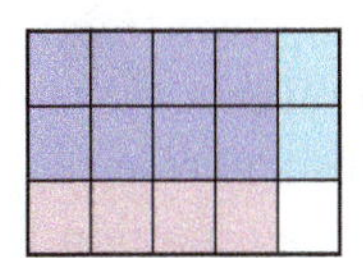

Note that we divided the whole into 15 parts and that our product, containing 8 of the 15 small squares, represents the double-shaded region. The answer is, therefore, $\frac{8}{15}$ of the original whole, which we can compute as follows.

$$\frac{2}{3} \times \frac{4}{5} = \frac{8}{15}$$

The numerator and denominator of the answer are the products of the original numerators and denominators.

To Multiply Fractions

- Multiply the numerators.
- Multiply the denominators.
- Write the answer in simplest form.

EXAMPLE 1

Multiply: $\frac{7}{8} \cdot \frac{9}{10}$

Solution

Multiply the numerators.

$$\frac{7}{8} \cdot \frac{9}{10} = \frac{7 \cdot 9}{8 \cdot 10} = \frac{63}{80}$$

Multiply the denominators.

PRACTICE 1

Find the product of $\frac{3}{4}$ and $\frac{5}{7}$.

EXAMPLE 2

Calculate: $\left(\frac{4}{5}\right)^2$

Solution

$$\left(\frac{4}{5}\right)^2 = \frac{4}{5} \cdot \frac{4}{5} = \frac{4 \cdot 4}{5 \cdot 5} = \frac{16}{25}$$

PRACTICE 2

Square $\frac{9}{10}$.

EXAMPLE 3

What is $\frac{3}{8}$ of 10?

Solution Finding $\frac{3}{8}$ of 10 means multiplying $\frac{3}{8}$ by 10.

$$\frac{3}{8} \times 10 = \frac{3}{8} \times \frac{10}{1} = \frac{3 \times 10}{8 \times 1} = \frac{30}{8} = \frac{15}{4}, \text{ or } 3\frac{3}{4}$$

PRACTICE 3

What is $\frac{2}{3}$ of 30?

In Example 3, we multiplied the two fractions first and then simplified the answer. It is preferable, however, to reverse these steps: Simplify first and then multiply. By first simplifying, sometimes referred to as *canceling*, we divide *any* numerator and *any* denominator by a common factor. Canceling before multiplying allows us to work with smaller numbers and still gives us the same answer.

EXAMPLE 4

Find the product of $\frac{4}{9}$ and $\frac{5}{8}$.

Solution

$$\frac{4}{9} \times \frac{5}{8} = \frac{\overset{1}{\cancel{4}}}{9} \times \frac{5}{\underset{2}{\cancel{8}}}$$

Divide the numerator 4 and the denominator 8 by 4.

$$= \frac{1 \times 5}{9 \times 2}$$

Multiply the resulting fractions.

$$= \frac{5}{18}$$

PRACTICE 4

Multiply: $\frac{7}{10} \cdot \frac{5}{11}$

EXAMPLE 5

Multiply: $\frac{9}{8} \times \frac{6}{5} \times \frac{7}{9}$

Solution We simplify and then multiply.

$$\frac{9}{8} \times \frac{6}{5} \times \frac{7}{9} = \frac{\overset{1}{\cancel{9}}}{\underset{4}{\cancel{8}}} \times \frac{\overset{3}{\cancel{6}}}{5} \times \frac{7}{\underset{1}{\cancel{9}}}$$

Divide the numerator 9 and the denominator 9 by 9. Divide the numerator 6 and the denominator 8 by 2.

$$= \frac{21}{20}, \text{ or } 1\frac{1}{20}$$

PRACTICE 5

Multiply: $\frac{7}{27} \cdot \frac{9}{4} \cdot \frac{8}{21}$

EXAMPLE 6

At a college, $\frac{3}{5}$ of the students take a math course. Of these students, $\frac{1}{6}$ take elementary algebra. What fraction of the students in the college take elementary algebra?

Solution We must find $\frac{1}{6}$ of $\frac{3}{5}$.

$$\frac{1}{6} \times \frac{3}{5} = \frac{1}{\underset{2}{\cancel{6}}} \times \frac{\overset{1}{\cancel{3}}}{5} = \frac{1 \times 1}{2 \times 5} = \frac{1}{10}$$

One-tenth of the students in the college take elementary algebra.

PRACTICE 6

A flight from New York to Los Angeles took 7 hours. With the help of the jet stream, the return trip took $\frac{3}{4}$ the time. How long did the trip from Los Angeles to New York take?

EXAMPLE 7

Of the 639 employees at a company, $\frac{4}{9}$ responded to a voluntary survey distributed by the human resources department. How many employees did not respond to the survey?

Solution Apply the strategy of breaking the problem into two parts.

- First, find $\frac{4}{9}$ of 639.
- Then, subtract the result from 639.

In short, we can solve this problem by computing $639 - \left(\frac{4}{9} \times 639\right)$.

$$639 - \left(\frac{4}{9} \times 639\right) = 639 - \left(\frac{4}{\underset{1}{\cancel{9}}} \times \frac{\overset{71}{\cancel{639}}}{1}\right) = 639 - 284 = 355$$

So 355 employees did not respond to the survey.

PRACTICE 7

The state sales tax on a car in Wisconsin is $\frac{1}{20}$ of the price of the car. What is the total amount a consumer would pay for a \$19,780 car?

Multiplying Mixed Numbers

Some situations require us to multiply mixed numbers. For instance, suppose that your regular hourly wage is $\$7\frac{1}{2}$ and that you make time-and-a-half for working overtime. To find your overtime hourly wage, you need to multiply $1\frac{1}{2}$ by $7\frac{1}{2}$. The key here is to first rewrite each mixed number as an improper fraction.

$$1\frac{1}{2} \times 7\frac{1}{2} = \frac{3}{2} \times \frac{15}{2} = \frac{45}{4}, \text{ or } 11\frac{1}{4}$$

So you make $\$11\frac{1}{4}$ per hour overtime.

To Multiply Mixed Numbers

- Write the mixed numbers as improper fractions.
- Multiply the fractions.
- Write the answer in simplest form.

EXAMPLE 8

Multiply $2\frac{1}{5}$ by $1\frac{1}{4}$.

Solution $2\frac{1}{5} \times 1\frac{1}{4} = \frac{11}{5} \times \frac{5}{4}$ Write each mixed number as an improper fraction.

$= \frac{11 \times \overset{1}{\cancel{5}}}{\underset{1}{\cancel{5}} \times 4}$ Simplify and multiply.

$= \frac{11}{4}$, or $2\frac{3}{4}$

PRACTICE 8

Find the product of $3\frac{3}{4}$ and $2\frac{1}{10}$.

EXAMPLE 9

Multiply: $\left(4\frac{3}{8}\right)(4)\left(2\frac{2}{5}\right)$

Solution

$$\left(4\frac{3}{8}\right)(4)\left(2\frac{2}{5}\right) = \left(\frac{35}{8}\right)\left(\frac{4}{1}\right)\left(\frac{12}{5}\right)$$

$$= \left(\frac{\overset{7}{\cancel{35}}}{\underset{\underset{1}{\cancel{2}}}{\cancel{8}}}\right)\left(\frac{\overset{1}{\cancel{4}}}{1}\right)\left(\frac{\overset{6}{\cancel{12}}}{\underset{1}{\cancel{5}}}\right) = 42$$

Note in this problem that, although there are several ways to simplify, the answer always comes out the same.

PRACTICE 9

Multiply: $\left(1\frac{3}{4}\right)\left(5\frac{1}{3}\right)(3)$

EXAMPLE 10

A lawn surrounding a garden is to be installed, as depicted in the following drawing.

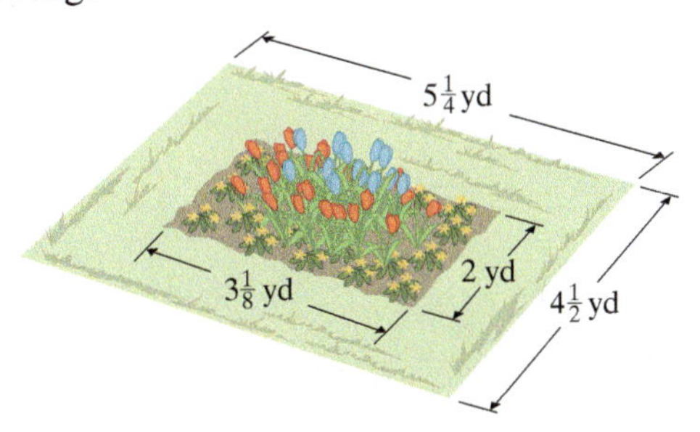

How many square yards of turf will we need to cover the lawn?

Solution Let's break this problem into three steps. First, we find the area of the rectangle with dimensions $5\frac{1}{4}$ yards and $4\frac{1}{2}$ yards. Then, we find the area of the small rectangle whose length and width are $3\frac{1}{8}$ yards and 2 yards, respectively.

$$5\frac{1}{4} \times 4\frac{1}{2} = \frac{21}{4} \times \frac{9}{2}$$

$$= \frac{189}{8}, \text{ or } 23\frac{5}{8}$$ The area of the large rectangle is $23\frac{5}{8}$ square yards.

$$3\frac{1}{8} \times 2 = \frac{25}{8} \times \frac{2}{1}$$

$$= \frac{25}{4}, \text{ or } 6\frac{1}{4}$$ The area of the small rectangle is $6\frac{1}{4}$ square yards.

Finally, we subtract the area of the small rectangle from the area of the large rectangle.

$$\begin{array}{rcr} 23\frac{5}{8} & = & 23\frac{5}{8} \\ -6\frac{1}{4} & = & -6\frac{2}{8} \\ \hline & & 17\frac{3}{8} \end{array}$$

The area of the lawn is, therefore, $17\frac{3}{8}$ square yards. So we will need $17\frac{3}{8}$ square yards of turf for the lawn.

PRACTICE 10

How much greater is the area of a sheet of legal-size paper than a sheet of letter-size paper?

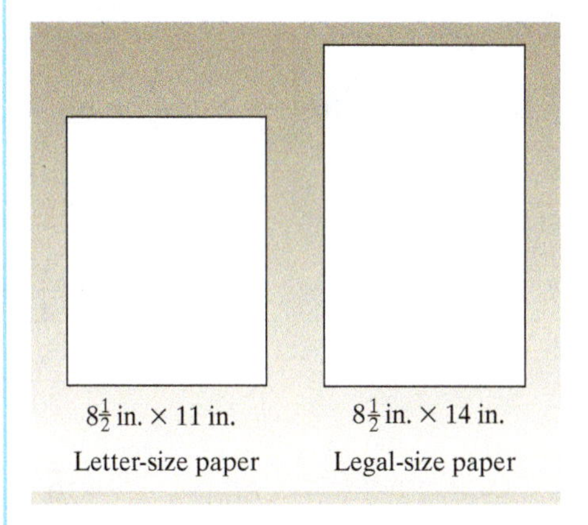

EXAMPLE 11

Simplify: $16\frac{1}{4} - 2 \cdot 4\frac{3}{5}$

Solution We use the order of operations rule, multiplying before subtracting.

$$\begin{aligned} 16\frac{1}{4} - 2 \cdot 4\frac{3}{5} &= 16\frac{1}{4} - \frac{2}{1} \cdot \frac{23}{5} \\ &= 16\frac{1}{4} - \frac{46}{5} \\ &= 16\frac{1}{4} - 9\frac{1}{5} \\ &= 16\frac{5}{20} - 9\frac{4}{20} \\ &= 7\frac{1}{20} \end{aligned}$$

PRACTICE 11

Calculate: $6 + \left(3\frac{1}{2}\right)^2$

Dividing Fractions

We now turn to quotients, beginning with dividing a fraction by a whole number. Suppose, for instance, that you want to share $\frac{1}{3}$ of a pizza with a friend, that is, to divide the $\frac{1}{3}$ into two equal parts. What part of the whole pizza will each of you receive?

This diagram shows $\frac{1}{3}$ of a pizza.

If we split each third into two equal parts, each part is $\frac{1}{6}$ of the pizza.

You and your friend will each get $\frac{1}{6}$ of the whole pizza, which you can compute as follows.

$$\frac{1}{3} \div 2 = \frac{1}{6}$$

Note that dividing a number by 2 is the same as taking $\frac{1}{2}$ of it. This equivalence suggests the procedure for dividing fractions shown next.

Divisor

$$\frac{1}{3} \div 2 = \frac{1}{3} \div \frac{2}{1} = \frac{1}{3} \times \frac{1}{2} = \frac{1 \times 1}{3 \times 2} = \frac{1}{6}$$

$\frac{2}{1}$ and $\frac{1}{2}$ are reciprocals.

This procedure involves *inverting*, or finding the *reciprocal* of the divisor. The reciprocal is found by switching the numerator and denominator.

To Divide Fractions

- Change the divisor to its reciprocal, and multiply the resulting fractions.
- Write the answer in simplest form.

EXAMPLE 12

Divide: $\frac{4}{5} \div \frac{3}{10}$

Solution $\frac{4}{5} \div \frac{3}{10} = \frac{4}{5} \times \frac{10}{3}$ Change the divisor to its reciprocal and multiply.

$= \frac{4}{\cancel{5}_1} \times \frac{\cancel{10}^2}{3}$ Divide the numerator 10 and the denominator 5 by 5.

$= \frac{4 \times 2}{1 \times 3}$ Multiply the fractions.

$= \frac{8}{3}$ Simplify.

$= 2\frac{2}{3}$

As in any division problem, we can check our answer by multiplying it by the divisor. $\frac{\cancel{8}^4}{\cancel{3}_1} \times \frac{\cancel{3}^1}{\cancel{10}_5} = \frac{4}{5}$

Because $\frac{4}{5}$ is the dividend, we have confirmed our answer.

PRACTICE 12

Divide: $\frac{3}{4} \div \frac{1}{8}$

Tip In a division problem, the fraction to the right of the division sign is the divisor. Always invert the divisor (the second fraction) and not the dividend (the first fraction).

EXAMPLE 13

What is $\frac{4}{7}$ divided by 20?

Solution $\frac{4}{7} \div 20 = \frac{4}{7} \times \frac{1}{20}$ — Invert $\frac{20}{1}$ and multiply.

$= \frac{\overset{1}{\cancel{4}}}{7} \times \frac{1}{\underset{5}{\cancel{20}}}$ — Divide the numerator 4 and the denominator 20 by 4.

$= \frac{1 \times 1}{7 \times 5} = \frac{1}{35}$

PRACTICE 13

Compute the following quotient:

$5 \div \frac{5}{8}$

EXAMPLE 14

To stop the developing process, photographers use a chemical called stop bath. Suppose that a photographer needs $\frac{1}{4}$ bottle of stop bath for each roll of film. If the photographer has $\frac{2}{3}$ bottle of stop bath left, can he develop three rolls of film?

Solution We want to find out how many $\frac{1}{4}$'s there are in $\frac{2}{3}$, that is, to compute $\frac{2}{3} \div \frac{1}{4}$.

$$\frac{2}{3} \div \frac{1}{4} = \frac{2}{3} \times \frac{4}{1} = \frac{8}{3} \text{ or } 2\frac{2}{3}$$

Find the reciprocal of the divisor, $\frac{1}{4}$, and then multiply.

So the photographer cannot develop three rolls of film.

PRACTICE 14

A house is built on ground that is sinking $\frac{3}{4}$ inch per year. How many years will it take the house to sink 2 inches?

Dividing Mixed Numbers

Dividing mixed numbers is similar to dividing fractions, except that there is an additional step.

To Divide Mixed Numbers

- Write the mixed numbers as improper fractions.
- Divide the fractions.
- Write the answer in simplest form.

EXAMPLE 15

Find: $9 \div 2\frac{7}{10}$.

Solution $9 \div 2\frac{7}{10} = \frac{9}{1} \div \frac{27}{10}$ Write the whole number and the mixed number as improper fractions.

$= \frac{\overset{1}{\cancel{9}}}{1} \times \frac{10}{\underset{3}{\cancel{27}}}$ Invert and multiply.

$= \frac{10}{3}$, or $3\frac{1}{3}$

PRACTICE 15

Divide: $6 \div 3\frac{3}{4}$

EXAMPLE 16

What is $2\frac{1}{2} \div 4\frac{1}{2}$?

Solution $2\frac{1}{2} \div 4\frac{1}{2} = \frac{5}{2} \div \frac{9}{2} = \frac{5}{\underset{1}{\cancel{2}}} \times \frac{\overset{1}{\cancel{2}}}{9} = \frac{5}{9}$

Invert and multiply.

PRACTICE 16

Divide $2\frac{3}{8}$ by $5\frac{3}{7}$.

EXAMPLE 17

There are $6\frac{3}{4}$ yards of silk in a roll. If it takes $\frac{3}{4}$ yards to make one designer tie, how many ties can be made from the roll?

Solution The question is: How many $\frac{3}{4}$'s fit into $6\frac{3}{4}$? It tells us that we must divide.

$$6\frac{3}{4} \div \frac{3}{4} = \frac{\overset{9}{\cancel{27}}}{\underset{1}{\cancel{4}}} \times \frac{\overset{1}{\cancel{4}}}{\underset{1}{\cancel{3}}} = 9$$

So nine ties can be made from the roll of silk.

PRACTICE 17

According to a newspaper advertisement for a "diet shake," a man lost 33 pounds in $5\frac{1}{2}$ months. How much weight did he lose per month?

Estimating Products and Quotients of Mixed Numbers

As with adding or subtracting mixed numbers, it is important to check our answers when multiplying or dividing. We can check a product or a quotient of mixed numbers by estimating the answer and then confirming that our estimate and answer are reasonably close.

Checking a Product by Estimating

$$2\frac{1}{5} \times 7\frac{2}{3} = \frac{11}{5} \times \frac{23}{3} = \frac{253}{15}, \text{ or } 16\frac{13}{15}$$

$$\downarrow \quad \downarrow$$

$$2 \times 8 = 16$$

Our answer, $16\frac{13}{15}$, is close to 16, the product of the rounded factors.

Because $16\frac{13}{15}$ is near 16, $16\frac{13}{15}$ is a reasonable answer.

Checking a Quotient by Estimating

$$6\frac{1}{4} \div 2\frac{7}{10} = \frac{25}{4} \div \frac{27}{10} = \frac{25}{\underset{2}{\cancel{4}}} \times \frac{\overset{5}{\cancel{10}}}{27} = \frac{125}{54}, \text{ or } 2\frac{17}{54}$$

$$\downarrow \quad \downarrow$$

$$6 \div 3 = 2$$

Our answer, $2\frac{17}{54}$, is close to 2, the quotient of the rounded dividend and divisor.

Because 2 is near $2\frac{17}{54}$, $2\frac{17}{54}$ is a reasonable answer.

EXAMPLE 18

Simplify and check: $3\frac{3}{4} \times 5\frac{1}{3} \div 2\frac{7}{9}$

Solution Following the order of operations rule, we work from left to right, multiplying the first two mixed numbers.

$$3\frac{3}{4} \times 5\frac{1}{3} = \frac{\overset{5}{\cancel{15}}}{\underset{1}{\cancel{4}}} \times \frac{\overset{4}{\cancel{16}}}{\underset{1}{\cancel{3}}} = 20$$

Then we divide 20 by $2\frac{7}{9}$ to get the answer.

$$20 \div 2\frac{7}{9} = \frac{20}{1} \div \frac{25}{9} = \frac{\overset{4}{\cancel{20}}}{1} \times \frac{9}{\underset{5}{\cancel{25}}} = \frac{36}{5}, \text{ or } 7\frac{1}{5}$$

Now let's check by estimating.

$$3\frac{3}{4} \times 5\frac{1}{3} \div 2\frac{7}{9}$$

$$\downarrow \quad \downarrow \quad \downarrow$$

$$4 \times 5 \div 3 = 20 \div 3 \approx 7$$

The answer, $7\frac{1}{5}$, and the estimate, 7, are reasonably close, confirming the answer.

PRACTICE 18

Compute and check:

$5\frac{3}{5} \div 2\frac{1}{10} \times 2\frac{1}{4}$

EXAMPLE 19

Calculate and check: $12 \div 1\frac{2}{3} + 5 \cdot 2\frac{9}{10}$

Solution According to the order of operations rule, we divide and multiply before adding.

$$12 \div 1\frac{2}{3} + 5 \cdot 2\frac{9}{10} = \frac{12}{1} \div \frac{5}{3} + \frac{5}{1} \cdot \frac{29}{10}$$

$$= \frac{12}{1} \cdot \frac{3}{5} + \frac{5}{1} \cdot \frac{29}{10}$$

$$= \frac{12}{1} \cdot \frac{3}{5} + \frac{\overset{1}{\cancel{5}}}{1} \cdot \frac{29}{\underset{2}{\cancel{10}}}$$

$$= \frac{36}{5} + \frac{29}{2}$$

$$= 7\frac{1}{5} + 14\frac{1}{2}$$

$$= 7\frac{2}{10} + 14\frac{5}{10}$$

$$= 21\frac{7}{10}$$

Now we estimate the answer in order to check.

$$12 \div 1\frac{2}{3} + 5 \cdot 2\frac{9}{10}$$

$$\downarrow \quad \downarrow \quad \downarrow \quad \downarrow$$

$$12 \div 2 + 5 \cdot 3 \approx 21$$

The estimate and the answer are close, confirming the answer.

PRACTICE 19

Compute and check:

$14\frac{1}{3} \div 2 - 6 \div 2\frac{1}{4}$

2.4 Exercises

FOR EXTRA HELP

PRACTICE WATCH DOWNLOAD READ REVIEW

Mathematically Speaking

Fill in each blank with the most approppriate term or phrase from the given list.

reverse	proper fraction	multiply
divide	simplify	reciprocal
		improper fraction

1. To find the product of the fractions $\frac{1}{7}$ and $\frac{5}{8}$, _______ 1 and 5, and 7 and 8.

2. To multiply mixed numbers, change each mixed number to its equivalent _______.

3. The fraction $\frac{2}{3}$ is said to be the _______ of the fraction $\frac{3}{2}$.

4. To _______ fractions, change the divisor to its reciprocal, and multiply the resulting fractions.

Multiply.

5. $\frac{1}{3} \times \frac{2}{5}$

6. $\frac{7}{8} \times \frac{1}{2}$

7. $\left(\frac{5}{8}\right)\left(\frac{2}{3}\right)$

8. $\left(\frac{3}{10}\right)\left(\frac{1}{4}\right)$

9. $\left(\frac{3}{4}\right)^2$

10. $\left(\frac{1}{8}\right)^2$

11. $\frac{4}{5} \times \frac{2}{5}$

12. $\frac{1}{2} \times \frac{3}{2}$

13. $\frac{7}{8} \times \frac{5}{4}$

14. $\frac{20}{3} \times \frac{2}{7}$

15. $\frac{5}{2} \cdot \frac{9}{8}$

16. $\frac{11}{10} \cdot \frac{9}{5}$

17. $\left(\frac{2}{5}\right)\left(\frac{5}{9}\right)$

18. $\left(\frac{4}{5}\right)\left(\frac{1}{4}\right)$

19. $\frac{7}{9} \times \frac{3}{4}$

20. $\frac{4}{5} \times \frac{1}{2}$

21. $\left(\frac{1}{8}\right)\left(\frac{6}{10}\right)$

22. $\left(\frac{4}{6}\right)\left(\frac{3}{8}\right)$

23. $\frac{10}{9} \times \frac{93}{100}$

24. $\frac{12}{5} \times \frac{15}{4}$

25. $\frac{2}{3} \times 20$

26. $\frac{5}{6} \times 5$

27. $\left(\frac{10}{3}\right)(4)$

28. $\frac{5}{3} \times 7$

29. $\frac{2}{3} \times 24$

30. $\frac{3}{4} \times 12$

31. $\frac{2}{3} \cdot 6$

32. $100 \cdot \frac{2}{5}$

33. $18 \cdot \frac{2}{9}$

34. $20 \cdot \frac{4}{5}$

35. $\frac{7}{8} \times 10$

36. $\frac{5}{8} \times 12$

37. $\left(\frac{7}{8}\right)\left(1\frac{1}{2}\right)$

38. $\left(4\frac{1}{3}\right)\left(\frac{1}{5}\right)$

39. $\frac{1}{4} \cdot 8\frac{1}{2}$

40. $\frac{1}{3} \cdot 2\frac{1}{5}$

41. $\left(\frac{5}{6}\right)\left(1\frac{1}{9}\right)$

42. $\left(\frac{9}{10}\right)\left(2\frac{1}{7}\right)$

43. $\frac{1}{2} \times 5\frac{1}{3}$

44. $4\frac{1}{2} \times \frac{2}{3}$

45. $\frac{4}{5} \cdot 1\frac{1}{4}$

46. $\frac{3}{8} \cdot 5\frac{1}{3}$

47. $\left(\frac{3}{16}\right)\left(4\frac{2}{3}\right)$

48. $\left(\frac{7}{9}\right)\left(2\frac{1}{4}\right)$

49. $1\frac{1}{7} \times 1\frac{1}{5}$

50. $2\frac{1}{3} \times 1\frac{1}{2}$

51. $\left(2\frac{1}{10}\right)^2$

52. $\left(1\frac{1}{2}\right)^2$

53. $3\frac{9}{10} \cdot 2$

54. $5 \cdot 1\frac{1}{2}$

55. $100 \times 3\frac{3}{4}$

56. $1\frac{5}{6} \times 20$

57. $1\frac{1}{2} \times 5\frac{1}{3}$

58. $5\frac{1}{4} \times 1\frac{1}{9}$

59. $\left(2\frac{1}{2}\right)\left(1\frac{1}{5}\right)$

60. $\left(1\frac{3}{10}\right)\left(2\frac{4}{9}\right)$

61. $12\frac{1}{2} \cdot 3\frac{1}{3}$

62. $5\frac{1}{10} \cdot 1\frac{2}{3}$

63. $66\frac{2}{3} \times 1\frac{7}{10}$

64. $37\frac{1}{2} \times 1\frac{3}{5}$

65. $1\frac{5}{9} \times \frac{3}{8} \times 2$

66. $\frac{1}{8} \times 2\frac{1}{4} \times 6$

67. $\left(\frac{1}{2}\right)^2\left(2\frac{1}{3}\right)$

68. $\left(1\frac{1}{4}\right)^2\left(\frac{1}{5}\right)$

69. $\frac{4}{5} \times \frac{7}{8} \times 1\frac{1}{10}$

70. $8\frac{1}{3} \times \frac{3}{10} \times \frac{5}{6}$

71. $\left(1\frac{1}{2}\right)^3$

72. $\left(2\frac{1}{2}\right)^3$

Divide.

73. $\frac{3}{5} \div \frac{2}{3}$

74. $\frac{2}{3} \div \frac{3}{5}$

75. $\frac{4}{5} \div \frac{7}{8}$

76. $\frac{7}{8} \div \frac{4}{5}$

77. $\frac{1}{2} \div \frac{1}{7}$

78. $\frac{1}{7} \div \frac{1}{2}$

79. $\frac{5}{9} \div \frac{1}{8}$

80. $\frac{1}{8} \div \frac{5}{9}$

81. $\frac{4}{5} \div \frac{8}{15}$

82. $\frac{3}{10} \div \frac{6}{5}$

83. $\frac{7}{8} \div \frac{3}{8}$

84. $\frac{10}{3} \div \frac{5}{6}$

85. $\frac{9}{10} \div \frac{3}{4}$

86. $\frac{5}{6} \div \frac{1}{3}$

87. $\frac{1}{10} \div \frac{2}{5}$

88. $\frac{3}{4} \div \frac{6}{5}$

89. $\frac{2}{3} \div 7$

90. $\frac{7}{10} \div 10$

91. $\frac{2}{3} \div 6$

92. $\frac{1}{20} \div 2$

93. $8 \div \frac{1}{5}$

94. $8 \div \frac{2}{9}$

95. $7 \div \frac{3}{7}$

96. $10 \div \frac{2}{5}$

97. $4 \div \frac{3}{10}$

98. $10 \div \frac{2}{3}$

99. $1 \div \frac{1}{7}$

100. $3 \div \frac{1}{8}$

101. $2\frac{5}{6} \div \frac{3}{7}$

102. $5\frac{1}{9} \div \frac{2}{3}$

103. $1\frac{1}{3} \div \frac{4}{5}$

104. $7\frac{1}{10} \div \frac{1}{2}$

105. $8\frac{5}{6} \div \frac{9}{10}$

106. $6\frac{1}{2} \div \frac{1}{2}$

107. $20\frac{1}{10} \div \frac{1}{5}$

108. $15\frac{2}{3} \div \frac{5}{6}$

109. $\frac{1}{6} \div 2\frac{1}{7}$

110. $\frac{2}{7} \div 1\frac{1}{3}$

111. $\frac{1}{2} \div 2\frac{3}{5}$

112. $\frac{3}{4} \div 3\frac{1}{9}$

113. $4 \div 1\frac{1}{4}$

114. $7 \div 1\frac{9}{10}$

115. $2\frac{1}{10} \div 20$

116. $5\frac{6}{7} \div 14$

117. $2\frac{1}{2} \div 3\frac{1}{7}$

118. $3\frac{1}{7} \div 2\frac{1}{2}$

119. $8\frac{1}{10} \div 5\frac{3}{4}$

120. $1\frac{7}{10} \div 5\frac{1}{8}$

121. $2\frac{1}{3} \div 4\frac{1}{2}$

122. $8\frac{1}{6} \div 2\frac{1}{2}$

123. $6\frac{3}{8} \div 2\frac{5}{6}$

124. $1\frac{2}{3} \div 1\frac{2}{5}$

Simplify.

125. $\frac{1}{2} + \frac{2}{3} \times 1\frac{1}{3}$

126. $\frac{9}{10} + \frac{4}{5} \cdot 8$

127. $5 - \frac{1}{3} \times \frac{2}{5}$

128. $3 \div \frac{2}{5} - 2\frac{1}{3}$

129. $2\frac{3}{4} \times \frac{1}{8} + \frac{1}{5}$

130. $\frac{3}{8} \cdot \frac{1}{2} - \frac{1}{10}$

131. $4 - \frac{2}{9} \div \frac{3}{4}$

132. $6 \div 5 \times \frac{1}{4}$

133. $3\frac{1}{2} \times 6 \div 5$

134. $4 \cdot \frac{2}{3} - 1\frac{1}{8}$

135. $10 \times \frac{1}{8} \times 2\frac{1}{2}$

136. $\frac{1}{3} \div \frac{1}{6} \times \frac{2}{3}$

137. $8 \div 1\frac{1}{5} + 3 \cdot 1\frac{1}{2}$

138. $3\frac{1}{8} \div 5 + 4 \div 2\frac{1}{2}$

139. $\left(1\frac{1}{2} \div \frac{1}{3}\right)^2 + \left(1 - \frac{1}{4}\right)^2$

140. $\left(3\frac{1}{2}\right)^2 + 2\left(1\frac{1}{2} - 1\frac{1}{3}\right)$

Mixed Practice

141. Divide $6\frac{1}{8}$ by $2\frac{3}{4}$.

142. Compute: $14 - 3 \div \left(\frac{4}{5}\right)^2$

143. Find the product of $\frac{3}{5}$ and $\frac{7}{8}$.

144. Find the quotient of $\frac{9}{10}$ and $\frac{2}{5}$.

145. Multiply $\frac{2}{3}$ by 12.

146. Calculate: $\left(4\frac{1}{2}\right)\left(6\frac{2}{3}\right)$

Applications

Solve. Write the answer in simplest form.

147. In a local town, $\frac{5}{6}$ of the voting-age population is registered to vote. If $\frac{7}{10}$ of the registered voters voted in the election for mayor, what fraction of the voting-age population voted?

148. Last year, $\frac{1}{8}$ of the emergency room visits at a hospital were injury related. Of these, $\frac{2}{5}$ were due to motor vehicle accidents. What fraction of the emergency room visits were due to motor vehicle accidents?

149. The house that a couple wants to buy is selling for \$240,000. They need to put $\frac{1}{20}$ of the selling price down, and take out a mortgage for the rest. How much money do they need to put down?

150. There is a rule of thumb that no one should spend more than $\frac{1}{4}$ of their income on rent. If someone makes \$24,000 a year, what is the most he or she should spend per month on rent according to this rule?

151. Students in an astronomy course learn that a first-magnitude star is $2\frac{1}{2}$ times as bright as a second-magnitude star, which in turn is $2\frac{1}{2}$ times as bright as a third-magnitude star. How many times as bright as a third-magnitude star is a first-magnitude star?

152. Which of these rooms has the larger area?

16 ft
$11\frac{1}{2}$ ft

$15\frac{1}{2}$ ft
12 ft

153. Find the cost of buying carpeting at $\$7\frac{1}{2}$ per square-foot for the hallway shown.

154. Some people believe that gasohol is superior to gasoline as an automotive fuel. Gasohol is a mixture of gasoline $\left(\frac{9}{10}\right)$ and ethyl alcohol $\left(\frac{1}{10}\right)$. How much more gasoline than ethyl alcohol is there in $10\frac{1}{2}$ gallons of gasohol?

155. Because of evaporation, a pond loses $\frac{1}{4}$ of its remaining water each month of summer. If it is full at the beginning of summer, what fraction of the original amount will the pond contain after three summer months?

156. A scientist is investigating the effects of cold on human skin. In one of the scientist's experiments, the temperature starts at 70°F and drops by $\frac{1}{10}°$ every 2 minutes. What is the temperature after 6 minutes?

157. A trip to a nearby island takes $3\frac{1}{2}$ hours by boat and $\frac{1}{2}$ hour by airplane. How many times as fast as the boat is the plane?

158. Each dose of aspirin weighs $\frac{3}{4}$ grain. If a nurse has 9 grains of aspirin on hand, how many doses can he administer?

159. A store sells two types of candles. The scented candle is 8 inches tall and burns $\frac{1}{2}$ inch per hour, whereas the unscented candle is 10 inches tall and burns $\frac{1}{3}$ inch per hour.

a. In an hour, which candle will burn more?

b. Which candle will last longer?

160. A college-wide fund-raising campaign collected \$3 million in $1\frac{1}{2}$ years for student scholarships.

a. What was the average amount collected per year?

b. By how much would this average increase if an additional \$1 million were collected?

• *Check your answers on page A-4.*

MINDSTRETCHERS

Writing

1. Every number except 0 has a reciprocal. Explain why 0 does not have a reciprocal.

Groupwork

2. In the following magic square, the *product* of every row, column, and diagonal is 1. Working with a partner, complete the square.

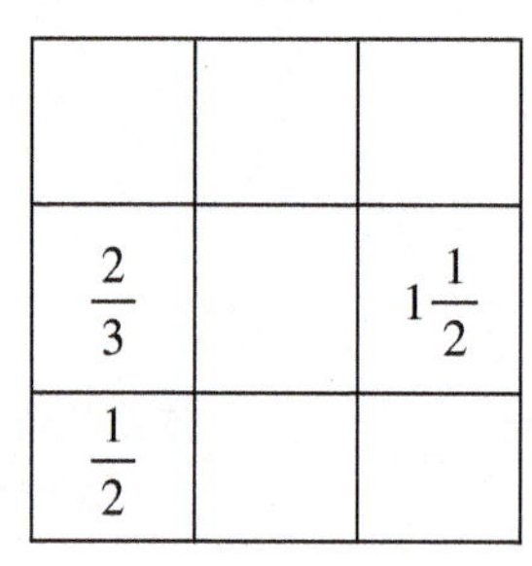

Patterns

3. Find the product: $1\frac{1}{2} \cdot 1\frac{1}{3} \cdot 1\frac{1}{4} \cdot \cdots \cdot 1\frac{1}{99} \cdot 1\frac{1}{100}$

KEY CONCEPTS AND SKILLS

CONCEPT SKILL

Concept/Skill	Description	Example
[2.1] **Prime number**	A whole number that has exactly two different factors: itself and 1.	2, 3, 5
[2.1] **Composite number**	A whole number that has more than two factors.	4, 8, 9
[2.1] **Prime factorization of a whole number**	The number written as the product of its prime factors.	$30 = 2 \cdot 3 \cdot 5$
[2.1] **Least common multiple (LCM) of two or more whole numbers**	The smallest nonzero whole number that is a multiple of each number.	The LCM of 30 and 45 is 90.
[2.1] **To compute the least common multiple (LCM)**	• Find the prime factorization of each number. • Identify the prime factors that appear in each factorization. • Multiply these prime factors, using each factor the greatest number of times that it occurs in any of the factorizations.	$20 = 2 \cdot 2 \cdot 5$ $= 2^2 \cdot 5$ $30 = 2 \cdot 3 \cdot 5$ The LCM of 20 and 30 is $2^2 \cdot 3 \cdot 5$, or 60.
[2.2] **Fraction**	Any number that can be written in the form $\frac{a}{b}$, where a and b are whole numbers and b is nonzero.	$\frac{3}{11}, \frac{9}{5}$
[2.2] **Proper fraction**	A fraction whose numerator is smaller than its denominator.	$\frac{2}{7}, \frac{1}{2}$
[2.2] **Mixed number**	A number with a whole-number part and a proper fraction part.	$5\frac{1}{3}, 4\frac{5}{6}$
[2.2] **Improper fraction**	A fraction whose numerator is greater than or equal to its denominator.	$\frac{9}{4}, \frac{5}{5}$
[2.2] **To change a mixed number to an improper fraction**	• Multiply the denominator of the fraction by the whole-number part of the mixed number. • Add the numerator of the fraction to this product. • Write this sum over the denominator to form the improper fraction.	$4\frac{2}{3} = \frac{3 \times 4 + 2}{3}$ $= \frac{14}{3}$
[2.2] **To change an improper fraction to a mixed number**	• Divide the numerator by the denominator. • If there is a remainder, write it over the denominator.	$\frac{14}{3} = 4\frac{2}{3}$
[2.2] **To find an equivalent fraction**	Multiply the numerator and denominator of $\frac{a}{b}$ by the same whole number; that is, $\frac{a}{b} = \frac{a \cdot n}{b \cdot n}$, where both b and n are nonzero.	$\frac{3}{4} = \frac{3 \cdot 2}{4 \cdot 2} = \frac{6}{8}$
[2.2] **To simplify (reduce) a fraction**	Divide the numerator and denominator of $\frac{a}{b}$ by the same whole number n; that is, $\frac{a}{b} = \frac{a \div n}{b \div n}$, where both b and n are nonzero.	$\frac{6}{8} = \frac{6 \div 2}{8 \div 2} = \frac{3}{4}$

continued

CONCEPT SKILL

Concept/Skill	Description	Example
[2.2] **Like fractions**	Fractions with the same denominator.	$\frac{2}{5}, \frac{3}{5}$
[2.2] **Unlike fractions**	Fractions with different denominators.	$\frac{3}{5}, \frac{3}{10}$
[2.2] **To compare fractions**	• If the fractions are like, compare their numerators. • If the fractions are unlike, write them as equivalent fractions with the same denominator and then compare their numerators.	$\frac{6}{8}, \frac{7}{8}$ $6 < 7$, so $\frac{6}{8} < \frac{7}{8}$ $\frac{2}{3}, \frac{12}{15}$ or $\frac{10}{15}, \frac{12}{15}$ $12 > 10$, so $\frac{12}{15} > \frac{2}{3}$
[2.2] **Least common denominator (LCD) of two or more fractions**	The least common multiple of their denominators.	The LCD of $\frac{11}{30}$ and $\frac{7}{45}$ is 90.
[2.3] **To add (or subtract) like fractions**	• Add (or subtract) the numerators. • Use the given denominator. • Write the answer in simplest form.	$\frac{1}{8} + \frac{1}{8} = \frac{2}{8} = \frac{1}{4}$ $\frac{3}{8} - \frac{1}{8} = \frac{2}{8} = \frac{1}{4}$
[2.3] **To add (or subtract) unlike fractions**	• Write the fractions as equivalent fractions with the same denominator, usually the LCD. • Add (or subtract) the numerators, keeping the same denominator. • Write the answer in simplest form.	$\frac{2}{3} + \frac{1}{2} = \frac{4}{6} + \frac{3}{6}$ $= \frac{7}{6}$, or $1\frac{1}{6}$ $\frac{5}{12} - \frac{1}{6}$ $= \frac{5}{12} - \frac{2}{12} = \frac{3}{12}$, or $\frac{1}{4}$
[2.3] **To add mixed numbers**	• Write the fractions as equivalent fractions with the same denominator, usually the LCD. • Add the fractions. • Add the whole numbers. • Write the answer in simplest form.	$4\frac{1}{2} = 4\frac{3}{6}$ $+6\frac{2}{3} = +6\frac{4}{6}$ $10\frac{7}{6} = 11\frac{1}{6}$
[2.3] **To subtract mixed numbers**	• Write the fractions as equivalent fractions with the same denominator, usually the LCD. • Rename (or borrow from) the whole number on top if the fraction on the bottom is larger than the fraction on top. • Subtract the fractions. • Subtract the whole numbers. • Write the answer in simplest form.	$4\frac{1}{5} = 3\frac{6}{5}$ $-1\frac{2}{5} = -1\frac{2}{5}$ $2\frac{4}{5}$
[2.4] **To multiply fractions**	• Multiply the numerators. • Multiply the denominators. • Write the answer in simplest form.	$\frac{1}{2} \cdot \frac{3}{5} = \frac{3}{10}$

continued

Concept/Skill	Description	Example
[2.4] To multiply mixed numbers	• Write the mixed numbers as improper fractions. • Multiply the fractions. • Write the answer in simplest form.	$2\frac{1}{2} \cdot 1\frac{2}{3} = \frac{5}{2} \cdot \frac{5}{3}$ $= \frac{25}{6}$, or $4\frac{1}{6}$
[2.4] Reciprocal of $\frac{a}{b}$	The fraction $\frac{b}{a}$ formed by switching the numerator and denominator.	The reciprocal of $\frac{4}{3}$ is $\frac{3}{4}$.
[2.4] To divide fractions	• Change the divisor to its reciprocal, and multiply the resulting fractions. • Write the answer in simplest form.	$\frac{2}{5} \div \frac{3}{7} = \frac{2}{5} \cdot \frac{7}{3}$ $= \frac{14}{15}$
[2.4] To divide mixed numbers	• Write the mixed numbers as improper fractions. • Divide the fractions. • Write the answer in simplest form.	$2\frac{1}{2} \div 1\frac{1}{3} =$ $\frac{5}{2} \div \frac{4}{3} =$ $\frac{5}{2} \cdot \frac{3}{4} = \frac{15}{8}$, or $1\frac{7}{8}$

CULTURAL NOTE

I/IIII Dieſſe figur iſt vñ bedeüt ain fiertel von ainẽ ganꝛen/alſo mag man auch ain fünfftail/ayn ſechſtail/ain ſybentail oder zwai ſechſtail ꝛc. vnd alle ander brüch beſchreiben/Als I/V | I/VI | I/VII | II/VI ꝛc.

VI/VIII Diß ſein Sechs achtail/das ſein ſechstail der acht ain gantz machen.

IX/XI Diß Figur bezaigt ann newn ayilfftail das ſeyn IX tail/der XI ain gantz machen.

XX/XXXI Diß Figur bezaichet/zwentzigk ainundreyſigk tail/das ſein zwentzigk tail der ainsundreiſſigk ain gantz machen.

IIC/IIIIC.LX Diß ſein zwaihundert tail/der Fierhundert vnd ſechtzigk ain gantz machen.

Source: David Eugene Smith and Jekuthiel Ginsburg, *Numbers and Numerals, a Story Book for Young and Old* (New York: Bureau of Publications, Teachers College, Columbia University, 1937).

In societies throughout the world and across the centuries, people have written fractions in strikingly different ways. In ancient Greece, for example, the fraction $\frac{1}{4}$ was written Δ″ where Δ (read "delta") is the fourth letter of the Greek alphabet.

At one time, people wrote the numerator and denominator of fractions in Roman numerals, as shown at the left in a page from a sixteenth-century German book. In today's notation, the last fraction is $\frac{200}{460}$.

Chapter 2 Review Exercises

To help you review this chapter, solve these problems.

[2.1] *Find all the factors of each number.*

1. 150 **2.** 180 **3.** 57 **4.** 70

Indicate whether each number is prime or composite.

5. 23 **6.** 33 **7.** 87 **8.** 67

Write the prime factorization of each number, using exponents.

9. 36 **10.** 75 **11.** 99 **12.** 54

Find the LCM.

13. 6 and 14 **14.** 5 and 10 **15.** 18, 24, and 36 **16.** 10, 15, and 20

[2.2] *Identify the fraction or mixed number represented by the shaded portion of each figure.*

17. 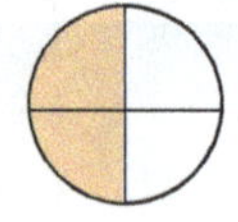

18. 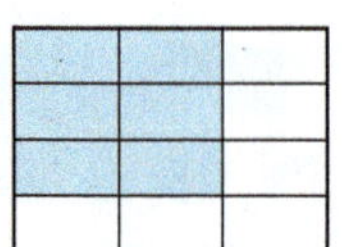

19. 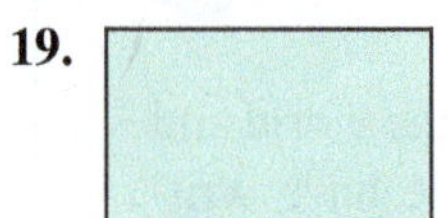 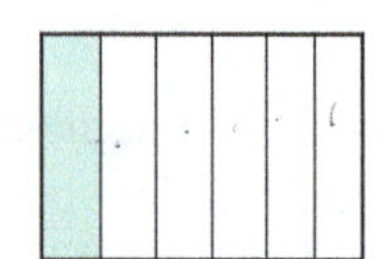

20.  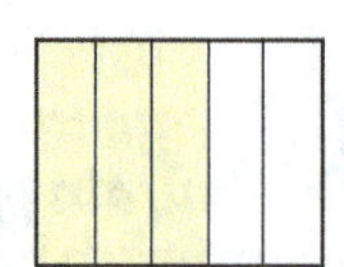

Indicate whether each number is a proper fraction, an improper fraction, or a mixed number.

21. $4\frac{1}{8}$ **22.** $\frac{5}{6}$ **23.** $\frac{3}{2}$ **24.** $\frac{7}{1}$

Write each mixed number as an improper fraction.

25. $7\frac{2}{3}$ **26.** $1\frac{4}{5}$ **27.** $9\frac{1}{10}$ **28.** $8\frac{3}{7}$

Write each fraction as a mixed number or a whole number.

29. $\frac{13}{2}$ **30.** $\frac{14}{3}$ **31.** $\frac{11}{4}$ **32.** $\frac{12}{12}$

Write an equivalent fraction with the given denominator.

33. $7 = \frac{}{12}$ **34.** $\frac{2}{7} = \frac{}{14}$ **35.** $\frac{1}{2} = \frac{}{10}$ **36.** $\frac{9}{10} = \frac{}{30}$

Simplify.

37. $\frac{14}{28}$ **38.** $\frac{15}{21}$ **39.** $\frac{30}{45}$ **40.** $\frac{54}{72}$

41. $5\frac{2}{4}$ **42.** $8\frac{10}{15}$ **43.** $6\frac{12}{42}$ **44.** $8\frac{45}{63}$

Insert the appropriate sign: <, =, or >.

45. $\frac{5}{8}\quad\frac{3}{8}$ **46.** $\frac{5}{6}\quad\frac{1}{6}$ **47.** $\frac{2}{3}\quad\frac{4}{5}$ **48.** $\frac{9}{10}\quad\frac{7}{8}$

49. $\frac{3}{4}\quad\frac{5}{8}$ **50.** $\frac{7}{10}\quad\frac{5}{9}$ **51.** $3\frac{1}{5}\quad1\frac{9}{10}$ **52.** $5\frac{1}{8}\quad5\frac{1}{9}$

Arrange in increasing order.

53. $\frac{2}{7}, \frac{3}{8}, \frac{1}{2}$ **54.** $\frac{1}{5}, \frac{1}{3}, \frac{2}{15}$ **55.** $\frac{4}{5}, \frac{9}{10}, \frac{3}{4}$ **56.** $\frac{7}{8}, \frac{7}{9}, \frac{13}{18}$

[2.3] *Add and simplify.*

57. $\frac{2}{5} + \frac{4}{5}$ **58.** $\frac{7}{20} + \frac{8}{20}$ **59.** $\frac{5}{8} + \frac{7}{8} + \frac{3}{8}$ **60.** $\frac{3}{10} + \frac{1}{10} + \frac{2}{10}$

61. $\frac{1}{3} + \frac{2}{5}$ **62.** $\frac{7}{8} + \frac{5}{6}$ **63.** $\frac{9}{10} + \frac{1}{2} + \frac{2}{5}$ **64.** $\frac{3}{8} + \frac{4}{5} + \frac{3}{4}$

65. $2 + 3\frac{7}{8}$ **66.** $6\frac{1}{4} + 3\frac{1}{4}$ **67.** $8\frac{7}{10} + 1\frac{9}{10}$ **68.** $5\frac{5}{6} + 2\frac{1}{6}$

69. $2\frac{1}{3} + 4\frac{1}{3} + 5\frac{2}{3}$ **70.** $1\frac{3}{10} + \frac{9}{10} + 2\frac{1}{10}$ **71.** $5\frac{2}{5} + \frac{3}{10}$ **72.** $9\frac{1}{6} + 8\frac{3}{8}$

73. $10\frac{2}{3} + 12\frac{3}{4}$ **74.** $20\frac{1}{2} + 25\frac{7}{8}$ **75.** $10\frac{3}{5} + 7\frac{9}{10} + 2\frac{1}{4}$ **76.** $20\frac{7}{8} + 30\frac{5}{6} + 4\frac{1}{3}$

Subtract and simplify.

77. $\frac{3}{8} - \frac{1}{8}$ **78.** $\frac{7}{9} - \frac{1}{9}$ **79.** $\frac{5}{3} - \frac{2}{3}$ **80.** $\frac{4}{6} - \frac{4}{6}$

81. $\frac{3}{10} - \frac{1}{20}$ **82.** $\frac{1}{2} - \frac{1}{8}$ **83.** $\frac{3}{5} - \frac{1}{4}$ **84.** $\frac{1}{3} - \frac{1}{10}$

85. $12\frac{1}{2} - 5$ **86.** $4\frac{3}{10} - 2$ **87.** $8\frac{7}{8} - 5\frac{1}{8}$ **88.** $20\frac{3}{4} - 2\frac{1}{4}$

89. $12 - 5\frac{1}{2}$ **90.** $4 - 2\frac{3}{10}$ **91.** $7 - 4\frac{1}{3}$ **92.** $1 - \frac{4}{5}$

93. $6\frac{1}{10} - 4\frac{3}{10}$ **94.** $2\frac{5}{8} - 1\frac{7}{8}$ **95.** $5\frac{1}{4} - 2\frac{3}{4}$ **96.** $7\frac{1}{6} - 3\frac{5}{6}$

97. $3\frac{1}{10} - 2\frac{4}{5}$ **98.** $7\frac{1}{2} - 4\frac{5}{8}$ **99.** $5\frac{1}{12} - 4\frac{1}{2}$ **100.** $6\frac{2}{9} - 2\frac{1}{3}$

101. $\frac{1}{3} + \frac{5}{6} - \frac{1}{2}$ **102.** $7\frac{9}{10} - 1\frac{1}{5} + 2\frac{3}{4}$

[2.4] *Multiply and simplify.*

103. $\frac{3}{4} \times \frac{1}{4}$ **104.** $\frac{1}{2} \times \frac{7}{8}$ **105.** $\left(\frac{5}{6}\right)\left(\frac{3}{4}\right)$ **106.** $\left(\frac{2}{3}\right)\left(\frac{1}{4}\right)$

107. $\frac{2}{3} \cdot 8$

108. $\frac{1}{10} \cdot 7$

109. $\left(\frac{1}{5}\right)^3$

110. $\left(\frac{2}{3}\right)^3$

111. $\frac{1}{2} \times \frac{2}{3} \times \frac{3}{4}$

112. $\frac{7}{8} \times \frac{2}{5} \times \frac{1}{6}$

113. $\frac{4}{5} \times 1\frac{1}{5}$

114. $\frac{2}{3} \times 2\frac{1}{3}$

115. $5\frac{1}{3} \cdot \frac{1}{2}$

116. $\frac{1}{10} \cdot 6\frac{2}{3}$

117. $1\frac{1}{3} \cdot 4\frac{1}{2}$

118. $3\frac{1}{4} \cdot 5\frac{2}{3}$

119. $6\frac{3}{4} \times 1\frac{1}{4}$

120. $8\frac{1}{2} \times 2\frac{1}{2}$

121. $\frac{7}{8} \times 1\frac{1}{5} \times \frac{3}{7}$

122. $1\frac{3}{8} \times \frac{10}{11} \times 1\frac{1}{4}$

123. $\left(3\frac{1}{3}\right)^3$

124. $\left(1\frac{1}{2}\right)^3$

125. $\frac{5}{8} + \frac{1}{2} \cdot 5$

126. $1\frac{9}{10} - \left(\frac{2}{3}\right)^2$

127. $4\left(\frac{2}{5}\right) + 3\left(\frac{1}{6}\right)$

128. $6\left(1\frac{1}{2} - \frac{3}{10}\right)$

Find the reciprocal.

129. $\frac{2}{3}$

130. $1\frac{1}{2}$

131. 8

132. $\frac{1}{4}$

Divide and simplify.

133. $\frac{7}{8} \div 5$

134. $\frac{5}{9} \div 9$

135. $\frac{2}{3} \div 5$

136. $\frac{1}{100} \div 2$

137. $\frac{1}{2} \div \frac{2}{3}$

138. $\frac{2}{3} \div \frac{1}{2}$

139. $6 \div \frac{1}{5}$

140. $7 \div \frac{4}{5}$

141. $\frac{7}{8} \div \frac{3}{4}$

142. $\frac{9}{10} \div \frac{1}{2}$

143. $\frac{3}{5} \div \frac{3}{10}$

144. $\frac{2}{3} \div \frac{1}{6}$

145. $3\frac{1}{2} \div 2$

146. $2 \div 3\frac{1}{2}$

147. $6\frac{1}{3} \div 4$

148. $4 \div 6\frac{1}{3}$

149. $8\frac{1}{4} \div 1\frac{1}{2}$

150. $3\frac{2}{5} \div 1\frac{1}{3}$

151. $4\frac{1}{2} \div 2\frac{1}{4}$

152. $7\frac{1}{5} \div 2\frac{2}{5}$

153. $\left(5 - \frac{2}{3}\right) \div \frac{4}{9}$

154. $6\frac{1}{2} \div \left(\frac{1}{2} + 4\frac{1}{2}\right)$

155. $7 \div 2\frac{1}{4} + 5 \div \left(1\frac{1}{2}\right)^2$

156. $\left(1\frac{2}{3}\right)^2 \times 2 + 9 \div 4\frac{1}{2}$

Mixed Applications

Solve. Write the answer in simplest form.

157. The Summer Olympic Games are held during each year divisible by 4. Were the Olympic Games held in 1990?

158. What is the smallest amount of money that you can pay in both all quarters and all dimes?

159. Eight of the 32 human teeth are incisors. What fraction of human teeth are incisors? (*Source:* Ilsa Goldsmith, *Human Anatomy for Children*)

160. The planets in the solar system (including the "dwarf planet" Pluto) consist of Earth, two planets closer to the Sun than Earth, and six planets farther from the Sun than Earth. What fraction of the planets in the solar system are closer than Earth to the Sun? (*Source:* Patrick Moore, *Astronomy for the Beginner*)

161. A Filmworks camera has a shutter speed of $\frac{1}{8{,}000}$ second and a Lensmax camera has a shutter speed of $\frac{1}{6{,}000}$ second. Which shutter is faster? (*Hint:* The faster shutter has the smaller shutter speed.)

162. Of the approximately 12,000 women that started the 2006 New York City Marathon, only about 300 did not finish the race. What fraction of the women did finish the race? (*Source:* www.nycmarathon.org)

163. An insurance company reimbursed a patient \$275 on a dental bill of \$700. Did the patient get more or less than $\frac{1}{3}$ of the money paid back? Explain.

164. A union goes on strike if at least $\frac{2}{3}$ of the workers voting support the strike call. If 23 of the 32 voting workers support a strike, should a strike be declared? Explain.

165. A grand jury has 23 jurors. Sixteen jurors are needed for a quorum, and a vote of 12 jurors is needed to indict.

a. What fraction of the full jury is needed to indict?

b. Suppose that 16 jurors are present. What fraction of those present is needed to indict?

166. In a tennis match, Lisa Gregory went to the net 12 times, winning the point 7 times. By contrast, Monica Yates won the point 4 of the 6 times that she went to the net.

a. Which player went to the net more often?

b. Which player had a better rate of winning points at the net?

167. In a math course, $\frac{3}{5}$ of a student's grade is based on four in-class exams, and $\frac{3}{20}$ of the grade is based on homework. What fraction of a student's grade is based on in-class exams and homework?

168. A metal alloy is made by combining $\frac{1}{4}$ ounce of copper with $\frac{2}{3}$ ounce of tin. Find the alloy's total weight.

169. The weight of a diamond is measured in carats. What is the difference in weight between a $\frac{1}{2}$-carat and a $\frac{3}{4}$-carat diamond?

170. During a sale, the price of a sweater was marked $\frac{1}{4}$ off the original price of \$45. Using a coupon, a customer received an additional $\frac{1}{5}$ off the sale price. What fraction of the original price was the final sale price?

171. In a math class, $\frac{3}{8}$ of the students are chemistry majors and $\frac{2}{3}$ of those students are women. If there are 48 students in the math class, how many women are chemistry majors?

172. Three-eighths of the undergraduate students at a two-year college receive financial aid. If the college has 4,296 undergraduate students, how many undergraduate students do not receive financial aid?

173. An investor bought $1,000 worth of a technology stock. At the beginning of last year, it had increased in value by $\frac{2}{5}$. During the year, the value of the stock declined by $\frac{1}{4}$. What was the value of the stock at the end of last year?

174. A sea otter eats about $\frac{1}{5}$ of its body weight each day. How much will a 35-pound otter eat in a day? (***Source:*** Karl W. Kenyon, *The Sea Otter in the Eastern Pacific Ocean*)

175. The regular price of roses is $27 a dozen. What is the cost of the roses on sale?

176. In Roseville, 40 of every 1,000 people who want to work are unemployed, in contrast to 8 of every 100 people in Georgetown. How many times as great as the unemployment rate in Roseville is the unemployment rate in Georgetown?

177. A brother and sister want to buy as many goldfish as possible for their new fish tank. A rule of thumb is that the total length of fish, in inches, should be less than the capacity of the tank in gallons. If they have a 10-gallon tank and goldfish average $\frac{1}{2}$ inch in length, how many fish should they buy?

178. A commuter is driving to the city of Denver 15 miles away. If he has already driven $3\frac{1}{4}$ miles, how far is he from Denver?

179. The wingspan of a Boeing 717 jet is $93\frac{1}{4}$ feet, whereas the wingspan of a Boeing 767 jet is $156\frac{1}{12}$ feet. How much longer is the wingspan of a Boeing 767 jet?

180. The Acela Express train from Boston to New York took $3\frac{5}{12}$ hours, whereas the Regional Service train took $4\frac{1}{4}$ hours. How much faster than the Regional Service train was the Acela Express train? (***Source:*** Amtrak).

181. An airplane is flying $1\frac{1}{2}$ times the speed of sound. If sound travels at about 1,000 feet per second, at what speed is the plane flying?

182. When you stand upright, the pressure per square inch on your hip joint is about $2\frac{1}{2}$ times your body weight. If you weigh 200 lb, what is that pressure?

183. A cubic foot of water weighs approximately $62\frac{1}{2}$ pounds. If a basin contains $4\frac{1}{2}$ cubic feet of water, how much does the water weigh?

184. During a housing boom, the market value of a home was $2\frac{1}{2}$ times its original value. If its market value is \$115,000, what was its original value?

185. It took the space shuttle Endeavor $1\frac{1}{2}$ hours to orbit Earth. How many orbits did the Endeavor make in 12 hours? (*Source:* NASA)

186. What is the area of the highway billboard sign shown below?

187. The following chart is a record of the amount of time (in hours) that two employees spent working the past weekend. Complete the chart.

Employee	Saturday	Sunday	Total
L. Chavis	$7\frac{1}{2}$	$4\frac{1}{4}$	
R. Young	$5\frac{3}{4}$	$6\frac{1}{2}$	
Total			

188. Complete the following chart.

Worker	Hours per Day	Days Worked	Total Hours	Wage per Hour	Gross Pay
Maya	5	3		\$7	
Noel	$7\frac{1}{4}$	4		\$10	
Alisa	$4\frac{1}{2}$	$5\frac{1}{2}$		\$9	

189. According to a newspaper advertisement, a man on a diet lost 60 pounds in $5\frac{1}{2}$ months. On the average, how much weight did he lose per month?

190. According to the nutrition label on a box of cereal, one serving is $1\frac{1}{4}$ cups. If the box contains 18 servings of cereal, how many cups of cereal does it contain?

• *Check your answers on page A-4.*

Chapter 2 POSTTEST

FOR EXTRA HELP

Test solutions are found on the enclosed CD.

To see if you have mastered the topics in this chapter, take this test.

1. List all the factors of 63.

2. Write 54 as the product of prime factors.

3. What fraction of the diagram is shaded?

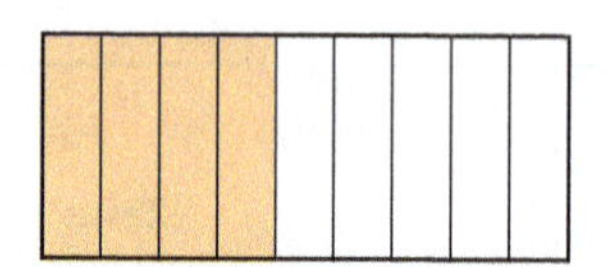

4. Write 12 as an improper fraction.

5. Express $\frac{41}{4}$ as a mixed number.

6. Write $\frac{875}{1{,}000}$ in simplest form

7. Which is smaller, $\frac{2}{3}$ or $\frac{5}{10}$?

8. What is the LCD for $\frac{3}{8}$ and $\frac{1}{12}$?

Add.

9. $\frac{2}{3} + \frac{1}{8} + \frac{3}{4}$

10. $6\frac{7}{8} + 1\frac{3}{10}$

Subtract.

11. $6 - 1\frac{5}{7}$

12. $10\frac{1}{6} - 4\frac{2}{5}$

Multiply.

13. $\left(\frac{1}{9}\right)^2$

14. $2\frac{2}{3} \times 4\frac{1}{2}$

15. Divide: $2\frac{1}{3} \div 3$

16. Calculate: $14\frac{1}{2} - 5 \cdot 1\frac{1}{3}$

Solve. Write your answer in simplest form.

17. Seven of the 42 men who have served as president of the United States were born in Ohio. What fraction of these men were *not* born in Ohio? (**Source:** *The World Almanac 2006*)

18. In an Ironman triathlon, an athlete completed the 112-mile bike ride in $5\frac{5}{6}$ hours. What was the average number of miles she bicycled each hour?

19. Find the area of the rectangular floor pictured.

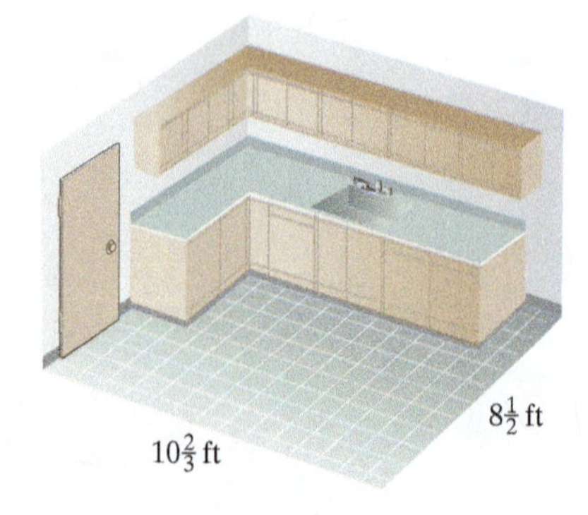

20. Find the distance across the hubcap of the tire shown in the diagram.

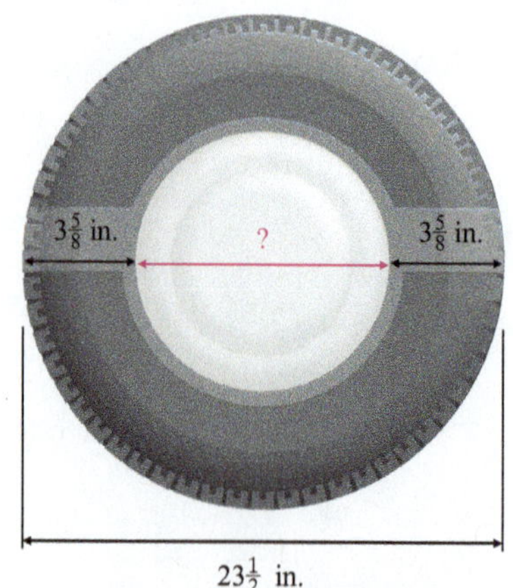

• Check your answers on page A-5.

Cumulative Review Exercises

To help you review, solve the following:

1. Write in words: 5,000,315

2. Multiply: $5{,}814 \times 100$

3. Find the quotient: $89\overline{)80{,}812}$

4. Write $\frac{75}{100}$ in simplest form.

5. Subtract: $8 - 1\frac{3}{5}$

6. Find the product: $1\frac{1}{2} \cdot 4\frac{2}{3}$

7. Which is larger, $\frac{1}{4}$ or $\frac{3}{8}$?

8. A jury decided on punishments in an oil spill case. The jury ordered the captain of the oil barge to pay \$5,000 in punitive damages and the oil company to pay \$5 billion. The amount that the company had to pay is how many times the amount that the captain had to pay?

9. The counter on a photocopier keeps track of the number of copies made. How many copies did you make if the counter showed 23,459 copies before you started copying and 24,008 after you finished?

10. A homeowner wants to refinish his basement over three weekends. He completes $\frac{1}{4}$ of the job the first weekend and $\frac{5}{12}$ of the job the second weekend. What fraction of the job remains to be completed?

• *Check your answers on page A-5.*

CHAPTER 3

Decimals

Decimals and Blood Tests

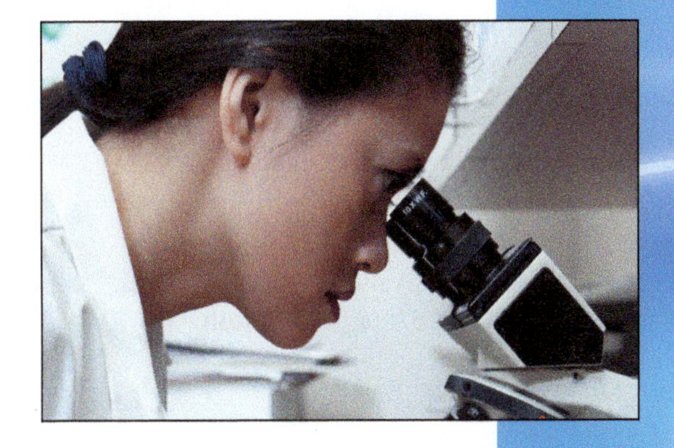

Blood tests reveal a great deal about a person's health—whether to reduce the cholesterol level to lower the risk of heart disease, or raise the red blood cell count to prevent anemia. And blood tests identify a variety of diseases, for example, AIDS and mononucleosis.

Blood analyses are typically carried out in clinical laboratories. Technicians in these labs operate giant machines that perform thousands of blood tests per hour.

What these blood tests, known as "chemistries," actually do is to analyze blood for a variety of substances, such as creatinine and calcium.

In any blood test, doctors look for abnormal levels of the substance being measured. For instance, the normal range on the creatinine test is typically from 0.7 to 1.5 milligrams (mg) per unit of blood. A high level may mean kidney disease; a low level, muscular dystrophy.

The normal range on the calcium test may be 9.0 to 10.5 mg per unit of blood. A result outside this range is a clue for any of several diseases.

(***Source:*** Dixie Farley, "Top 10 Laboratory Tests: Blood Will Tell," *FDA Consumer*, Vol. 23)

Chapter 3 PRETEST

To see if you have already mastered the topics in this chapter, take this test.

1. In the number 27.081, what place does the 8 occupy?
2. Write in words: 4.012
3. Round 3.079 to the nearest tenth.
4. Which is largest: 0.00212, 0.0029, or 0.000888?

Perform the indicated operations.

5. $7.02 + 3.5 + 11$
6. $2.37 + 5.0038$
7. $13.79 - 2.1$
8. $9 - 2.7 + 3.51$
9. $8.3 \times 1{,}000$
10. 8.01×2.3
11. $(0.12)^2$
12. $5 + 3 \times 0.7$
13. $6.05 \div 1{,}000$
14. $\dfrac{9.81}{0.3}$

Express as a decimal.

15. $\dfrac{7}{8}$
16. $2\dfrac{5}{6}$, rounded to the nearest hundredth

Solve.

17. In a science course, a student learns that an acid is stronger if it has a lower pH value. Which is stronger, an acid with a pH value of 3.7 or an acid with a pH value of 2.95?
18. In 2005, Microsoft Corporation had quarterly revenues of $9.189 billion, $10.818 billion, $9.62 billion, and $10.161 billion. What was Microsoft's total revenue that year? (**Source:** Microsoft Corporation, 2005 Annual Report)
19. A serving of iceberg lettuce contains 3.6 milligrams (mg) of vitamin C, whereas romaine lettuce contains 11.9 mg of vitamin C. Romaine lettuce is how many times as rich in vitamin C as iceberg lettuce? Round the answer to the nearest whole number. (**Source:** *The Concise Encyclopedia of Foods and Nutrition*)
20. Suppose that a long-distance telephone call costs $0.85 for the first 3 minutes and $0.17 for each additional minute. What is the cost of a 20-minute call?

• *Check your answers on page A-5.*

3.1 Introduction to Decimals

OBJECTIVES

- To read and write decimals
- To find the fraction equivalent to a decimal
- To compare decimals
- To round decimals
- To solve word problems involving decimals

What Decimals Are and Why They Are Important

Decimal notation is in common use. When we say that the price of a book is $32.75, that the length of a table is 1.8 meters, or that the answer displayed on a calculator is 5.007, we are using decimals.

A number written as a **decimal** has

- a whole-number part, which *precedes* the decimal point, and
- a fractional part, which *follows* the decimal point.

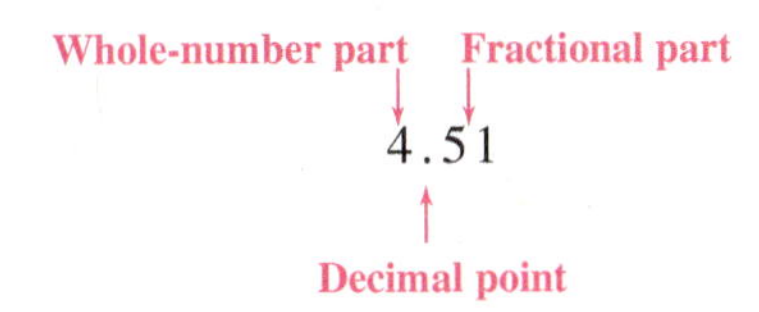

A decimal without a decimal point shown is understood to have one at the right end and is the same as a whole number. For instance, 3 and 3. are the same number.

The fractional part of any decimal has as its denominator a power of 10, such as 10, 100, or 1,000. The use of the word *decimal* reminds us of the importance of the number 10 in this notation, just as decade means 10 years or December meant the 10th month of the year (which it was for the early Romans).

In many problems, we can choose to work with either decimals or fractions. Therefore, we need to know how to work with both if we are to use the easier approach to solve a particular problem.

Decimal Places

Each digit in a decimal has a place value. The place value system for decimals is an extension of the place value system for whole numbers.

The places to the right of the decimal point are called **decimal places**. For instance, the number 64.149 is said to have three decimal places.

Recall that, for a whole number, place values are powers of 10: 1, 10, 100, ... By contrast, each place value for the fractional part of a decimal is the reciprocal of a power of 10: $\frac{1}{10}, \frac{1}{100}, \frac{1}{1{,}000}, \ldots$.

The first decimal place after the decimal point is the tenths place. Working to the right, the next decimal places are the hundredths place, the thousandths place, and so on.

The following table shows the place values in the decimals 0.54 and 0.30716.

Ones	.	Tenths	Hundredths	Thousandths	Ten-thousandths	Hundred-thousandths
0	.	5	4			
0	.	3	0	7	1	6

The next table shows the place values for the decimals 7,204.5 and 513.285.

Thousands	Hundreds	Tens	One	.	Tenths	Hundredths	Thousandths
7	2	0	4	.	5		
	5	1	3	.	2	8	5

EXAMPLE 1

In each number, identify the place that the digit 3 occupies.

a. 0.134 **b.** 92.388 **c.** 0.600437

Solution

a. The hundredths place

b. The tenths place

c. The hundred-thousandths place

PRACTICE 1

What place does the digit 1 occupy in each number?

a. 566.184

b. 43.57219

c. 0.921

Changing Decimals to Fractions

Knowing the place value system is the key to understanding what decimals mean, how to read them, and how to write them.

- The decimal 0.9, or .9, is another way of writing $(0 \times 1) + \left(9 \times \frac{1}{10}\right)$, or $\frac{9}{10}$.
  This decimal is read the same as the equivalent fraction: "nine tenths."
- The decimal 0.21 represents 2 tenths + 1 hundredth. This expression simplifies to the following:

$$\left(2 \times \frac{1}{10}\right) + \left(1 \times \frac{1}{100}\right) = \frac{2}{10} + \frac{1}{100} = \frac{20}{100} + \frac{1}{100}, \text{ or } \frac{21}{100}$$

  So 0.21 is read "twenty-one hundredths."
- The decimal 0.149 stands for $\frac{149}{1,000}$.

$$\left(1 \times \frac{1}{10}\right) + \left(4 \times \frac{1}{100}\right) + \left(9 \times \frac{1}{1,000}\right) = \frac{149}{1,000}$$

  So 0.149 is read "one hundred forty-nine thousandths."

Let's summarize these examples.

Decimal	Equivalent Fraction	Read as
0.9	$\frac{9}{10}$	Nine tenths
0.21	$\frac{21}{100}$	Twenty-one hundredths
0.149	$\frac{149}{1,000}$	One hundred forty-nine thousandths

Note that in each of these decimals, the fractional part is the same as the numerator of the equivalent fraction: $0.149 = \frac{149}{1{,}000}$.

We can use the following rule to rewrite any decimal as a fraction or a mixed number.

To Change a Decimal to the Equivalent Fraction or Mixed Number

- Copy the nonzero whole-number part of the decimal and drop the decimal point.
- Place the fractional part of the decimal in the numerator of the equivalent fraction.
- Make the denominator of the equivalent fraction 1 followed by as many zeros as the decimal has decimal places.
- Simplify the resulting fraction, if possible.

EXAMPLE 2

Express 0.75 in fractional form and simplify.

Solution The whole-number part of the decimal is 0. We drop the decimal point. The fractional part (75) of the decimal becomes the numerator of the equivalent fraction. Since the decimal has two decimal places, we make the denominator of the equivalent fraction 1 followed by two zeros (100). So we can write 0.75 as $\frac{75}{100}$, which simplifies to $\frac{3}{4}$.

PRACTICE 2

Write 0.875 as a fraction reduced to lowest terms.

EXAMPLE 3

Express 1.87 as a mixed number.

Solution This decimal is equivalent to a mixed number whose whole-number part is 1. The fractional part (87) of the decimal is the numerator of the equivalent fraction. The decimal has two decimal places, so the fraction's denominator has two zeros (that is, it is 100).

$$1.87 = 1\frac{87}{100}$$

Do you see that the answer can also be written as $\frac{187}{100}$?

PRACTICE 3

The decimal 2.03 is equivalent to what mixed number?

EXAMPLE 4

Find the equivalent fraction of each decimal.

a. 3.2 **b.** 3.200

Solution

a. 3.2 represents $3\frac{2}{10}$, or $3\frac{1}{5}$.

b. 3.200 equals $3\frac{200}{1{,}000}$, or $3\frac{1}{5}$.

PRACTICE 4

Express each decimal in fractional form.

a. 5.6 **b.** 5.6000

Tip Adding zeros in the rightmost decimal places does not change a decimal's value. However, generally decimals can be written without these extra zeros.

EXAMPLE 5

Write each decimal as a mixed number.

a. 1.309 **b.** 1.39

Solution

a. $1.309 = 1\frac{309}{1{,}000}$

b. $1.39 = 1\frac{39}{100}$

PRACTICE 5

What mixed number is equivalent to each decimal?

a. 7.003 **b.** 4.1

Knowing how to change a decimal to its equivalent fraction also helps us read the decimal.

EXAMPLE 6

Express each decimal in words.

a. 0.319 **b.** 2.71 **c.** 0.08

Solution

a. $0.319 = \frac{319}{1{,}000}$ We read the decimal as "three hundred nineteen thousandths."

b. $2.71 = 2\frac{71}{100}$ The decimal point is read as "and." We read the decimal as "two and seventy-one hundredths."

c. $0.08 = \frac{8}{100}$ We read the decimal as "eight hundredths." Note that we *do not simplify* the equivalent fraction when reading the decimal.

PRACTICE 6

Express each decimal in words.

a. 0.61

b. 4.923

c. 7.05

EXAMPLE 7

Write each number in decimal notation.

a. Seven tenths **b.** Five and thirty-two thousandths

Solution

a. Since 7 is in the tenths place, the decimal is written as 0.7.

b.

The whole number preceding *and* is in the ones place.

The last digit of 32 is in the thousandths place.

5.0 3 2

We replace *and* with the decimal point.

We need a 0 to hold the tenths place.

The answer is 5.032.

PRACTICE 7

Write each number in decimal notation.

a. Forty-three thousandths

b. Ten and twenty-six hundredths

EXAMPLE 8

For hay fever, an allergy sufferer takes a decongestant pill that has a tablet strength of three hundredths of a gram. Write the equivalent decimal.

Solution "Three hundredths" is written 0.03, with the digit 3 in the hundredths place.

PRACTICE 8

The number pi (usually written π) is approximately three and fourteen hundredths. Write this approximation as a decimal.

Comparing Decimals

Suppose that we want to compare two decimals—say, 0.6 and 0.7. The key is to rethink the problem in terms of fractions.

$$0.6 = \frac{6}{10} \qquad 0.7 = \frac{7}{10}$$

Because $\frac{7}{10} > \frac{6}{10}$, $0.7 > 0.6$.

The following procedure provides another way to compare decimals that is faster than converting the decimals to fractions.

To Compare Decimals

- Rewrite the numbers vertically, lining up the decimal points.
- Working from left to right, compare the digits that have the same place value. The decimal which has the largest digit with this place value is the largest decimal.

EXAMPLE 9

Which is larger, 0.729 or 0.75?

Solution First let's line up the decimal point.

0.729
0.75

We see that both decimals have a 0 in the ones place. We next compare the digits in the tenths place and see that, again, they are the same. Looking to the right in the hundredths place, we see that $5 > 2$. Therefore, $0.75 > 0.729$. Note that the decimal with more digits is not necessarily the larger decimal.

PRACTICE 9

Which is smaller, 0.83 or 0.8297?

EXAMPLE 10

Rank from smallest to largest: 2.17, 2.1, and 0.99

Solution First, we position the decimals so that the decimal points are aligned.

2.17
2.1
0.99

PRACTICE 10

Rewrite in decreasing order: 3.5, 3.51, and 3.496

Working from left to right, we see that in the ones place, the first two decimals have a 2 and the third decimal has a 0, so the third decimal is the smallest of the three. To decide which of the first two decimals is smaller, we compare the digits in the tenths place. Since both of these decimals have a 1 in the tenths place, we proceed to the hundredths place. A 0 is understood to the right of the 1 in 2.1, so we compare 0 and 7.

$$\begin{array}{l} \downarrow \\ 2.17 \\ 2.10 \\ 0.99 \\ \uparrow \end{array}$$

Since $0 < 7$, we conclude that $2.10 < 2.17$. Therefore, the three decimals from smallest to largest are 0.99, 2.1, and 2.17.

EXAMPLE 11

Plastic garbage bags come in three thicknesses (or gauges): 0.003 inch, 0.0025 inch, and 0.002 inch. The three gauges are called lightweight, regular weight, and heavyweight. Which is the lightweight gauge?

Solution To find the smallest of the decimals, we first line up the decimal points.

$$\begin{array}{l} \downarrow \\ 0.003 \\ 0.0025 \\ 0.002 \\ \uparrow \end{array}$$

Working from left to right, we see that the three decimals have the same digits until the thousandths place, where $3 > 2$. Therefore, 0.003 must be the heavyweight gauge. To compare 0.0025 and 0.002, we look at the ten-thousandths place. The 5 is greater than the 0 that is understood to be there. So 0.0025 inch must be the regular-weight gauge, and 0.002 the lightweight gauge.

PRACTICE 11

The higher the energy efficiency rating (EER) of an air conditioner, the more efficiently it uses electricity. Which of the following air conditioners is least efficient with EER ratings shown to be 8.2, 9, and 8.1?

(*Source: Consumer Guide*)

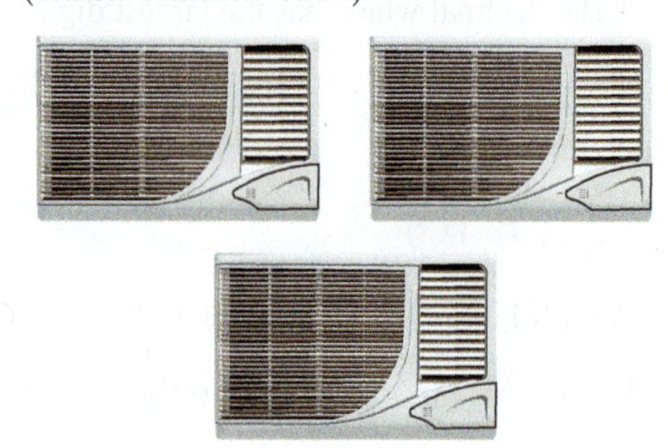

Rounding Decimals

As with whole numbers, we can round decimals to a given place value. For instance, suppose that we want to round the decimal 1.38 to the nearest tenth. The decimal 1.38 lies between 1.3 and 1.4, so one of these two numbers will be our answer—but which? To decide, let's take a look at a number line.

1.31 1.33 1.35 1.37 1.39

1.3 1.32 1.34 1.36 1.38 1.4

Do you see from this diagram that 1.38 is closer to 1.4 than to 1.3?

$$1.38 \approx 1.4$$

Tenths place

Rounding a decimal to the nearest tenth means that the last digit lies in the tenths place.

The following table shows the relationship between the place to which we are rounding and the number of decimal places in our answer.

Rounding to the Nearest	Means That the Rounded Decimal Has
tenth $\left(\frac{1}{10}\right)$	one decimal place.
hundredth $\left(\frac{1}{100}\right)$	two decimal places.
thousandth $\left(\frac{1}{1,000}\right)$	three decimal places.
ten-thousandth $\left(\frac{1}{10,000}\right)$	four decimal places.

Note that the number of decimal places is the same as the number of zeros in the corresponding denominator.

The following rule can be used to round decimals.

To Round a Decimal to a Given Decimal Place

- Underline the digit in the place to which the number is being rounded.
- Look at the digit to the right of the underlined digit—the critical digit. If this digit is 5 or more, add 1 to the underlined digit; if it is less than 5, leave the underlined digit unchanged.
- Drop all decimal places to the right of the underlined digit.

Let's apply this rule to the problem that we just considered—namely, rounding 1.38 to the nearest tenth.

Tenths place

$1.\underline{3}8$ — Underline the digit 3, which occupies the tenths place.

$1.\underline{3}8 \approx 1.4$ — The critical digit, 8, is 5 or more, so add 1 to the 3 and then drop all digits to its right.

Critical digit

The following examples illustrate this method of rounding.

EXAMPLE 12

Round 94.735 to

a. the nearest tenth.

b. two decimal places.

c. the nearest thousandth.

d. the nearest ten.

e. the nearest whole number.

PRACTICE 12

Round 748.0772 to

a. the nearest tenth.

b. the nearest hundredth.

c. three decimal places.

d. the nearest whole number.

e. the nearest hundred.

Solution

a. First, we underline the digit 7 in the tenths place: 94.<u>7</u>35. The critical digit, 3, is less than 5, so we do not add 1 to the underlined digit. Dropping all digits to the right of the 7, we get 94.7. Note that our answer has only one decimal place because we are rounding to the nearest tenth.

b. We need to round 94.735 to two decimal places (to the nearest hundredth).

$94.7\underline{3}5 \approx 94.74$

The critical digit is 5 or more. Add 1 to the underlined digit and drop the decimal place to the right.

c. $94.73\underline{5} \approx 94.735$ because the critical digit to the right of the 5 is understood to be 0.

d. We are rounding 94.735 to the nearest 10 (*not tenth*), which is a whole-number place.

$\underline{9}4.735 \approx 90$

Because 4 < 5, keep 9 in the tens place, insert 0 in the ones place and drop all decimal places.

e. Rounding to the nearest whole number means rounding to the nearest 1.

$9\underline{4}.735 \approx 95$

Because 7 > 5, change the 4 to 5 and drop all decimal places.

EXAMPLE 13

Round 3.982 to the nearest tenth.

Solution First, we underline the digit 9 in the tenths place and identify the critical digit: 3.<u>9</u>82. The critical digit, 8, is more than 5, so we add 1 to the 9, get 10, and write down the 0. We add the carried 1 to 3 getting 4, and drop the 8 and 2.

$$3.\underline{9}82 \approx 4.0$$

Drop

The answer is 4.0. Note that we do not drop the 0 in the tenths place of the answer to indicate that we have rounded to that place.

PRACTICE 13

Round 7.2962 to two decimal places.

EXAMPLE 14

On the commodities market, prices are often quoted in terms of thousandths or even ten-thousandths of a dollar. Suppose that the commodities market price of a pound of coffee is \$2.0883. What is this price to the nearest cent?

Solution A cent is one-hundredth of a dollar. Therefore, we need to round 2.0883 to the nearest hundredth.

$$2.0\underline{8}83 \approx 2.09$$

The price to the nearest cent is \$2.09.

PRACTICE 14

Mount Waialeale on the Hawaiian island of Kauai is one of the world's wettest places, with an average annual rainfall of 11.43 meters. What is the amount of rainfall to the nearest tenth of a meter?

3.1 Exercises

FOR EXTRA HELP

Mathematically Speaking

Fill in each blank with the most appropriate term or phrase from the given list.

less	greater	increasing	left
decreasing	ten	hundredths	multiple
thousandths	power	right	tenth

1. A decimal place is a place to the _______ of the decimal point.

2. The fractional part of a decimal has as its denominator a _______ of 10.

3. The decimal 0.17 is equivalent to the fraction seventeen _______.

4. The decimal 209.95 rounded to the nearest _______ is 210.0.

5. The decimal 0.371 is _______ than the decimal 0.3499.

6. The decimals 0.48, 0.4, and 0.371 are written in _______ order.

Underline the digit that occupies the given place.

7. 2.78 Tenths place

8. 9.01 Hundredths place

9. 2.00175 Ten-thousandths place

10. 6.835 Tenths place

11. 358.02 Tens place

12. 823.001 Thousandths place

13. 0.772 Hundredths place

14. 135.83 Hundreds place

Identify the place occupied by the underlined digit.

15. $25.\underline{7}1$

16. $3.00\underline{2}$

17. $8.1\underline{8}3$

18. $\underline{4}9.771$

19. $1{,}077.04\underline{2}$

20. $2.8371\underline{0}7$

21. $\$25\underline{3}.72$

22. $\$7{,}571.3\underline{9}$

For each decimal, find the equivalent fraction or mixed number, reduced to lowest terms.

23. 0.6

24. 0.8

25. 0.39

26. 0.27

27. 1.5

28. 9.8

29. 8.000

30. 6.700

31. 5.012

32. 20.304

Write each decimal in words.

33. 0.53

34. 0.72

35. 0.305

36. 0.849

37. 0.6

38. 0.3

39. 5.72

40. 3.89

41. 24.002

42. 370.081

Write each number in decimal notation.

43. Eight tenths

44. Eleven hundredths

45. One and forty-one thousandths

46. Five and sixty-three hundredths

47. Sixty and one hundredth

48. Eighteen and four thousandths

49. Four and one hundred seven thousandths

50. Ninety-two and seven hundredths

51. Three and two tenths meters

52. Ninety-eight and six tenths degrees

Between each pair of numbers, insert the appropriate sign, <, =, or >, to make a true statement.

53. 3.21 2.5

54. 8.66 4.952

55. 0.71 0.8

56. 1.2 1.38

57. 9.123 9.11

58. 0.5 0.52

59. 4 4.000

60. 7.60 7.6

61. 8.125 feet 8.2 feet

62. 2.45 pounds 2.5 pounds

Rearrange each group of numbers from smallest to largest.

63. 7.1, 7, 7.07

64. 0.002, 0.2, 0.02

65. 5.001, 4.9, 5.2

66. 3.85, 3.911, 2

67. 9.6 miles, 9.1 miles, 9.38 miles

68. 2.7 seconds, 2.15 seconds, 2 seconds

Round as indicated.

69. 17.36 to the nearest tenth

70. 8.009 to two decimal places

71. 3.5905 to the nearest thousandth

72. 3.5902 to the nearest thousandth

73. 37.08 to one decimal place

74. 3.08 to the nearest whole number

75. 0.396 to the nearest hundredth

76. 0.978 to the nearest tenth

77. 7.0571 to two decimal places

78. 3.038 to one decimal place

79. 8.7 miles to the nearest mile

80. $35.75 to the nearest dollar

Round to the indicated place.

81.

To the Nearest	8.0714	0.9916
Tenth		
Hundredth		
Ten		

82.

To the Nearest	0.8166	72.3591
Tenth		
Hundredth		
Ten		

Mixed Practice

Solve.

83. In the decimal 0.024, underline the digit in the tenths place.

84. What is the equivalent fraction of 3.8?

85. Round 870.062 to the nearest hundredth.

86. Write four and thirty-one thousandths in decimal notation.

87. Write in increasing order: 2.14 meters, 2.4 meters, and 2.04 meters.

88. Write 0.05 in words.

Applications

The following statements involve decimals. Write all decimals in words.

89. It takes the Earth 23.934 hours to rotate once about its axis. (*Source:* NASA)

90. A chemistry text gives 55.847 as the atomic weight of iron.

91. Over two years, the average score on a college admissions exam increased from 18.7 to 18.8.

92. Male Rufous hummingbirds weigh an average of 0.113 ounces and have a wingspan of 4.25 inches. (*Source:* Lanny Chambers, *Facts about Hummingbirds*)

93. The following table shows the number of people per square mile living on various continents in a recent year.

Continent	Number of People per Square Mile
Asia	301.3
Africa	55.9
Europe	268.2
North America	46.6
South America	43.6

94. The coefficient of friction is a measure of the amount of friction produced when one surface rubs against another. The following table gives these coefficients for various surfaces.

Materials	Coefficient of Friction
Wood on wood	0.3
Steel on steel	0.15
Steel on wood	0.5
A rubber tire on a dry concrete road	0.7
A rubber tire on a wet concrete road	0.5

(*Source: CRC Handbook of Chemistry and Physics*)

95. Bacteria are single-celled organisms that typically measure from 0.00001 inch to 0.00008 inch across.

96. In one month, the consumer confidence index rose from 71.9 points to 80.2 points.

Write each number in decimal notation.

97. The area of a plot of land is one and two tenths acres.

98. The lead in many mechanical pencils is seven tenths millimeter thick.

99. At the first Indianapolis 500 auto race in 1911, the winning speed was seventy-four and fifty-nine hundredths miles per hour. (***Source:*** Jack Fox, *The Indianapolis 500*)

100. According to the owner's manual, the voltage produced by a camcorder battery is nine and six tenths volts (V).

101. At sea level, the air pressure on each square inch of surface area is fourteen and seven tenths pounds.

102. A doctor prescribed a dosage of one hundred twenty-five thousandths milligram of Prolixin.

103. In 1796, there was a U.S. coin in circulation, the half cent, worth five thousandths of a dollar.

104. In preparing an injection, a nurse measured out one and eight tenths milliliters of sterile water.

105. The electrical usage in a tenant's apartment last month amounted to three hundred fifty-two and one tenth kilowatt hours (kWh).

106. In one day, the Dow Jones Industrial Average fell by three and sixty-three hundredths points.

Solve.

107. A jury awarded a plaintiff $1.85 million. Was this award more or less than the $2.1 million that the plaintiff had demanded?

108. The more powerful an earthquake is, the higher its magnitude is on the Richter scale. Great earthquakes, such as the 1906 San Francisco earthquake, have magnitudes of 8.0 or higher. Is an earthquake with magnitude 7.8 considered to be a great earthquake? (***Source:*** *The New Encyclopedia Britannica*)

109. Last winter, a homeowner's average daily heating bill was for 8.75 units of electricity. This winter, it was for 8.5 units. During which winter was the average higher?

110. Suppose that in order to qualify for the dean's list at a community college, a student's grade point average (GPA) must be 3.5 or above. Did a student with a GPA of 3.475 make the dean's list?

111. A person with reasonably good vision can see objects as small as 0.0004 inch long. Can such a person see a mite—a tiny bug that is 0.003 inch long?

112. At the 2006 Winter Olympics, the two top skaters in the 500-meter women's speed skating event finished in 76.57 seconds and 76.78 seconds. Which time was better? (***Source:*** www.nbcolympics.com)

113. As part of her annual checkup, a patient had a blood test. The normal range for a particular substance is 1.1 to 2.3. If she scored 0.95, was her blood in the normal range?

114. Last year, an electronics factory released 1.5 million pounds of toxic gas into the air. During the same time, a food factory and a chemical factory released 1.4 and 1.48 million pounds of toxic emissions, respectively. Which of the three factories was the worst polluter?

Round to the indicated place.

115. A bank pays interest on all its accounts to the nearest cent. If the interest on an account is $57.0285, how much interest does the bank pay?

116. A city's sales tax rate, expressed as a decimal, is 0.0825. What is this rate to the nearest hundredth?

117. According to the organizers of a lottery, the probability of winning the lottery is 0.0008. Round this probability to three decimal places.

118. One day last week, a particular foreign currency was worth 0.7574 U.S. dollars ($US). How much is this currency worth to the nearest tenth of a dollar?

119. According to a recent survey, the cost of medical care is 1.77 times what it was a decade ago. Express this decimal to the nearest tenth.

120. The length of the Panama Canal is 50.7 miles. Round this length to the nearest mile. (***Source:*** *The New Encyclopedia Britannica*)

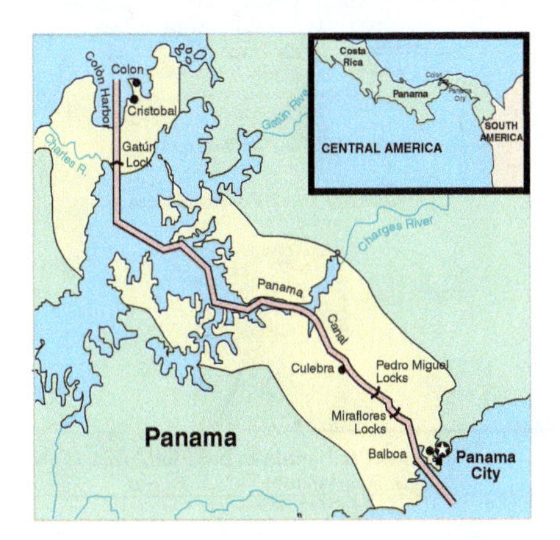

• Check your answers on page A-5.

MINDSTRETCHERS

Critical Thinking

1. For each question, either give the answer or explain why there is none.

a. Find the *smallest* decimal that when rounded to the nearest tenth is 7.5.

b. Find the *largest* decimal that when rounded to the nearest tenth is 7.5.

Writing

2. The next whole number after 7 is 8. What is the next decimal after 0.7? Explain.

Groupwork

3. Working with a partner, list fifteen numbers between 2.5 and 2.6.

CULTURAL NOTE

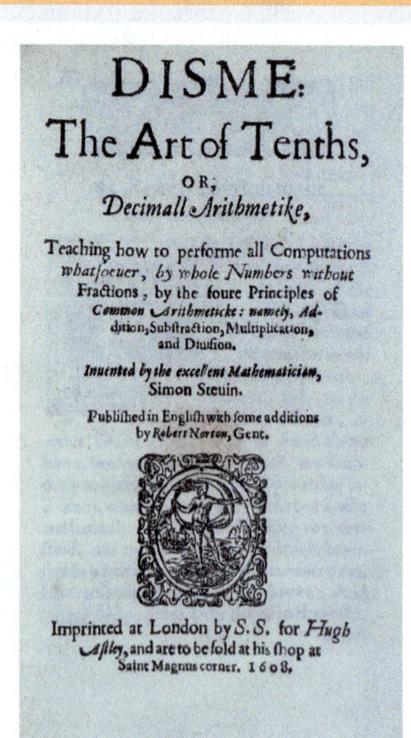

DISME:
The Art of Tenths,
OR;
Decimall Arithmetike,

Teaching how to performe all Computations *whatsoeuer, by whole Numbers without* Fractions, by the foure Principles of *Common Arithmeticke: namely,* Addition, Substraction, Multiplication, and Diuision.

Inuented by the excellent Mathematician, Simon Steuin.

Published in English with some additions by *Robert Norton,* Gent.

Imprinted at London by *S. S.* for *Hugh Astley*, and are to be sold at his shop at Saint Magnus corner. 1608.

In 1585, Simon Stevin, a Dutch engineer, published a book entitled *The Art of Tenths* (*La Disme* in French) in which he presented a thorough account of decimals. Stevin sought to teach everyone "with an ease unheard of, all computations necessary between men by integers without fractions."

Stevin did not invent decimals; their history dates back thousands of years to ancient China, medieval Arabia, and Renaissance Europe. However, Stevin's writings popularized decimals and also supported the notion of decimal coinage—as in American currency, where there are 10 dimes to the dollar.

Source: Morris Kline, *Mathematics, a Cultural Approach* (Reading, Massachusetts: Addison-Wesley Publishing Company, 1962), p. 614.

3.2 Adding and Subtracting Decimals

OBJECTIVES

- To add and subtract decimals
- To estimate the sum and difference of decimals
- To solve word problems involving the addition or subtraction of decimals

In Section 3.1 we discussed the meaning of decimals and how to compare and round them. Now we turn our attention to computing with decimals, starting with addition and subtraction.

Adding Decimals

Adding decimals is similar to adding whole numbers: We add the digits in each place value position, carrying when necessary. Suppose that we want to find the sum of two decimals: 1.2 + 3.5. First, we rewrite the problem vertically, lining up the decimal points in the addends. Then we add as usual, inserting the decimal point below the other decimal points.

$$\begin{array}{r} 1.2 \\ +3.5 \\ \hline 4.7 \end{array} \quad \text{This addition is equivalent to} \quad \begin{array}{r} 1\frac{2}{10} \\ +3\frac{5}{10} \\ \hline 4\frac{7}{10} \end{array}$$

Note that, when we added the mixed numbers corresponding to the decimals, we got $4\frac{7}{10}$, which is equivalent to 4.7. This example suggests the following rule.

To Add Decimals

- Rewrite the numbers vertically, lining up the decimal points.
- Add.
- Insert a decimal point in the sum below the other decimal points.

EXAMPLE 1

Find the sum: 2.7 + 80.13 + 5.036

Solution

$$\begin{array}{r} 2.7 \\ 80.13 \\ +5.036 \\ \hline 87.866 \end{array}$$

Rewrite the addends with decimal points lined up vertically.

Add.

Insert the decimal point in the sum.

PRACTICE 1

Add: 5.12 + 4.967 + 0.3

EXAMPLE 2

Compute: 2.367 + 5 + 0.143

Solution Recall that 5 and 5. are equivalent.

$$\begin{array}{r} 2.367 \\ 5. \\ +\ 0.143 \\ \hline 7.510 \end{array} = 7.51$$

Line up the decimal points and add.

Insert the decimal point in the sum.

We can drop the extra 0 at the right end.

PRACTICE 2

What is the sum of 7.31, 8, and 23.99?

EXAMPLE 3

A runner's time was 0.06 second longer than the world record of 21.71 seconds. What was the runner's time?

Solution We need to compute the sum of the two numbers. The runner's time was 21.77 seconds.

$$\begin{array}{r} 0.06 \\ +\ 21.71 \\ \hline 21.77 \end{array}$$

PRACTICE 3

A child has the flu. This morning, his body temperature was 99.4°F. What was his temperature after it went up by 2.7°?

Subtracting Decimals

Now let's discuss subtracting decimals. As with addition, subtracting decimals is similar to subtracting whole numbers. To compute the difference between 12.83 and 4.2, we rewrite the problem vertically, lining up the decimal points. Then we subtract as usual, inserting a decimal point below the other decimal points.

$$\begin{array}{r} 12.83 \\ -4.2 \\ \hline 8.63 \end{array} \quad \text{is equivalent to} \quad \begin{array}{rcr} 12\frac{83}{100} & = & 12\frac{83}{100} \\ -4\frac{2}{10} & = & -4\frac{20}{100} \\ \hline & & 8\frac{63}{100}, \end{array} \quad \text{or} \quad 8.63$$

Again, note that when we subtracted the equivalent mixed numbers, we got the same difference.

As in any subtraction problem, we can check this answer by adding the subtrahend (4.2) to the difference (8.63), confirming that we get the original minuend (12.83). This example suggests the following rule.

$$\begin{array}{r} 8.63 \\ +4.2 \\ \hline 12.83 \end{array}$$

To Subtract Decimals

- Rewrite the numbers vertically, lining up the decimal points.
- Subtract, inserting extra zeros in the minuend if necessary for borrowing.
- Insert a decimal point in the difference, below the other decimal points.

EXAMPLE 4

Subtract and check: 5.038 − 2.11

Solution

$$\begin{array}{r} 5.038 \\ -2.11 \\ \hline 2.928 \end{array}$$

Rewrite the problem with decimal points lined up vertically.

Subtract. Borrow when necessary.

Insert the decimal point in the answer.

Check To verify that our difference is correct, we check by addition.

$$\begin{array}{r} 2.928 \\ +2.11 \\ \hline 5.038 \end{array}$$

PRACTICE 4

Find the difference and check: 71.3825 − 25.17

EXAMPLE 5

65 is how much larger than 2.04?

Solution Recall that 65 and 65. are equivalent.

Insert zeros needed for borrowing.

$$\begin{array}{r} 65.00 \\ -2.04 \\ \hline 62.96 \end{array}$$

Line up the decimal points.
Subtract.
Insert the decimal point in the answer.

Check

$$\begin{array}{r} 62.96 \\ +\ 2.04 \\ \hline 65.00 \end{array} = 65$$

PRACTICE 5

How much greater is \$735 than \$249.57?

EXAMPLE 6

A McDonald's hamburger contains 0.53 grams of sodium, whereas a McDonald's cheeseburger contains 0.74 grams. How much more sodium does the cheeseburger contain? (***Source:*** McDonald's)

Solution We need to find the difference between 0.74 and 0.53.

$$\begin{array}{r} 0.74 \\ -0.53 \\ \hline 0.21 \end{array}$$

So a McDonald's cheeseburger contains 0.21 grams more sodium.

PRACTICE 6

A swimmer is competing in the 28.5 mile swim around the island of Manhattan. After swimming 15 miles, how much farther does she have to go?

EXAMPLE 7

Suppose that a part-time employee's salary is \$350 a week, less deductions. The following table shows these deductions.

Deduction	Amount
Federal, state, and city taxes	\$100.80
Social Security	13.50
Union dues	8.88

What is the employee's take-home pay?

Solution Let's use the strategy of breaking the question into two simpler questions.

- *How much money is deducted per week?* The weekly deductions (\$100.80, \$13.50, and \$8.88) add up to \$123.18.
- *How much of the salary is left after subtracting the total deductions?* The difference between \$350 and \$123.18 is \$226.82, which is the employee's take-home pay.

PRACTICE 7

A sales rep, working in Ohio, wants to drive from Circleville to Columbus. How much shorter is it to drive directly to Columbus instead of going by way of Lancaster?

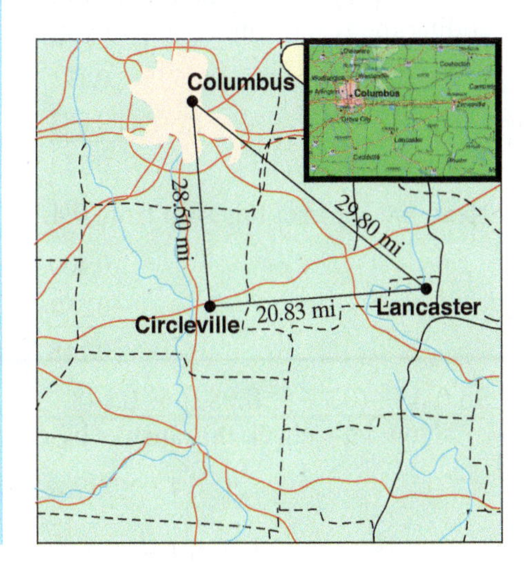

Estimating Sums and Differences

Being able to estimate in your head the sum or difference between two decimals is a useful skill, if only for checking an exact answer. To estimate, simply round the numbers to be added or subtracted and then carry out the operation on the rounded numbers.

EXAMPLE 8

Compute the sum 0.17 + 0.4 + 0.083. Use estimation to check.

Solution First, we add. Then, to check, we round the addends—say, to the nearest tenth—and add the rounded numbers.

$$\begin{array}{rcr} 0.17 & \approx & 0.2 \\ 0.4 & \approx & 0.4 \\ +0.083 & \approx & +0.1 \\ \hline 0.653 & & 0.7 \end{array}$$

Exact sum → 0.653   0.7 ← Estimated sum

Our exact sum is close to our estimated sum, and in fact, rounds to it. So we can have confidence that our answer, 0.653, is correct.

PRACTICE 8

Add 0.093, 0.008, and 0.762. Then check by estimating.

EXAMPLE 9

Subtract 0.713 − 0.082. Then check by estimating.

Solution First we find the exact answer and then round the given numbers to get an estimate.

$$\begin{array}{rcr} 0.713 & \approx & 0.7 \\ -0.082 & \approx & -0.1 \\ \hline 0.631 & & 0.6 \end{array}$$

Exact difference → 0.631   0.6 ← Estimated difference

Our exact answer, 0.631, is close to 0.6, so we can feel confident that it is correct.

PRACTICE 9

Compute: 0.17 − 0.091. Use estimation to check.

EXAMPLE 10

Combine and check: 0.4 − (0.17 + 0.082)

Solution Following the order of operations rule, we begin by adding the two decimals in parentheses.

$$\begin{array}{r} 0.17 \\ +0.082 \\ \hline 0.252 \end{array}$$

Next, we subtract this sum from 0.4.

$$\begin{array}{r} 0.400 \\ -0.252 \\ \hline 0.148 \end{array}$$

So 0.4 − (0.17 + 0.082) = 0.148.

Now let's check this answer by estimating:

$$0.4 - (0.17 + 0.082)$$
$$\downarrow \quad \downarrow \quad \downarrow$$
$$0.4 - (0.2 + 0.1) = 0.4 - 0.3 = 0.1$$

The estimate, 0.1, is sufficiently close to 0.148 to confirm our answer.

PRACTICE 10

Calculate and check:
0.813 − (0.29 − 0.0514)

EXAMPLE 11

A movie budgeted at \$7.25 million ended up costing \$1.655 million more. Estimate the final cost of the movie.

Solution Let's round each number to the nearest million dollars.

$$\begin{array}{lr} 1.655 \approx & 2 \text{ million} \\ 7.25 \approx & \underline{+7 \text{ million}} \\ & 9 \text{ million} \end{array}$$

Adding the rounded numbers, we see that the movie cost approximately \$9 million.

PRACTICE 11

From the deposit ticket shown below, estimate the total amount deposited.

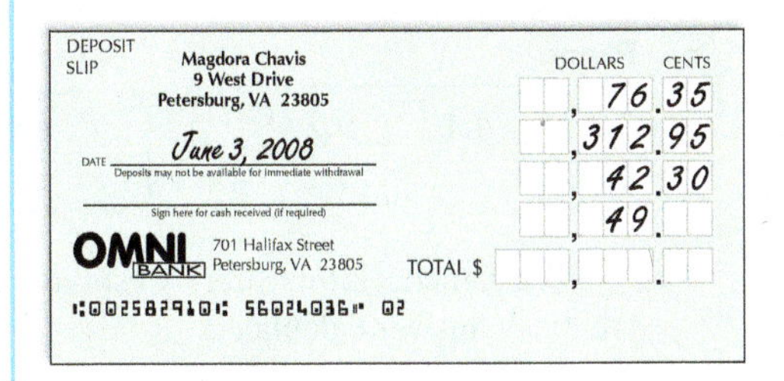

DEPOSIT SLIP

Magdora Chavis
9 West Drive
Petersburg, VA 23805

DATE June 3, 2008

Deposits may not be available for immediate withdrawal

Sign here for cash received (if required)

OMNI BANK
701 Halifax Street
Petersburg, VA 23805

DOLLARS	CENTS
76	35
312	95
42	30
49	
TOTAL $	

Estimate: ____________

EXAMPLE 12

When the underwater tunnel connecting the United Kingdom and France was built, French and British construction workers dug from their respective countries. They met at the point shown on the map.

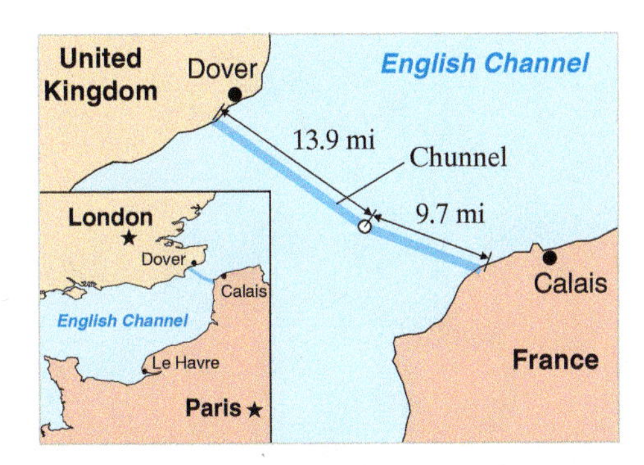

Estimate how much farther the British workers had dug than the French workers. (***Source:*** *The New York Times*)

Solution We can round 13.9 to 14 and 9.7 to 10. The difference between 14 and 10 is 4, so the British workers dug about 4 miles farther than the French workers.

PRACTICE 12

An art collector bought a painting for \$2.3 million. A year later, she sold the painting for \$4.1 million. Estimate her profit on the sale.

Adding and Subtracting Decimals on a Calculator

When adding or subtracting decimals, press the [.] key to enter the decimal point. If a sum or difference ends with a 0 in the rightmost decimal place, does your calculator drop the 0? If a sum or difference has no whole-number part, does your calculator insert a 0?

EXAMPLE 13

Compute: 2.7 + 4.1 + 9.2

Solution

Press	Display
2.7 [+] 4.1 [+] 9.2 [ENTER]	2.7 + 4.1 + 9.2 16.

We can check this sum by estimating: 3 + 4 + 9 = 16, which is the exact answer calculated.

PRACTICE 13

Find the sum: 3.82 + 9.17 + 66.24

EXAMPLE 14

Find the difference: 83.71 − 83.70002

Solution

Press	Display
83.71 [−] 83.70002 [ENTER]	83.71 - 83.720002 0.00998

We can check this difference by adding: 0.00998 + 83.70002 = 83.71.

PRACTICE 14

Compute: 5.00003 − 5.00001

3.2 Exercises

FOR EXTRA HELP

Mathematically Speaking

Fill in each blank with the most appropriate term or phrase from the given list.

sum	decimal points	difference
any number	rightmost digits	zeros

1. When adding decimals, rewrite the numbers vertically, lining up the _______.
2. Inserting _______ at the right end of a decimal does not change its value.
3. To estimate the _______ of 0.31 and 0.108, add 0.3 and 0.1.
4. To estimate the _______ between 0.31 and 0.108, subtract 0.1 from 0.3.

Find the sum. Check by estimating.

5. 3.89 + 5.44
6. 2.17 + 4.29
7. 0.6 + 0.3
8. 12.7 + 3.9
9. 6.03 + 2.1
10. 0.4 + 3.96
11. 13.05 + 8.4
12. 3.922 + 5.1
13. 2.67 + 5
14. 8 + 4.99
15. $74 + $3.21
16. $8.77 + $6
17. 0.49023 + 0.5997
18. 1.002 + 0.20013
19. 8.01 + 6.7 + 9.45
20. 9.73 + 5.99 + 3.688
21. 34.7 + 5.84 + 3 + 0.882
22. 75.285 + 2 + 3.871 + 0.5
23. 7 millimeters + 3.5 millimeters + 9.82 millimeters
24. 10.35 inches + 32 inches + 54.9 inches
25. 4.7 kilograms + 2.98 kilograms + 9.002 kilograms
26. 0.85 second + 1.72 seconds + 3.009 seconds
27. 3.861 + 2.89 + 3.775 + 9.00813 + 3.77182
28. $8.99 + $3.99 + $17.83 + $15 + $201.75

Find the difference. Check either by estimating or by adding.

29. 0.8 − 0.1
30. 12.98 − 5.73
31. 20.72 − 3.92
32. 0.68 − 0.59
33. 23.81 − 5.4
34. 17.49 − 10.2
35. 80.2 − 4.57
36. 9.71 − 3.225
37. 25.99 − 3.666
38. 80.2 − 3.51
39. 0.27 − 0.1
40. 4.92 − 1.01
41. 1.032 − 0.9178
42. 0.01 − 0.0001
43. 13.2 − 7
44. 9.662 − 4
45. 20 − 4.63
46. 8 − 2.55
47. 10 − 4.1
48. 13 − 7.2
49. 8 − 1.79
50. 20 − 4.63
51. 3.2 pounds − 1.35 pounds
52. 23.5 seconds − 2.8 seconds
53. 103.7°F − 98.8°F
54. 32.5 grams − 19.27 grams

Compute.

55. $35.2 - 2.86 + 9.07 - 1.658$

56. $10 - 2.38 - 4.92 + 6.02$

57. 30 milligrams − 0.5 milligram − 1.6 milligrams

58. \$20.93 + \$1.07 − \$19.58

59. $5.21 - (1.03 + 0.975)$

60. $6.953 - (4.09 - 0.008)$

61. $41.075 - 2.87104 - 17.005$

62. $0.00661 + 1.997 - 0.05321$

In each group of three computations, one answer is wrong. Use estimation to identify which answer is incorrect.

63. **a.** $\begin{array}{r} 0.059 \\ 0.00234 \\ +0.036 \\ \hline 0.09734 \end{array}$ **b.** $\begin{array}{r} 0.1903 \\ 0.074 \\ +0.2051 \\ \hline 0.4694 \end{array}$ **c.** $\begin{array}{r} 0.00441 \\ 0.06882 \\ +0.0103 \\ \hline 0.8353 \end{array}$

64. **a.** $\begin{array}{r} \$32.71 \\ 43.09 \\ +\ \ 8.27 \\ \hline \$74.07 \end{array}$ **b.** $\begin{array}{r} \$19.37 \\ 2. \\ +\ \ 7.22 \\ \hline \$28.59 \end{array}$ **c.** $\begin{array}{r} \$139.26 \\ 82.87 \\ +\ \ 3.01 \\ \hline \$225.14 \end{array}$

65. **a.** $\begin{array}{r} 0.35 \\ -0.1007 \\ \hline 0.2493 \end{array}$ **b.** $\begin{array}{r} 0.072 \\ -0.0056 \\ \hline 0.664 \end{array}$ **c.** $\begin{array}{r} 0.03 \\ -0.008 \\ \hline 0.022 \end{array}$

66. **a.** $\begin{array}{r} 8.551 \\ -2.9995 \\ \hline 5.5515 \end{array}$ **b.** $\begin{array}{r} 78.328 \\ -\ \ 5.5 \\ \hline 7.2828 \end{array}$ **c.** $\begin{array}{r} 65 \\ -\ \ 2.778 \\ \hline 62.222 \end{array}$

Mixed Practice

Solve.

67. Calculate: $4.78 + 13 - 10.009$

68. Find the difference between 90.1 and 12.58.

69. Add: 0.5 pound + 3 pounds + 4.25 pounds

70. Subtract: \$20 − \$6.95

71. Compute: $8 - 2.4 + 6.0013$

72. What is the sum of 1.265, 7, and 0.14?

Applications

Solve and check.

73. A paperback book that normally sells for \$13 is now on sale for \$11.97. What is the discount in dollars and cents?

74. During a drought, the mayor of a city attempted to reduce daily water consumption to 3.1 million gallons. If daily water consumption fell to 1.948 million gallons above that goal, estimate the city's consumption.

75. A skeleton was found at an archaeological dig. Radiocarbon dating—a technique used for estimating age—indicated that the skeleton was 56 centuries old, plus or minus 0.8 centuries. According to this estimate, what is the greatest possible age of the skeleton?

76. A college launched a campaign to collect \$3 million to build a new technology complex. If \$1.316 million has been collected so far, how much more money, to the nearest million dollars, is needed?

77. As an investment, a couple bought an apartment house for \$2.3 million. Two years later, they sold the apartment house for \$4 million. What was their profit?

78. A woman sued her business partner and was awarded \$1.5 million. On appeal, however, the award was reduced to \$0.75 million. By how much was the award reduced?

p. 184. 55, 57, 63, 73, 77.

e margins of
rts of an inch.
ng is each

ompact
minutes.
nd

scores in
four separate events: vault (VT), uneven bar (UB), balance bar (BB), and floor exercises (FX). The total of these four event scores is called the all-around score (AA). The following chart shows the results for two Indiana high school gymnasts at a 2006 competition:

Gymnast	VT	UB	BB	FX	AA
Madeline Whiteman	9.2	9.275	8.6	8.05	
Jordyn Stengel	9	9	8.65	8.45	

(***Source:*** http://www.sigsgym.org/Results/2005-2006/State_Championships_5.asp)

a. Calculate the all-around scores for these two competitors.

b. Which of the competitors had the higher all-around score?

80. In the boat pictured below, what is the distance between the bottom of the rudder and the bottom of the lake?

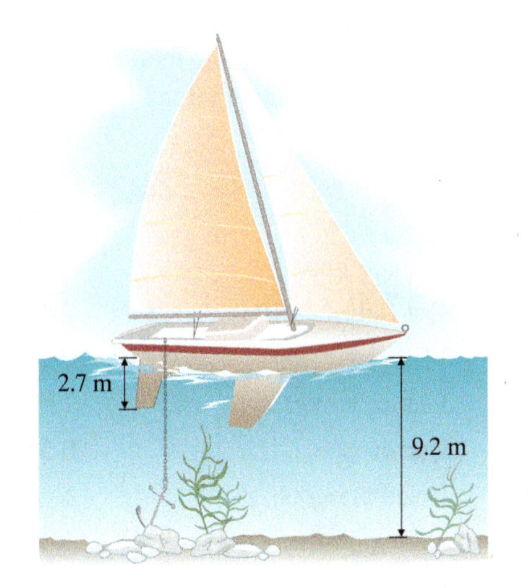

82. A shopper plans to buy three items that cost \$4.99, \$7.99, and \$2.99 each. If she has \$15 with her, will she have enough money to pay for all three items? Explain.

84. The graph shown gives the number of households that watched TV programs in a recent week according to the Nielsen Top 20 ratings. (***Source:*** http://tv.yahoo.com/nielsen/)

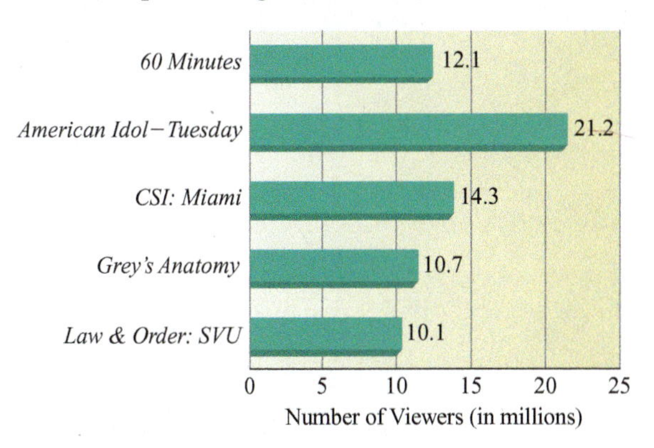

a. How many more households watched *American Idol—Tuesday* than *CSI: Miami*?

b. How many households altogether watched the five programs?

Use a calculator to solve each problem, giving (a) the operation(s) carried out in the solution, (b) the exact answer, and (c) an estimate of the answer.

85. To prevent anemia, a doctor advises his patient to take at least 18 milligrams of iron each day. The following table shows the amount of iron in the food that the patient ate yesterday. Did she get enough iron? If not, how much more does she need?

Food	Iron (mg)
Tomato juice	1.1
Oat flakes	2.4
Milk	0.1
Peanut butter sandwich	1.8
Carrot	0.5
Dried apricots	1
Oatmeal raisin cookies	1.4
Fish	1.1
Baked potato	1.1
Green peas	1.5
Romaine lettuce	0.8
Chocolate pudding	0.4

86. When filling a prescription, buying a generic drug rather than a brand-name drug can often save money. The following table shows the prices of various brand-name and generic drugs.

Drug	Brand-Name Price	Generic Price
Metformin	$36.89	$27.19
Allopurinal	$47	$18.55
Imipramine	$43.90	$10.15
Propranolol	$27.60	$10.55
Forosemide	$13.99	$10.29

How much money will a shopper save if he buys all five generic drugs rather than the brand-name drugs?

• *Check your answers on page A-6.*

MINDSTRETCHERS

Groupwork

1. Working with a partner, find the missing entries in the following magic square, in which 3.75 is the sum of every row, column, and diagonal.

0.75	1.25	
2		

Mathematical Reasoning

2. Suppose that a spider is sitting at point A on the rectangular web shown. If the spider wants to crawl along the web horizontally and vertically to munch on the delicious fly caught at point B, how long is the shortest route that the spider can take?

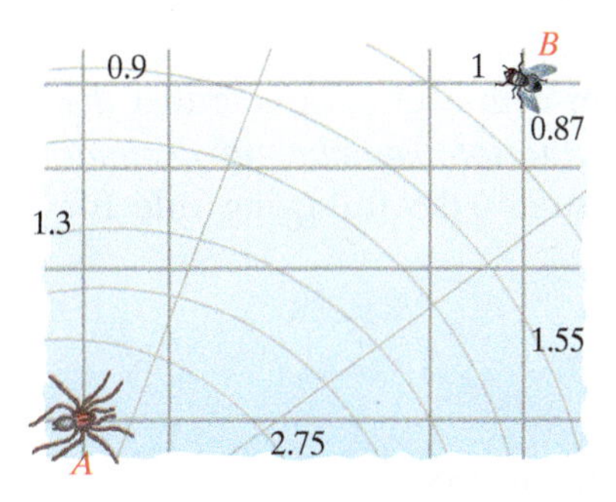

Writing

3. a. How many pairs of whole numbers are there whose sum is 7?

b. How many pairs of decimals are there whose sum is 0.7?

c. Explain why (a) and (b) have different answers.

3.3 Multiplying Decimals

OBJECTIVES

- To multiply decimals
- To estimate the product of decimals
- To solve word problems involving the multiplication of decimals

In this section, we discuss how to multiply two or more decimals, finding both the exact product and an estimated product.

Multiplying Decimals

To find the product of two decimals—say, 1.02 and 0.3—we multiply the same way we multiply whole numbers. But with decimals we need to know where the decimal point goes in the product. To find out, let's change each decimal to its fractional equivalent.

$$\begin{array}{r} 1.02 \\ \times\ 0.3 \\ \hline 306 \end{array} \quad \text{is equivalent to} \longrightarrow \quad 1\frac{2}{100} \times \frac{3}{10} = \frac{102}{100} \times \frac{3}{10} = \frac{306}{1{,}000}, \text{ or } 0.306$$

Where should we place the decimal point?

The product is in thousandths, so it has three decimal places.

Looking at the multiplication problem with decimals, note that *the product has as many decimal places as the total number of decimal places in the factors*. This example illustrates the following rule for multiplying decimals.

$$\begin{array}{r} 1.02 \\ 0.3 \\ \hline 0.306 \end{array}$$

To Multiply Decimals

- Multiply the factors as if they were whole numbers.
- Find the total number of decimal places in the factors.
- Count that many places from the right end of the product, and insert a decimal point.

EXAMPLE 1

Multiply: 6.1 × 3.7

Solution First, multiply 61 by 37, ignoring the decimal points.

$$\begin{array}{r} 6.1 \\ \times\ 3.7 \\ \hline 427 \\ 183 \\ \hline 2257 \end{array}$$

Then count the total number of decimal places in the factors.

$$\begin{array}{r} 6.1 \\ \times\ 3.7 \\ \hline 427 \\ 183 \\ \hline 22.57 \end{array} \quad \begin{array}{l} \leftarrow \text{One decimal place} \\ \leftarrow \text{One decimal place} \\ \\ \\ \leftarrow \text{Two decimal places in the product} \end{array}$$

Insert the decimal point two places from the right end. So 22.57 is the product.

PRACTICE 1

Find the product: 2.81 × 3.5

EXAMPLE 2

Find the product of 0.75 and 4.

Solution Let's multiply 0.75 by 4, ignoring the decimal point.

$$\begin{array}{r} 0.75 \\ \times\ \ 4 \\ \hline 300 \end{array}$$

Count the total number of decimal places.

$$\begin{array}{rl} 0.75 & \leftarrow \text{Two decimal places.} \\ \times\ \ 4 & \leftarrow \text{Zero decimal places (4 is a whole number)} \\ \hline 3.00 & \leftarrow \text{Two decimal places in the product} \end{array}$$

So the product is 3.00, which simplifies to 3.

PRACTICE 2

Multiply: 0.28×5

EXAMPLE 3

Multiply 0.03 and 0.25, rounding the answer to the nearest thousandth.

Solution

$$\begin{array}{r} 0.2\,5 \\ \times\ \ 0.0\,3 \\ \hline 0\,7\,5 \\ 0\,0\,0\ \ \\ \hline 0.0\,0\,7\,5 \end{array}$$

Rounding to the nearest thousandth, we get 0.008.

PRACTICE 3

What is the product of 0.44 and 0.03, rounded to the nearest hundredth?

EXAMPLE 4

Multiply: $(1.1)(3.5)(0.8)$

Solution To find the product of three factors, we can first multiply the two left factors and then multiply this product by the third factor:

$$\underbrace{(1.1)(3.5)}(0.8) =$$
$$(3.85)(0.8) = 3.08$$

So 3.08 is the final product.

PRACTICE 4

Evaluate: $(0.2)(0.3)(0.4)$

EXAMPLE 5

Simplify: $3 + (1.2)^2$

Solution Recall that, according to the order of operations rule, we first must find $(1.2)^2$ and then add 3.

$$3 + (1.2)^2 =$$
$$3 + 1.44 = 4.44$$

PRACTICE 5

Evaluate: $10 - (0.3)^2$

EXAMPLE 6

Multiply: 8.274 × 100

Solution

```
  8.274  ← Three decimal places
 ×100    ← Zero decimal places
827.400  ← Three decimal places in the product
```

So the product is 827.400, or 827.4 after we drop the extra zeros.

Note that the second factor (100) is a power of 10 ending in **two** zeros and that the product is identical to the first factor except that the decimal point is moved to the right **two** places.

PRACTICE 6

Compute: 0.325 × 1,000

As Example 6 illustrates, a shortcut for multiplying a decimal by a power of 10 is to *move the decimal point to the right the same number of places as the power of 10 has zeros.* Let's apply this shortcut in the next example.

EXAMPLE 7

Find the product: 1,000 × 2.89

Solution We see that 1,000 is a power of 10 and has three zeros. So to multiply 1,000 by 2.89, we simply move the decimal point in 2.89 to the right three places.

Add a 0 to move three places. 2.890 = 2890. = 2,890

So the product is 2,890, with the 0 serving as a placeholder.

PRACTICE 7

Multiply 32.7 by 10,000.

EXAMPLE 8

The popularity of a television show is measured in ratings, where each rating point represents 900,000 homes in which the show is watched. After examining the table at the right, answer each question.

Show	Rating
1	17.3
2	14.25

a. In how many homes was show 1 watched?

b. In how many more homes was show 1 watched than show 2?

Solution

a. To find the number of homes in which show 1 was watched, we multiply its rating, 17.3, by 900,000, which gives us 15,570,000.

b. To compare the popularity of show 1 and show 2, we compute the number of homes in which show 2 was viewed: 14.25 × 900,000 = 12,825,000. The number of show 1 homes exceeds the number of show 2 homes by 15,570,000 − 12,825,000, or 2,745,000 homes.

PRACTICE 8

A chemistry student learns that a molecule is made up of atoms. For instance, the water molecule, H_2O, consists of two atoms of hydrogen, H, and one atom of oxygen, O. Each of these atoms has an atomic weight.

Atom	Atomic Weight
H	1.008
O	15.999

a. After examining the chart above, compute the weight of the water molecule.

b. Round this weight to the nearest whole number.

Estimating Products

A good way to estimate the product of decimals is to round each factor so that it has only one nonzero digit. Then multiply the rounded factors.

For instance, suppose that we want to estimate the product of the decimals 19.0382 and 0.061.

$$\begin{array}{rcr} 19.0382 & \approx & 20 \\ \times 0.061 & \approx & \times 0.06 \\ \hline & & 01.20 = 1.2 \end{array}$$

Before we dropped the extra 0, the estimated product, 1.20, has two decimal places—the total number of decimal places in the factors.

EXAMPLE 9

Multiply 0.703 by 0.087 and check the answer by estimating.

Solution First, we multiply the factors to find the exact product. Then, we round each factor and multiply them.

$$\begin{array}{rcrl} 0.703 & \approx & 0.7 & \leftarrow \text{Rounded to have one nonzero digit} \\ \times\ 0.087 & \approx & \times 0.09 & \leftarrow \text{Rounded to have one nonzero digit} \\ \hline \text{Exact product} \rightarrow 0.061161 & & 0.063 & \leftarrow \text{Estimated product} \end{array}$$

We see that the exact product and the estimated product are fairly close, as expected.

PRACTICE 9

Find the product of 0.0037×0.092, estimating to check.

EXAMPLE 10

Calculate and check: $(4.061)(0.72) + (0.91)(0.258)$

Solution Following the order of operations rule, we begin by finding the two products.

$$(4.061)(0.72) = 2.92392 \qquad (0.91)(0.258) = 0.23478$$

Then we add these two products.

$$2.92392 + 0.23478 = 3.1587$$

So $(4.061)(0.72) + (0.91)(0.258) = 3.1587$.

Now, let's check this answer by estimating.

$$\begin{array}{l} (4.061)(0.72) + (0.91)(0.258) \\ \downarrow \quad\ \downarrow \qquad\ \downarrow \quad\ \downarrow \\ (4)\ (0.7) + (0.9)\ (0.3) = 2.8 + 0.27 = 3.07 \approx 3 \end{array}$$

The estimate, 3, is sufficiently close to 3.1587 to confirm our answer.

PRACTICE 10

Compute and check:
$(0.488)(9.1) - (3.5)(0.227)$

EXAMPLE 11

The sound waves of an elephant call can travel through both the ground and the air. Through the air, the waves may travel 6.63 miles. If they travel 1.5 times as far through the ground, what is the estimated ground distance? (*Source:* http://www.abc.net.au/science/k2/moments/s434107.htm)

Solution We know that the waves may travel 6.63 miles through the air and 1.5 times as far through the ground. To find the estimated ground distance, we compute this product.

$$\begin{array}{r} 6.63 \approx 7 \\ \underline{\times 1.5} \approx \underline{\times 2} \\ 14 \end{array}$$

So the estimated ground distance of the sound waves of an elephant call is about 14 miles.

PRACTICE 11

The Earth travels through space at a speed of 18.6 miles per second. Estimate how far the Earth travels in 60 seconds. (*Source:* The Diagram Group, *Comparisons*)

Multiplying Decimals on a Calculator

Multiply decimals on a calculator by entering each decimal as you would enter a whole number, but insert a decimal point as needed. If there are too many decimal places in your answer to fit in the display, investigate how your calculator displays the answer.

EXAMPLE 12

Compute 8,278.55 × 0.875, rounding your answer to the nearest hundredth. Then check the answer by estimating.

Solution

Press	Display
8278.55 [×] 0.875 [ENTER]	8278.55 * 0.875 7243.73125

Now 7,243.73125 rounded to the nearest hundredth is 7,243.73. Checking by estimating, we get 8,000 × 0.9, or 7,200, which is close to our exact answer.

PRACTICE 12

Find the product of 2,471.66 and 0.33, rounding to the nearest tenth. Check the answer.

EXAMPLE 13

Find $(1.9)^2$

Solution

Press	Display
1.9 [^] 2 [ENTER]	1.9 ^ 2 3.61

Now let's check by estimating. Since 1.9 rounded to the nearest whole number is 2, $(1.9)^2$ should be close to 2^2, or 4, which is close to our exact answer, 3.61.

PRACTICE 13

Calculate: $(2.1)^3$

3.3 Exercises

FOR EXTRA HELP

Mathematically Speaking

Fill in each blank with the most appropriate term or phrase from the given list.

add	three	factors	five
first factor	four	multiplication	
square	two	division	

1. The operation understood in the expression (3.4) (8.9) is _______.
2. When multiplying decimals, the number of decimal places in the product is equal to the total number of decimal places in the _______.
3. To multiply a decimal by 100, move the decimal point _______ places to the right.
4. The product of 0.27 and 8.18 has _______ decimal places.
5. To compute the expression $(8.5)^2 + 2.1$, first _______.
6. To multiply a decimal by 1,000, move the decimal point _______ places to the right.

Insert a decimal point in each product. Check by estimating.

7. $2.356 \times 1.27 = 299212$
8. $97.26 \times 5.3 = 515478$
9. $3{,}144 \times 0.065 = 204360$
10. $837 \times 0.15 = 12555$
11. $71.2 \times 35 = 24920$
12. $0.002 \times 37 = 0074$
13. $0.0019 \times 0.051 = 969$
14. $0.0089 \times 0.0021 = 1869$
15. $2.87 \times 1{,}000 = 287000$
16. $492.31 \times 10 = 492310$
17. $\$4.25 \times 0.173 = \$73525$
18. 11.2 feet $\times$ 0.75 = 8400 feet

Find the product. Check by estimating.

19. 0.6×0.9
20. 0.8×0.7
21. 0.5×0.8
22. 0.6×0.8
23. 0.1×0.2
24. 0.09×0.5
25. 0.04×0.07
26. 0.03×0.01
27. 2.55×0.3
28. 80.7×0.6
29. 0.96×2.1
30. 0.043×0.02
31. 38.01×0.2
32. 1.22×0.09
33. 125×0.004
34. 0.003×0.7
35. 3.8×1.54
36. 9.51×0.7
37. 13.74×11
38. $1{,}245 \times 2.5$
39. 12.459×0.3
40. 72.558×0.2
41. $(0.675)(2.66)$
42. $(4.003)(0.59)$
43. 83.127×100
44. 4.9×10
45. $0.0023 \times 10{,}000$
46. $0.0135 \times 1{,}000$

47. $(1.5)(0.6)(0.1)$ **48.** $(12)(3.5)(0.2)$ **49.** $(0.03)(1.4)(25)$ **50.** $(2.6)(0.5)(0.9)$

51. $(0.001)^3$ **52.** $(0.1)^4$ **53.** 17 feet $\times$ 2.5 **54.** 5 hours $\times$ 0.75

55. 3.5 miles $\times$ 0.4 **56.** 9.1 meters $\times$ 1,000

57. 43.87×0.075 **58.** $18{,}275.33 \times 0.39$ **59.** $99{,}125 \times 2.75$ **60.** 3.512×1.47

Simplify.

61. 0.7×10^2 **62.** 0.6×10^4 **63.** $30 - 2.5 \times 1.7$ **64.** $8 + 4.1 \times 2$

65. $1 + (0.3)^2$ **66.** $6 - (1.2)^2$ **67.** $0.8(1.3 + 2.9) - 0.5$ **68.** $4 - 2.1(3.5 - 1.8)$

Complete each table.

69.

Input	Output
1	$3.8 \times \mathbf{1} - 0.2 =$
2	$3.8 \times \mathbf{2} - 0.2 =$
3	$3.8 \times \mathbf{3} - 0.2 =$
4	$3.8 \times \mathbf{4} - 0.2 =$

70.

Input	Output
1	$7.5 \times \mathbf{1} + 0.4 =$
2	$7.5 \times \mathbf{2} + 0.4 =$
3	$7.5 \times \mathbf{3} + 0.4 =$
4	$7.5 \times \mathbf{4} + 0.4 =$

Each product is rounded to the nearest hundredth. In each group of three products, one is wrong. Use estimation to explain which product is incorrect.

71. a. $51.6 \times 0.813 = 419.51$ **b.** $2.93 \times 7.283 = 21.34$ **c.** $(5.004)^2 = 25.04$

72. a. $0.004 \times 3.18 = 0.01$ **b.** $2.99 \times 0.287 = 0.86$ **c.** $(1.985)^3 = 10.82$

73. a. $4.913 \times 2.18 = 10.71$ **b.** $0.023 \times 0.71 = 0.16$ **c.** $(8.92)(1.0027) = 8.94$

74. a. \$138.28 $\times$ 0.075 = \$10.37 **b.** 0.19 $\times$ \$487.21 = \$92.57 **c.** 0.77 $\times$ \$6,005.79 = \$462.45

Mixed Practice

Solve.

75. Simplify: $9 - (0.5)^2$

76. Compute: $2.1 + 5 \times 0.6$

77. Multiply 0.75 and 0.09, rounding the answer to the nearest thousandth.

78. Multiply: $(2.3)(4.5)(0.6)$

79. Find the product of 0.56 and 8.

80. Find the product: $3.01 \times 1{,}000$

Applications

Solve. Check by estimating.

81. Sound travels at approximately 1,000 feet per second (fps). If a jet is flying at Mach 2.9 (that is, 2.9 times the speed of sound), what is its speed?

82. If insurance premiums of \$323.50 are paid yearly for 10 years for a life insurance policy, how much did the policy holder pay altogether in premiums?

83. The planet in the solar system closest to the Sun is Mercury. The average distance between these two bodies is 57.9 million kilometers. Express this distance in standard form. (***Source:*** Jeffrey Bennett et al., *The Cosmic Perspective*)

84. According to the first American census in 1790, the population of the United States was approximately 3.9 million. Write this number in standard form. (***Source:*** *The Statistical History of the United States*)

85. The area of this circle is approximately $3.14 \times (2.5)^2$ square feet. Find this area to the nearest tenth.

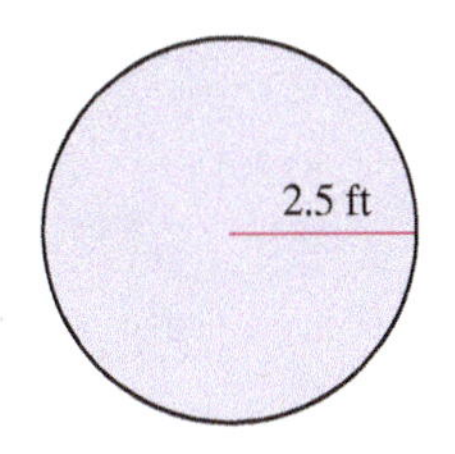

86. Find the area (in square meters) of the room pictured.

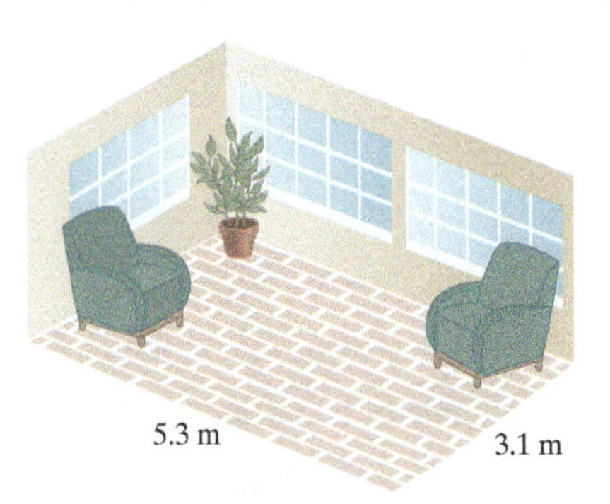

87. Over a 5-day period, a nurse administered 10 tablets to a patient. If each tablet contained 0.125 milligram of the drug Digoxin, how much Digoxin did the nurse administer?

88. Water weighs approximately 62.5 pounds per cubic foot (lb/ft^3). If a bathtub contains about 30 ft^3 of water, how much does the water in the bathtub weigh?

89. A tennis player weighing 180 pounds burns 10.9 calories per minute while playing singles tennis. How many calories would he burn in 2 hours? (***Source:*** http://www.caloriesperhour.com)

90. A plumber is paid $37.50 per hour for the first 40 hours worked. She gets time and a half, $56.25, for any time over her 40-hour week. If she works 49 hours in a week, how much is her pay?

91. The sales receipt for a shopper's purchases is as follows.

Purchase	Quantity	Unit Price	Price
Belt	1	$11.99	$___.__
Shirt	3	$16.95	$___.__
Total Price			$___.__

a. Complete the table.

b. If the shopper pays for these purchases with four $20 bills, how much change should he get?

92. On an electric bill, *usage* is the difference between the meter's *current reading* and the *previous reading* in kilowatt hours (kWh). The *amount due* is the product of the usage and the *rate per kWh*. Find the two missing quantities in the table, rounding to the nearest hundredth.

Previous Reading	750.07 kWh
Current Reading	1,115.14 kWh
Usage	_____ kWh
Rate per kWh	$0.10
Amount Due	$____

Use a calculator to solve each problem, giving (a) the operation(s) carried out in the solution, (b) the exact answer, and (c) an estimate.

93. Scientists have discovered a relationship between the length of a person's bones and the person's overall height. For instance, an adult male's height (in inches) can be predicted from the length of his femur bone by using the formula $(1.9 \times \text{femur}) + 32.0$. With this formula, estimate the height of the German giant Constantine, whose femur measured 29.9 inches. (***Source:*** *Guinness World Records 2007*)

94. In order to buy a $125,000 house, a couple puts down $25,000 and takes out a mortgage on the balance. To pay off the mortgage, they pay $877.57 per month for the following 360 months. How much more will they end up paying for the house than the original price of $125,000?

• *Check your answers on page A-6.*

MINDSTRETCHERS

Patterns

1. When $(0.001)^{100}$ is multiplied out, how many decimal places will it have?

Mathematical Reasoning

2. Give an example of two decimals

 a. whose sum is greater than their product, and

 b. whose product is greater than their sum.

Groupwork

3. In the product to the right, each letter stands for a different digit. Working with a partner, identify all the digits.

$$\begin{array}{r} A.B \\ \times\ B.A \\ \hline C\ D \end{array}$$

3.4 Dividing Decimals

OBJECTIVES

- To find the decimal equivalent to a fraction
- To divide decimals
- To estimate the quotient of decimals
- To solve word problems involving the division of decimals

In this section, we first consider changing a fraction to its decimal equivalent, which involves both division and decimals. Then we move on to our main concern—the division of decimals.

Changing a Fraction to the Equivalent Decimal

Earlier in this chapter, we discussed how to change a decimal to its equivalent fraction. Now let's consider the opposite problem—how to change a fraction to its equivalent decimal.

When the denominator of a fraction is already a power of 10, the problem is simple. For example, the decimal equivalent of $\frac{43}{100}$ is 0.43.

But what about the more difficult problem where the denominator is *not* a power of 10? A good strategy is to find an equivalent fraction that does have a power of 10 as its denominator. Consider, for instance, the fraction $\frac{3}{4}$. Since 4 is a factor of 100, which is a power of 10, we can easily find an equivalent fraction having a denominator of 100.

$$\frac{3}{4} = \frac{3 \cdot 25}{4 \cdot 25} = \frac{75}{100} = 0.75$$

So 0.75 is the decimal equivalent of $\frac{3}{4}$.

There is a faster way to show that $\frac{3}{4}$ is the same as 0.75, without having to find an equivalent fraction. Because $\frac{3}{4}$ can mean $3 \div 4$, we divide the numerator (3) by the denominator (4). Note that if we continue to divide to the hundredths place, there is no remainder.

```
    0.75     ← Insert the decimal point directly above the decimal point in the dividend.
 4)3.00      ← The decimal point is after the 3. Insert enough 0's to continue dividing as far as necessary.
   0
   -
   3 0
   2 8
   ---
     20
     20
     --
```

So this division also tells us that $\frac{3}{4}$ equals 0.75.

To Change a Fraction to the Equivalent Decimal

- Divide the denominator of the fraction into the numerator, inserting to its right both a decimal point and enough zeros to get an answer either without a remainder or rounded to a given decimal place.
- Place a decimal point in the quotient directly above the decimal point in the dividend.

EXAMPLE 1

Express $\frac{1}{2}$ as a decimal.

Solution To find the decimal equivalent, we divide the fraction's numerator by its denominator.

$$\begin{array}{r} 0.5 \\ 2\overline{)1.0} \\ \underline{0} \\ 1\,0 \\ \underline{1\,0} \end{array}$$

Add a decimal point and a 0 to the right of the 1.

So 0.5 is the decimal equivalent of $\frac{1}{2}$.

Check We verify that the fractional equivalent of 0.5 is $\frac{1}{2}$.

$$0.5 = \frac{5}{10} = \frac{1}{2}$$

The answer checks.

PRACTICE 1

Write the fraction $\frac{3}{8}$ as a decimal.

EXAMPLE 2

Convert $2\frac{3}{5}$ to a decimal.

Solution Let's first change this mixed number to an improper fraction: $2\frac{3}{5} = \frac{13}{5}$. We can then change this improper fraction to a decimal by dividing its numerator by its denominator.

$$\begin{array}{r} 2.6 \\ 5\overline{)13.0} \\ \underline{10} \\ 3\,0 \\ \underline{3\,0} \end{array}$$

So 2.6 is the decimal form of $2\frac{3}{5}$.

Check We convert this answer back from a decimal to its mixed number form.

$$2.6 = 2\frac{6}{10} = 2\frac{3}{5}$$

The answer checks.

PRACTICE 2

Write $7\frac{5}{8}$ as a decimal.

When converting some fractions to decimal notation, we keep getting a remainder as we divide. In this case, we round the answer to a given decimal place.

EXAMPLE 3

Convert $4\frac{8}{9}$ to a decimal, rounded to the nearest hundredth.

Solution First, we change this mixed number to an improper fraction. Then, we convert it to a decimal.

$$4\frac{8}{9} = \frac{44}{9} = 9\overline{)44.000} = 4.888$$

$$\begin{array}{r} 4.888 \\ 9\overline{)44.000} \\ \underline{36} \\ 8\,0 \\ \underline{7\,2} \\ 80 \\ \underline{72} \\ 80 \\ \underline{72} \end{array}$$

← In order to round to the nearest hundredth, we must continue to divide to the thousandths place. So we insert three 0's.

Finally, we round to the nearest hundredth: $4.8\underline{8}8 \approx 4.89$

PRACTICE 3

Express $83\frac{1}{3}$ as a decimal, rounded to the nearest tenth.

In Example 3, note that if instead of rounding we had continued to divide we would have gotten as our answer the *repeating decimal* 4.88888 . . . (also written $4.\overline{8}$). Can you think of any other fraction that is equivalent to a repeating decimal?

Let's now consider some word problems in which we need to convert fractions to decimals.

EXAMPLE 4

A share of stock sells for $5\frac{7}{8}$ dollars. Express this amount in dollars and cents, to the nearest cent.

Solution To solve, we must convert the mixed number $5\frac{7}{8}$ to a decimal.

$$5\frac{7}{8} = \frac{47}{8} = 8\overline{)47.000}$$

$$\begin{array}{r} 5.8\underline{7}5 \approx 5.88 \\ 8\overline{)47.000} \\ \underline{40} \\ 7\,0 \\ \underline{6\,4} \\ 60 \\ \underline{56} \\ 40 \\ \underline{40} \end{array}$$

So a share of this stock sells for \$5.88, to the nearest cent.

PRACTICE 4

The gas nitrogen makes up about $\frac{39}{50}$ of the air in the atmosphere. Express this fraction as a decimal, rounded to the nearest tenth. (***Source:*** *World of Scientific Discovery*)

Dividing Decimals

Before we turn our attention to dividing one decimal by another, let's consider simpler problems in which we are dividing a decimal by a whole number. An example of such a problem is $0.6 \div 2$. We can write this expression as the fraction $\frac{0.6}{2}$, which can be rewritten as the quotient of two whole numbers by multiplying the numerator and denominator by 10.

$$\frac{0.6}{2} = \frac{0.6 \times 10}{2 \times 10} = \frac{6}{20}$$

We then convert this fraction to the equivalent decimal, as we have previously discussed.

$$\begin{array}{r} 0.3 \\ 20\overline{)\,6.0} \\ \underline{0} \\ 6\,0 \\ \underline{6\,0} \end{array}$$

So $\dfrac{0.6}{2} = 0.3$

Note that we get the same quotient if we divide the number in the original problem as follows:

$$\begin{array}{r} 0.3 \leftarrow \text{Quotient} \\ \text{Divisor} \rightarrow 2\overline{)0.6} \leftarrow \text{Dividend} \\ \underline{0} \\ 6 \\ \underline{6} \end{array}$$

It is important to write the decimal point in the quotient directly above the decimal point in the dividend.

Next, let's consider a division problem where we are dividing one decimal by another: $0.006 \div 0.02$. Writing this expression as a fraction, we get $\dfrac{0.006}{0.02}$. Since we have already discussed how to divide a decimal by a whole number, the goal here is to find a fraction equivalent to $\dfrac{0.006}{0.02}$ where the denominator is a whole number. Multiplying the numerator and denominator by 100 will do just that.

$$0.006 \div 0.02 = \frac{0.006}{0.02} = \frac{0.006 \times 100}{0.02 \times 100} = \frac{0.006}{0.02} = \frac{0.6}{2}$$

We know from the previous problem that $\dfrac{0.6}{2} = 0.3$. Since $\dfrac{0.006}{0.02} = \dfrac{0.6}{2}$, we conclude $\dfrac{0.006}{0.02} = 0.3$.

A shortcut to multiply by 100 in both the divisor and the dividend is to move the decimal point two places to the right.

So $\begin{array}{r} 0.3 \\ 0.02\overline{)0.006} \end{array}$ is equivalent to $\begin{array}{r} 0.3 \\ 2\overline{)0.6} \end{array}$

As in any division problem, we can check our answer by confirming that the product of the quotient and the *original divisor* equals the *original dividend.*

Division Problem	**Check**
$0.6 \div 2 = 0.3$	$\begin{array}{r} 0.3 \\ \underline{\times 2} \\ 0.6 \end{array}$
$0.006 \div 0.02 = 0.3$	$\begin{array}{r} 0.3 \\ \underline{\times 0.02} \\ 0.006 \end{array}$

These examples suggest the following rule.

To Divide Decimals

- Move the decimal point in the divisor to the right end of the number.
- Move the decimal point in the dividend the same number of places to the right as in the divisor.
- Insert a decimal point in the quotient directly above the decimal point in the dividend.
- Divide the new dividend by the new divisor, inserting zeros at the right end of the dividend as necessary.

EXAMPLE 5

What is 0.035 divided by 0.25?

Solution Move the decimal point to the right end, making the divisor a whole number.

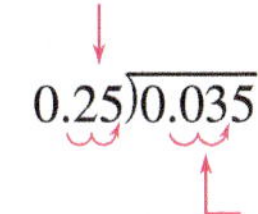

Move the decimal point in the dividend the same number of places.

Now, we divide 3.5 by 25, which gives us 0.14.

```
   0.14
25)3.50
   2 5
   1 00
   1 00
```

Check We see that the product of the quotient and the original divisor is equal to the original dividend.

```
     0.14
   × 0.25
     0 7 0
    0 2 8
  0.03 5 0 = 0.035
```

PRACTICE 5

Divide and check: $2.706 \div 0.15$

EXAMPLE 6

Find the quotient: $6 \div 0.0012$. Check the answer.

Solution The decimal point is moved four places to the right.

```
0.0012)6.0000
```

To move the decimal point four places to the right, we must add four 0's as placeholders.

```
   5,000
12)60,000
```

Check

```
     5,000
 × 0.0012
 0006.0000 = 6
```

The answer checks.

PRACTICE 6

Divide $\frac{8.2}{0.004}$ and then check.

EXAMPLE 7

Divide and round to the nearest hundredth: $0.7\overline{)40.2}$
Then check.

Solution $0.7\overline{)40.2}$

```
   57.428 ≈ 57.43 to the nearest hundredth
7)402.000
  35
   52
   49
    3 0
    2 8
      20
      14
       60
       56
        4
```

Check

$$\begin{array}{r} 57.43 \\ \times\ 0.7 \\ \hline 40.201 \end{array} \approx 40.2$$

Because we rounded our answer, the check gives us a product only approximately equal to the original dividend.

PRACTICE 7

Find the quotient of 8.07 and 0.11, rounded to the nearest tenth.

EXAMPLE 8

Compute and check: $8.319 \div 1{,}000$

Solution

```
         0.008319
1,000)8.319000
      8 000
        3190
        3000
         1900
         1000
          9000
          9000
```

Check

$$\begin{array}{r} 0.008319 \\ \times\ \ \ 1{,}000 \\ \hline 0008.319000 \end{array} = 8.319$$

Note that the divisor (1,000) is a power of 10 ending in three zeros, and that the quotient is identical to the dividend except that the decimal point is moved to the left three places.

$$\frac{8.319}{1{,}000} = 0.008319$$

PRACTICE 8

Divide: $100\overline{)3.41}$

As Example 8 illustrates, a shortcut for dividing a decimal by a power of 10 is to *move the decimal point to the left the same number of places as the power of 10 has zeros.* Can you explain the difference between the shortcuts for multiplying and for dividing by a power of 10?

EXAMPLE 9

Compute: $\dfrac{7.2}{100}$

Solution Since we are dividing by 100, a power of 10 with two zeros, we can find this quotient simply by moving the decimal point in 7.2 to the left two places.

$$\frac{7.2}{100} = .072, \quad \text{or} \quad 0.072$$

So the quotient is 0.072.

PRACTICE 9

Calculate: $0.86 \div 1{,}000$

Now let's try using these skills in some applications.

EXAMPLE 10

The following graph shows the number of people who attended a Broadway show in selected seasons.

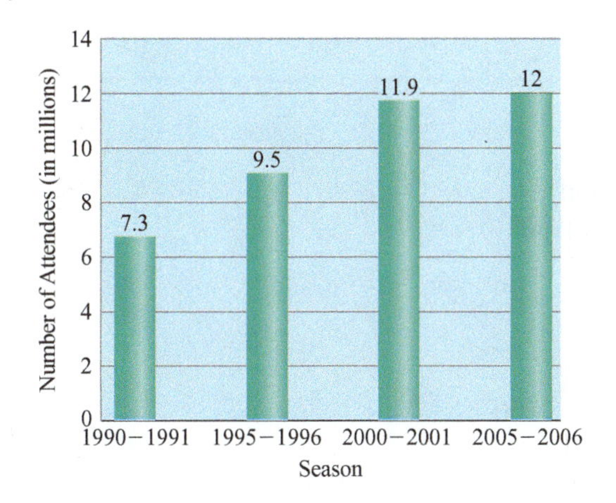

The number of Broadway attendees in the 2005–2006 season was how many times as great as the number of attendees 15 years earlier? Round to the nearest tenth. (*Source:* The League of American Theatres and Producers)

Solution The number of attendees in 2005–2006 (in millions) was 12, and the number in 1990–1991 was 7.3. To find how many times as great 12 is as compared to 7.3, we find their quotient.

$$7.3\overline{)12} = 7.3\overline{)12.0} = 73\overline{)120}$$

$$\begin{array}{r} 1.64 \\ 73\overline{)120.00} \\ \underline{73} \\ 47\,0 \\ \underline{43\,8} \\ 3\,20 \\ \underline{2\,92} \\ 28 \end{array}$$

Rounding to the nearest tenth, we conclude that the number of Broadway attendees in 2005–2006 was 1.6 times the corresponding figure in 1990–1991.

PRACTICE 10

The table gives the amount of selected foods consumed per capita in the United States in a recent year.

Food	Annual Per Capita Consumption (in pounds)
Red meat	112.0
Poultry	72.7
Fish and shellfish	16.5

The amount of red meat consumed was how many times as great as the amount of poultry, rounded to the nearest tenth? (*Source:* U.S. Department of Agriculture)

Estimating Quotients

As we have shown, one way to check the quotient of two decimals is by multiplying. Another way is by estimating.

To check a decimal quotient by estimating, we can round each decimal to one nonzero digit and then mentally divide the rounded numbers. But we must be careful to position the decimal point correctly in our estimate.

EXAMPLE 11

Divide and check by estimating: $3.36 \div 0.021$

Solution $0.021\overline{)3.360}$

We compute the exact answer.

$$\begin{array}{r} 160 \\ 21\overline{)3{,}360} \\ \underline{2\,1} \\ 1\,26 \\ \underline{1\,26} \\ 00 \\ \underline{00} \end{array}$$

So 160 is our quotient.

Now let's check by estimating. Because $3.36 \approx 3$ and $0.021 \approx 0.02$, $3.36 \div 0.021 \approx 3 \div 0.02$. We mentally divide to get the estimate.

$$\begin{array}{r} 150 \\ 0.02\overline{)3.00} \end{array}$$

Our estimate, 150, is reasonably close to our exact answer, 160, and so confirms it.

PRACTICE 11

Compute and check by estimating:
$8.229 \div 0.39$

EXAMPLE 12

Calculate and check: $(9.13) \div (0.2) + (4.6)^2$

Solution Following the order of operations rule, we begin by finding the square and then the quotient.

$$(4.6)^2 = 21.16$$
$$(9.13) \div (0.2) = 45.65$$

Then we add these two answers.

$$21.16 + 45.65 = 66.81$$

So $(9.13) \div (0.2) + (4.6)^2 = 66.81$.

Now, let's check this answer by estimating.

$$(9.13) \div (0.2) + (4.6)^2$$
$$\underbrace{9 \div 0.2}_{45} + 25 \approx 45 + 25 \approx 70$$

The estimate, 70, is sufficiently close to 66.81 to confirm our answer.

PRACTICE 12

Compute and check:
$13.07 + (8.4 \div 0.5)^2$

EXAMPLE 13

The water in a filled aquarium weighs 638.25 pounds. If 1 cubic foot of water weighs 62.5 pounds, estimate how many cubic feet of water there are in the aquarium.

Solution We know that the water in the aquarium weighs 638.25 pounds. Since 1 cubic foot of water weighs 62.5 pounds, we can estimate the number of cubic feet of water in the aquarium by computing the quotient $638.25 \div 62.5$, which is approximately $600 \div 60$, or 10. So a reasonable estimate for the amount of water in the aquarium is 10 cubic feet.

PRACTICE 13

The following graph shows the number of farms, in a recent year, in five states.

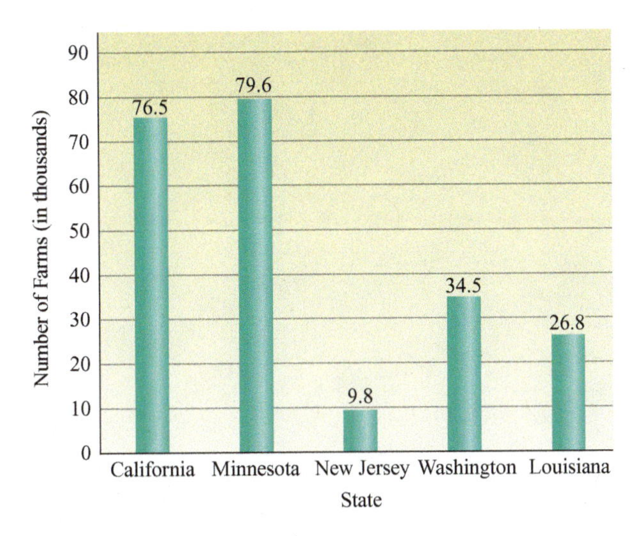

The number of farms in Minnesota is about how many times as great as the number in New Jersey? (*Source:* U.S. Department of Agriculture)

Dividing on a Calculator

When dividing decimals on a calculator, be careful to enter the dividend first and then the divisor. Note that when the dividend is larger than the divisor, the quotient is greater than 1, and when the dividend is smaller than the divisor, the quotient is less than 1.

EXAMPLE 14

Calculate $8.6 \div 1.6$ and round to the nearest tenth.

Solution

Press	**Display**
8.6 [÷] 1.6 [ENTER =]	8.6/1.6 5.375

The answer, when rounded to the nearest tenth, is 5.4. As expected, the answer is greater than 1, because $8.6 > 1.6$.

PRACTICE 14

Compute the quotient $8.6\overline{)1.6}$ and round to the nearest tenth.

EXAMPLE 15

Divide $0.3\overline{)0.07}$, rounding to the nearest hundredth.

Solution

Press	Display
0.07 [÷] 0.3 [ENTER =]	0.07/0.3 0.233333333

The answer, when rounded to the nearest hundredth, is 0.23. As expected, the answer is less than 1, because $0.07 < 0.3$.

PRACTICE 15

Find the quotient, rounding to the nearest hundredth: $0.3 \div 0.07$

3.4 Exercises

FOR EXTRA HELP

Mathematically Speaking

Fill in each blank with the most appropriate term or phrase from the given list.

quotient	three	divisor	decimal
dividend	right	terminating	four
fraction	product	left	repeating

1. To change a fraction to the equivalent _______, divide the numerator of the fraction by its denominator.
2. An example of a(n) _______ decimal is 0.3333 . . . .
3. When dividing decimals, move the decimal point in the divisor to the _______ end.
4. To divide a decimal by 1,000, move the decimal point _______ places to the left.
5. To estimate the _______ of 0.813 and 0.187, divide 0.8 by 0.2.
6. When dividing a decimal by a whole number, the decimal point in the quotient is placed above the decimal point in the _______.

Change to the equivalent decimal. Then check.

7. $\frac{1}{2}$ **8.** $\frac{3}{5}$ **9.** $\frac{1}{4}$ **10.** $\frac{1}{8}$

11. $\frac{37}{10}$ **12.** $\frac{517}{100}$ **13.** $1\frac{5}{8}$ **14.** $10\frac{3}{4}$

15. $2\frac{7}{8}$ **16.** $8\frac{2}{5}$ **17.** $21\frac{3}{100}$ **18.** $60\frac{17}{100}$

Divide and check.

19. $4\overline{)17}$ **20.** $2\overline{)35}$ **21.** $5\overline{)21}$ **22.** $6\overline{)33}$

23. $8\overline{)11}$ **24.** $6\overline{)9}$ **25.** $18\overline{)153}$ **26.** $14\overline{)217}$

Change to the equivalent decimal. Round to the nearest hundredth.

27. $\frac{2}{3}$ **28.** $\frac{5}{6}$ **29.** $\frac{7}{9}$ **30.** $\frac{1}{3}$

31. $3\frac{1}{9}$ **32.** $2\frac{4}{7}$ **33.** $5\frac{1}{16}$ **34.** $10\frac{11}{32}$

Divide. Express any remainder as a decimal rounded to the nearest thousandth.

35. $7\overline{)23}$ **36.** $9\overline{)41}$ **37.** $11\overline{)3}$ **38.** $13\overline{)2}$

39. $7\overline{)46}$ **40.** $6\overline{)82}$ **41.** $13\overline{)911}$ **42.** $12\overline{)208}$

Insert the decimal point in the appropriate place.

43. $\begin{array}{r}5\,8\,8\,2\\0.7\overline{)4\,1.1\,7\,4}\end{array}$ **44.** $\begin{array}{r}5\,7\\3\overline{)0.0\,1\,7\,1}\end{array}$ **45.** $\begin{array}{r}6\,6\,3\\0.5\,8\overline{)0.0\,3\,8\,4\,5\,4}\end{array}$ **46.** $\begin{array}{r}6\,9\\3.9\overline{)2\,6.9\,1}\end{array}$

Divide. Check, either by multiplying or by estimating.

47. $8\overline{)23.1}$ **48.** $2\overline{)0.0035}$ **49.** $7\overline{)2.002}$ **50.** $6\overline{)24.042}$

51. $\frac{17.2}{4}$ **52.** $\frac{0.75}{5}$ **53.** $\frac{0.003}{2}$ **54.** $\frac{1.04}{8}$

55. $8.65 \div 5$ **56.** $0.42 \div 3$ **57.** $11.5 \div 4$ **58.** $7.3 \div 2$

59. $0.2\overline{)0.8}$ **60.** $0.3\overline{)0.6}$ **61.** $0.05\overline{)3.52}$ **62.** $0.04\overline{)1.92}$

63. $\frac{47}{0.5}$ **64.** $\frac{86}{0.2}$ **65.** $\frac{5}{0.4}$ **66.** $\frac{9}{0.6}$

67. $0.03 \div 0.1$ **68.** $1.2 \div 0.01$ **69.** $0.38 \div 1.9$ **70.** $0.075 \div 0.25$

71. $95.2 \div 100$ **72.** $81.6 \div 10$ **73.** $0.082 \div 100$ **74.** $9.03 \div 1{,}000$

Divide, rounding to the nearest hundredth. Check, either by multiplying or by estimating.

75. $0.8\overline{)307.1}$ **76.** $0.4\overline{)81.9}$ **77.** $0.9\overline{)0.0057}$ **78.** $0.2\overline{)0.057}$

79. $\frac{3.69}{0.4}$ **80.** $\frac{3.995}{0.7}$ **81.** $\frac{87}{0.009}$ **82.** $\frac{23}{0.06}$

83. $41 \div 0.021$ **84.** $9.13 \div 0.007$ **85.** $35.77 \div 0.11$ **86.** $0.291 \div 0.17$

87. $49.071 \div 0.728$ **88.** $18.3 \div 7.96$ **89.** $3 \div 0.0721$ **90.** $100 \div 3.89$

Perform the indicated operations.

91. $\frac{10.71}{5} \cdot \frac{0.4}{5}$ **92.** $\frac{2.04}{3} + 1$ **93.** $\frac{51.3}{10} - 5$ **94.** $\frac{26.77 - 10.1}{0.4}$

95. $\frac{13.05}{7.27 - 7.02}$ **96.** $\frac{81.51}{3} - \frac{25.2}{9}$ **97.** $\frac{8.1 \times 0.2}{0.4}$

98. $(82.9 - 3.6) \div (0.21 - 0.01)$

Complete each table.

99.

Input	Output
1	$1 \div 5 - 0.2 =$
2	$2 \div 5 - 0.2 =$
3	$3 \div 5 - 0.2 =$
4	$4 \div 5 - 0.2 =$

100.

Input	Output
1	$1 \div 4 + 0.4 =$
2	$2 \div 4 + 0.4 =$
3	$3 \div 4 + 0.4 =$
4	$4 \div 4 + 0.4 =$

Each of the following quotients is rounded to the nearest hundredth. In each group of three quotients, one is wrong. Use estimation to identify which quotient is incorrect.

101. a. $5.7 \div 89 \approx 0.06$ **b.** $0.77 \div 0.0019 \approx 405.26$ **c.** $31.5 \div 0.61 \approx 516.39$

102. a. $\frac{9.83}{4.88} \approx 0.20$ **b.** $\frac{2.771}{0.452} \approx 6.13$ **c.** $\frac{389.224}{1.79} \approx 217.44$

103. a. $61.27 \div 0.057 \approx 1{,}074.91$ **b.** $0.614 \div 2.883 \approx 2.13$ **c.** $0.0035 \div 0.00481 \approx 0.73$

104. a. $\$365 \div \$4.89 \approx 7.46$ **b.** $\$17{,}358.27 \div \$365 \approx 47.56$ **c.** $\$3{,}000 \div \$2.54 \approx 1{,}181.10$

Mixed Practice

Solve.

105. Express $\frac{4}{5}$ as a decimal.

106. Divide $1.6\overline{)8.5}$ and round to the nearest tenth.

107. Change $1\frac{1}{6}$ to a decimal, rounded to the nearest hundredth.

108. Simplify: $81.5 - \frac{32}{0.4}$

109. What is 0.063 divided by 0.14?

110. Find the quotient: $9 \div 0.0072$

Applications

Solve and check.

111. A stalactite is an icicle-shaped mineral deposit that hangs from the roof of a cave. If it took a thousand years for a stalactite to grow to a length of 3.7 inches, how much did it grow per year?

112. In a strong earthquake, the damage to 100 houses was estimated at \$12.7 million. What was the average damage per house?

113. So far this season, the women's softball team has won 21 games and lost 14 games. The men's softball team has won 22 games and lost 18.

a. The women's team has won what fraction of the games that it played, expressed as a decimal?

b. The men's team has won what fraction of its games, expressed as a decimal?

c. Which team has a better record? Explain.

114. Yesterday, 0.08 inches of rain fell. Today, $\frac{1}{4}$ inch of rain fell.

a. How much rain fell today, expressed as a decimal?

b. Which day did more rain fall? Explain.

115. The table shown gives the best gasoline mileage for a road test of three SUVs.

SUVs	Distance Driven (in miles)	Gasoline Used (in gallons)	Miles per Gallon
A	17.4	1.2	
B	8.4	0.6	
C	23.4	1.2	

a. For each SUV, compute how many miles it gets per gallon.

b. Which SUV gives the best mileage?

116. The following table shows the number of assists that three basketball players handed out over the same three-year career period.

Player	No. of Games	No. of Assists	Average
Vince Carter	229	1,013	
Richard Hamilton	234	957	
Dwayne Wade	213	1,298	

a. Compute the average number of assists per game for each player, expressed as a decimal rounded to the nearest tenth.

b. Decide which player has the highest average.

117. If Rite Aid stock sells for $4.20 a share, how many shares can be bought for $8,400? (***Source:*** finance.yahoo.com, 2006)

118. A light microscope can distinguish two points separated by 0.0005 millimeters, whereas an electron microscope can distinguish two points separated by 0.0000005 millimeters. The electron microscope is how many times as powerful as the light microscope?

119. Typically, the heaviest organ in the body is the skin, weighing about 9 pounds. By contrast, the heart weighs approximately 0.7 pound. About how many times the weight of the heart is that of the skin? (***Source:*** *World of Scientific Discovery*)

120. At a community college, each student enrolled pays a $19.50 student fee per semester. In a given semester, if the college collected $39,000 in student fees, how many students were enrolled?

121. A shopper buys four organic chickens. The chickens weigh 3.2 pounds, 3.5 pounds, 2.9 pounds, and 3.6 pounds. How much less than the average weight of the four chickens was the weight of the lightest one?

122. A dieter joins a weight-loss club. Over a 5-month period, she loses 8 pounds, 7.8 pounds, 4 pounds, 1.5 pounds, and 0.8 pound. What was her average monthly weight loss, to the nearest tenth of a pound?

Use a calculator to solve the the following problems, giving (a) the operation(s) carried out in your solution, (b) the exact answer, and (c) an estimate of the answer.

123. Babe Ruth got 2,873 hits in 8,398 times at bat, resulting in a batting average of $\frac{2,873}{8,398}$, or approximately .342. Another great player, Ty Cobb, got 4,189 hits out of 11,434 times at bat. What was his batting average, expressed as a decimal rounded to the nearest thousandth? (Note that batting averages don't have a zero to the left of the decimal point because they can never be greater than 1.) (***Source:*** http://www.baseball-reference.com)

124. Light travels at a speed of 186,000 miles per second. If Earth is about 93,000,000 miles from the Sun, how many seconds, to the nearest tenth of a second, does it take for light to reach Earth from the Sun?

• *Check your answers on page A-6.*

MINDSTRETCHERS

Patterns

1. In the *repeating decimal* 0.142847142847142847... , identify the 994th digit to the right of the decimal point.

Groupwork

2. In the following magic square, 3.375 is the *product* of the numbers in every row, column, and diagonal. Working with a partner, fill in the missing numbers.

	0.25	3
		0.5

Writing

3. a. $0.5 \div 0.8 = ?$

   b. $0.8 \div 0.5 = ?$

   c. Find the product of your answers in parts (a) and (b). Explain how you could have predicted this product.

KEY CONCEPTS AND SKILLS

CONCEPT SKILL

Concept/Skill	Description	Example
[3.1] **Decimal**	A number written with two parts: a whole number, which precedes the decimal point, and a fractional part, which follows the decimal point.	Whole-number part → 3.721 ← Fractional part; ↑ Decimal point
[3.1] **Decimal place**	A place to the right of the decimal point.	Decimal places: 8.<u>035</u> — Tenths, Hundredths, Thousandths
[3.1] **To change a decimal to the equivalent fraction or mixed number**	• Copy the nonzero whole-number part of the decimal and drop the decimal point. • Place the fractional part of the decimal in the numerator of the equivalent fraction. • Make the denominator of the equivalent fraction 1 followed by as many zeros as the decimal has decimal places. • Simplify the resulting fraction, if possible.	The decimal 3.25 is equivalent to the mixed number $3\frac{25}{100}$ or $3\frac{1}{4}$.
[3.1] **To compare decimals**	• Rewrite the numbers vertically, lining up the decimal points. • Working from left to right, compare the digits that have the same place value. The decimal which has the largest digit with this place value is the largest decimal.	1.073 1.06999 In the ones place and the tenths place, the digits are the same. But in the hundredths place, $7 > 6$, so $1.073 > 1.06999$.
[3.1] **To round a decimal to a given decimal place**	• Underline the digit in the place to which the number is being rounded. • Look at the digit to the right of the underlined digit—the critical digit. If this digit is 5 or more, add 1 to the underlined digit; if it is less than 5, leave the underlined digit unchanged. • Drop all decimal places to the right of the underlined digit.	$23.9\underline{3}81 \approx 23.94$ Critical digit (8)
[3.2] **To add decimals**	• Rewrite the numbers vertically, lining up the decimal points. • Add. • Insert a decimal point in the sum below the other decimal points.	0.035 0.08 +0.00813 0.12313
[3.2] **To subtract decimals**	• Rewrite the numbers vertically, lining up the decimal points. • Subtract, inserting extra zeros in the minuend if necessary for borrowing. • Insert a decimal point in the difference below the other decimal points.	0.90370 −0.17052 0.73318

continued

Concept/Skill	Description	Example
[3.3] **To multiply decimals**	• Multiply the factors as if they were whole numbers. • Find the total number of decimal places in the factors. • Count that many places from the right end of the product, and insert a decimal point.	21.07 ← Two decimal places × 0.18 ← Two decimal places 3.7926 ← Four decimal places
[3.4] **To change a fraction to the equivalent decimal**	• Divide the denominator of the fraction into the numerator, inserting to its right both a decimal point and enough zeros to get an answer either without a remainder or rounded to a given decimal place. • Place a decimal point in the quotient directly above the decimal point in the dividend.	$\frac{7}{8} = 8\overline{)7.000}$ with quotient 0.875
[3.4] **To divide decimals**	• Move the decimal point in the divisor to the right end of the number. • Move the decimal point in the dividend the same number of places to the right as in the divisor. • Insert a decimal point in the quotient directly above the decimal point in the dividend. • Divide the new dividend by the new divisor, inserting zeros at the right end of the dividend as necessary.	$3.5\overline{)71.05} =$ 20.3 $35\overline{)710.5}$ 70 10 0 105 105

Chapter 3 Review Exercises

To help you review this chapter, solve these problems.

[3.1] *Name the place that each underlined digit occupies.*

1. $8.3\underline{5}9$ **2.** $13.\underline{0}05$ **3.** $8{,}024.\underline{5}$ **4.** $0.000\underline{3}$

Express each number as a fraction, mixed number, or whole number.

5. 0.35 **6.** 8.2 **7.** 4.007 **8.** 10.000

Write each decimal in words.

9. 0.72 **10.** 5.6

11. 3.0009 **12.** 510.036

Write each decimal in decimal notation.

13. Seven thousandths **14.** Two and one tenth

15. Nine hundredths **16.** Seven and forty-one thousandths

Between each pair of numbers, insert the appropriate sign, $<$, $=$, or $>$, to make a true statement.

17. 0.037 0.04 **18.** 2.031 2.0301 **19.** 5.12 4.71932 **20.** 2 1.8

Rearrange each group of numbers from largest to smallest.

21. 0.72, 0.8, 1.002 **22.** 0.003, 0.00057, 0.004

Round as indicated.

23. 7.31 to the nearest tenth **24.** 0.0387 to the nearest thousandth

25. 4.3868 to two decimal places **26.** $899.09 to the nearest dollar

[3.2] *Perform the indicated operations. Check by estimating.*

27. $8.2 + 3.91$ **28.** $50 + 2.7 + 0.05$ **29.** $\$8 + \$3.25 + \$12.88$ **30.** $8.4\text{ m} + 3.6\text{ m}$

31. $30.7 - 1.92$ **32.** $93 - 5.248$ **33.** $2.5 - (0.72 - 0.054)$ **34.** $54.17 - (8 - 2.731)$

35. $5.398 + 8.72 + 92.035 + 0.7723 - 3.714 - 5.008$ **36.** $\$87{,}259.39 + \$2{,}098.35 + \$1{,}387.92 + \$203.14$

[3.3] *Find the product. Check by estimating.*

37. 7.28×0.4 **38.** $(288)(3.5)$ **39.** 0.005×0.002 **40.** $(3.7)^2$

41. $2.71 \cdot 1{,}000$ **42.** 0.0034×10 **43.** $8 - (1.5)^2$ **44.** $3(2.4) + 7(0.9)$

45. $18{,}772.35 \times 0.0836$ **46.** $(74.862)(5.901)$

[3.4] *Change to the equivalent decimal.*

47. $\frac{5}{8}$

48. $90\frac{1}{5}$

49. $4\frac{1}{16}$

50. $\frac{45}{1,000}$

Express each fraction as a decimal. Round to the nearest hundredth.

51. $\frac{1}{6}$

52. $\frac{2}{7}$

53. $8\frac{1}{3}$

54. $11\frac{2}{9}$

Divide and check.

55. $2\overline{)1.3}$

56. $\frac{4.8}{3}$

57. $0.7 \div 4$

58. $\frac{2.77}{10}$

Divide. Round to the nearest tenth.

59. $4.67 \div 0.9$

60. $\frac{2.35}{0.73}$

61. $\frac{7.11}{0.3}$

62. $0.06\overline{)981.5}$

63. $18.74 \div 9.7$

64. $220 \div 0.61$

65. $81.37\overline{)247.062}$

66. $247.062\overline{)81.37}$

Simplify.

67. $\frac{(1.3)^2 - 1.1}{0.5}$

68. $\frac{2.5 - (0.4)^2}{0.02}$

69. $\frac{13.75}{9.6 - 9.2}$

70. $(2.5)(3.5) \div 6.25$

Mixed Applications

Solve.

71. The venom of a certain South American frog is so poisonous that 0.0000004 ounce of the venom can kill a person. How is this decimal read?

72. On a certain day, the closing price of one share of Apple Computer stock was \$59.58. If the closing price was \$1.72 higher than the opening price, what was the opening price? (*Source:* http://www.nasdaq.com)

73. Recently, a champion swimmer swam 50 meters in 25.2 seconds and then swam 100 meters in 29.29 seconds longer. How long did she take to swim the 150 meters?

74. Find the missing length.

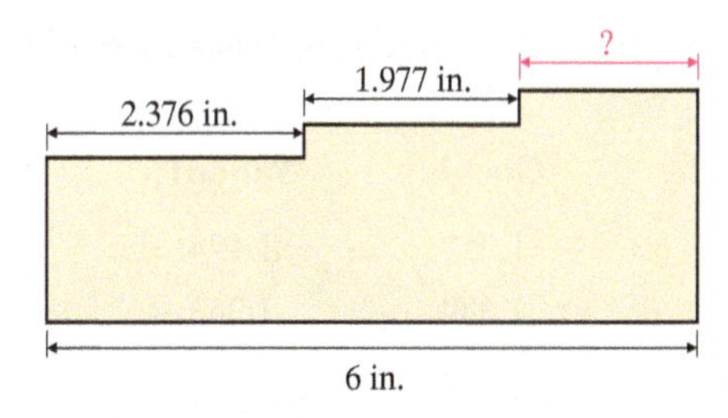

75. The fastest speed ever recorded for a spider was 1.17 miles per hour. Crawling at this speed for $\frac{1}{2}$ hour, could a spider reach a wall $\frac{3}{4}$ mile away?

(*Source: World Almanac and Book of Facts*)

76. In the United States Congress, there are 100 senators and 435 representatives. How many times as many representatives as senators are there?

77. A supermarket sells a 4-pound package of ground meat for \$5.20 and a 5-pound package of ground meat for \$6.20. What is the difference between the two prices per pound?

78. Most compact discs are sold in plastic boxes. Suppose that in a CD collection there are 29 boxes 0.4 inch thick and 3 boxes 0.94 inch thick. Estimate how many inches of shelf space is needed to house this collection.

79. In a chemistry lab, a student weighs a compound three times, getting 7.15 grams, 7.18 grams, and 7.23 grams. What is the average of these weights, to the nearest hundredth of a gram?

80. A team of geologists scaled a mountain. At the base of the mountain, the temperature had been 11°C. The temperature fell 0.75 degrees for every 100 meters the team climbed. After they climbed 1,000 meters, what was the temperature?

81. The following form was adapted from the *U.S. Individual Income Tax Return.* Find the total income in line 23.

Line	Description	Line	Amount
7	Wages, salaries, tips, etc.	7	28,774.71
8	Taxable interest income	8	
9	Dividend income	9	232.55
10	Taxable refunds, credits, or offsets of state and local income taxes	10	349.77
11	Alimony received	11	
12	Business income or (loss)	12	
13	Capital gain or (loss)	13	511.74
14	Capital gain distributions not reported on line 13	14	
15	Other gains or (losses)	15	5,052.71
16	Total IRA distributions: taxable amount	16	
17	Total pensions and annuities: taxable amount	17	
18	Rents, royalties, partnerships, estates, trusts, etc.	18	1,240.97
19	Farm income or (loss)	19	
20	Unemployment compensation	20	
21	Social Security benefits: taxable amount	21	
22	Other income	22	
23	Add the amounts shown in the far right column	23	

82. The following table shows the quarterly revenues, in billions of dollars, for Google and Yahoo! for 2005. (***Source:*** http://Google.com and http://Yahoo.com)

Quarter	Google	Yahoo!
1st	1.257	1.174
2nd	1.384	1.253
3rd	1.578	1.330
4th	1.919	1.501

How much more were Google's earnings than Yahoo's for the year?

• *Check your answers on page A-7.*

Chapter 3 POSTTEST

FOR EXTRA HELP

Test solutions are found on the enclosed CD.

To see if you have mastered the topics in this chapter, take this test.

1. In the number 0.79623, which digit occupies the thousandths place?

2. Write in words: 5.102

3. Round 320.1548 to the nearest hundredth.

4. Which is smallest, 0.04, 0.0009, or 0.00028?

5. Express 3.04 as a mixed number.

6. Write as a decimal: four thousandths

Perform the indicated operations.

7. $2.3 + 0.704 + 1.35$

8. $\$5.27 + \$9 - \$8.61$

9. 2.09×10

10. 5.2×1.1

11. $(0.1)^3$

12. $\frac{3.52}{2} + \frac{4.8}{3}$

13. $2.9 \div 1{,}000$

14. $\frac{9.81}{0.3}$

Express as a decimal.

15. $\frac{3}{8}$

16. $4\frac{1}{6}$, rounded to the nearest hundredth.

Solve.

17. The element hydrogen is so light that 1 cubic foot (ft^3) of hydrogen weighs only 0.005611 pound. Round this weight to the nearest hundredth of a pound.

18. Historically, a mile was the distance that a Roman soldier covered when he took 2,000 steps. If a mile is 5,280 feet, how many feet, to the nearest tenth of a foot, was a Roman's step?

19. The Triple Crown consists of three horse races—the Kentucky Derby (1.25 miles), the Preakness Stakes (1.1875 miles), and the Belmont Stakes (1.5 miles). Which race is longest? (***Source:*** *World Almanac*)

20. To compute the annual property tax on a house in Keller, Texas, multiply the house's assessed value by 0.02807207 and then round to the nearest cent. What is the tax on a house if its assessed value is $100,000? (***Source:*** http://www.cityofkeller.com)

• *Check your answers on page A-7.*

Cumulative Review Examples

To help you review, solve the following:

1. Round 591,622 to the nearest million.
2. Estimate: $7\frac{9}{10} \times 4\frac{1}{13}$
3. Which is larger, $1\frac{1}{2}$ or $1\frac{3}{8}$?
4. Multiply: (409)(67)
5. Subtract: $5 - 2\frac{1}{3}$
6. Calculate: $\frac{2}{5} + \frac{1}{3} \cdot \frac{1}{2}$
7. Divide: $29.89 \div 0.049$
8. A dating service advertises that it has been introducing thousands of singles for 20 years, resulting in 6,500 successful marriages. On the average, how many marriages is this per year?
9. A satellite orbiting the Earth travels at 16,000 miles per hour. An orbit takes 1.6 hours. How far will the satellite travel once around, to the nearest thousand miles?

10. An electric company charges $0.09693 per kilowatt hour. If a restaurant used 2,000 kilowatt hours of electricity in a certain week, what was its weekly cost?

• Check your answers on page A-7.

CHAPTER 4

Basic Algebra: Solving Simple Equations

Algebra and Physics

Physicists use algebra to describe the relationship between physical quantities. For instance, consider the situation in which two children are riding a seesaw. Physicists have observed that when the seesaw is balanced, the product of each child's weight and distance from the pivot are equal.

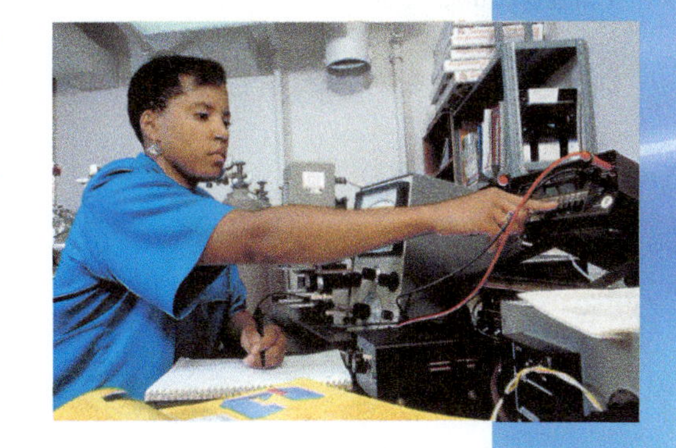

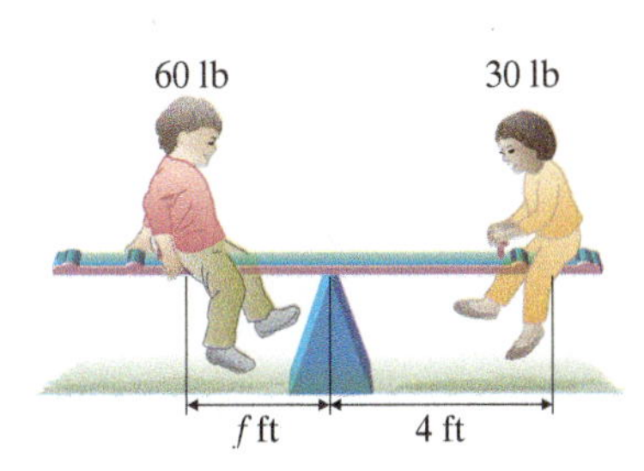

In the example pictured, if the 30-pound child is 4 feet from the pivot, we can conclude that

$$60 \cdot f = 30 \cdot 4,$$

where f represents the distance of the 60-pound child from the pivot.

Only one value of f will make the seesaw balance. To find that value, algebra can be used to solve the equation.

(***Source:*** W. Thomas Griffith, *The Physics of Everyday Phenomena,* Wm. C. Brown, 1992)

Chapter 4 PRETEST

To see if you have already mastered the topics in this chapter, take this test.

Write each algebraic expression in words.

1. $t - 4$

2. $\frac{y}{3}$

Translate each phrase to an algebraic expression.

3. 8 more than m

4. Twice n

Evaluate each algebraic expression.

5. $\frac{x}{4}$, for $x = 16$

6. $5 - y$, for $y = 3\frac{1}{2}$

Translate each sentence to an equation.

7. The sum of x and 3 equals 5.

8. The product of 4 and y is 12.

Solve and check.

9. $x + 4 = 10$

10. $t - 1 = 9$

11. $2n = 26$

12. $\frac{a}{4} = 3$

13. $8 = m + 1.9$

14. $15 = 0.5n$

15. $7m = 3\frac{1}{2}$

16. $\frac{n}{10} = 1.5$

Write an equation. Solve and check.

17. The planet Jupiter has 36 more moons than the planet Uranus. If Jupiter has 63 moons, how many does Uranus have? (***Source:*** NASA)

18. Tickets for all movies shown before 5:00 P.M. at a local movie theater qualify for the bargain matinee price, which is $2.75 less than the regular ticket price. If the bargain price is $6.75, what is the regular price?

19. In Michigan, about two-fifths of the area is covered with water. This portion of the state represents about 39,900 square miles. What is the area of Michigan? (***Source:*** *The New York Times Almanac, 2006*)

20. An 8-ounce cup of regular tea has about 40 milligrams of caffeine, which is 10 times the amount of caffeine in a cup of decaffeinated tea. How much caffeine is in a cup of decaffeinated tea?

• *Check your answers on page A-7.*

4.1 Introduction to Basic Algebra

OBJECTIVES

- To translate phrases to algebraic expressions and vice versa
- To evaluate an algebraic expression for a given value of the variable
- To solve word problems involving algebraic expressions

What Algebra Is and Why It Is Important

In this chapter, we discuss some of the basic ideas in algebra. These ideas will be important throughout the rest of this book.

In algebra, we use letters to represent unknown numbers. The expression $2 + 3$ is arithmetic, whereas the expression $x + y$ is algebraic, since x and y represent numbers whose values are not known. With *algebraic expressions*, such as $x + y$, we can make general statements about numbers and also find the value of unknown numbers.

We can think of algebra as a *language*: The idea of translating ordinary words to algebraic notation and vice versa is the key. Often, just writing a problem algebraically makes the problem much easier to solve. We present ample proof of this point repeatedly in the chapters that follow.

We begin our discussion of algebra by focusing on what algebraic expressions mean and how to translate and evaluate them.

Translating Phrases to Algebraic Expressions and Vice Versa

To apply mathematics to a real-world situation, we often need to be able to express that situation algebraically. Consider the following example of this kind of translation.

Suppose that you are enrolled in a college course that meets 50 minutes a day for 3 days a week. The course therefore meets for $50 \cdot 3$, or 150 minutes, a week.

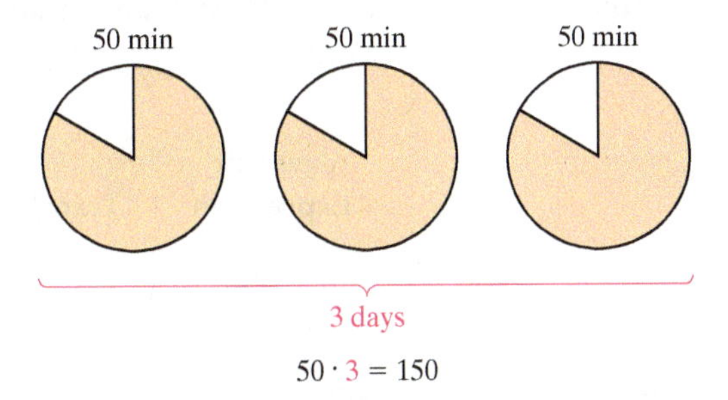

Now suppose that in a semester the 50-minute class meets d days but that we do not know what number the letter d represents. How many minutes per semester does the class meet? Do you see that we can express the answer as $50d$, that is, 50 times d min?

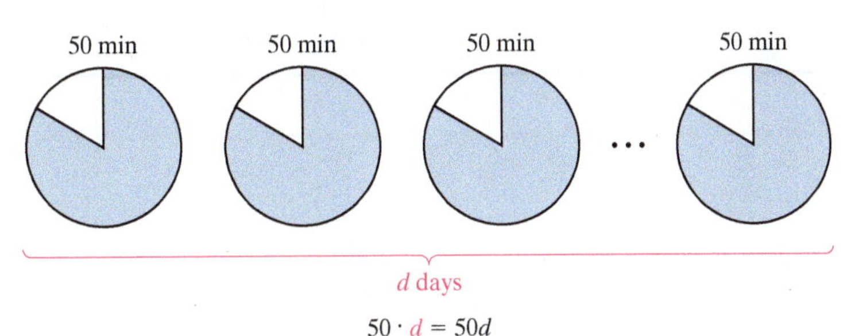

In algebra, a *variable* is a letter, or other symbol, used to represent an unknown number. In the algebraic expression $50d$, for instance, d is a variable and 50 is a *constant.* Note that in writing an algebraic expression, we usually omit any multiplication symbol: $50d$ means $50 \cdot d$.

Definitions

A **variable** is a letter that represents an unknown number.

A **constant** is a known number.

An **algebraic expression** is an expression that combines variables, constants, and arithmetic operations.

There are many translations of an algebraic expression to words, as the following table indicates.

$x + 4$ translates to	$n - 3$ translates to	$\frac{3}{4} \cdot y$ or $\frac{3}{4}y$ translates to	$z \div 5$ or $\frac{z}{5}$ translates to
• x plus 4 • x increased by 4 • the sum of x and 4 • 4 more than x • 4 added to x	• n minus 3 • n decreased by 3 • the difference between n and 3 • 3 less than n • 3 subtracted from n	• $\frac{3}{4}$ times y • the product of $\frac{3}{4}$ and y • $\frac{3}{4}$ of y	• z divided by 5 • the quotient of z and 5 • z over 5

EXAMPLE 1

Translate each algebraic expression in the table to words.

Solution

Algebraic Expression	Translation
a. $\frac{p}{3}$	p divided by 3
b. $x - 4$	4 less than x
c. $5f$	5 times f
d. $2 + y$	the sum of 2 and y
e. $\frac{2}{3}a$	$\frac{2}{3}$ of a

PRACTICE 1

Translate each algebraic expression to words.

Algebraic Expression	Translation
a. $\frac{1}{2}p$	
b. $5 - x$	
c. $y \div 4$	
d. $n + 3$	
e. $\frac{3}{5}b$	

EXAMPLE 2

Translate each word phrase in the table to an algebraic expression.

Solution

	Word Phrase	Translation
a.	16 more than m	$m + 16$
b.	the product of 5 and b	$5b$
c.	the quotient of 6 and z	$6 \div z$
d.	a decreased by 4	$a - 4$
e.	$\frac{3}{8}$ of t	$\frac{3}{8}t$

PRACTICE 2

Express each word phrase as an algebraic expression.

	Word Phrase	Translation
a.	x plus 9	
b.	10 times y	
c.	the difference between n and 7	
d.	p divided by 5	
e.	$\frac{2}{5}$ of v	

As we have seen, any letter or symbol can be used to represent a variable. For example, *five less than a number* can be translated to $n - 5$, where n represents the number.

Let's consider the following example.

EXAMPLE 3

Express each phrase as an algebraic expression.

Solution

	Word Phrase	Translation
a.	2 less than a number	$n - 2$, where n represents the number
b.	an amount divided by 10	$\frac{a}{10}$, where a represents the amount
c.	$\frac{3}{8}$ of a price	$\frac{3}{8}p$, where p represents the price

PRACTICE 3

Translate each word phrase to an algebraic expression.

	Word Phrase	Translation
a.	a quantity increased by 12	
b.	the quotient of 9 and an account balance	
c.	a cost multiplied by $\frac{2}{7}$	

Now let's look at word problems that involve translations.

EXAMPLE 4

Suppose that p partners share equally in the profits of a business. What is each partner's share if the profit was $2,000?

Solution Each partner should get the quotient of 2,000 and p, which can be written algebraically as $\frac{2{,}000}{p}$ dollars.

PRACTICE 4

Next weekend, a student wants to study for his four classes. If he has h hours to study in all and he wants to devote the same amount of time to each class, how much time will he study per class?

EXAMPLE 5

At registration, n out of 100 classes are closed. How many classes are not closed?

Solution Since n classes are closed, the remainder of the 100 classes are not closed. So we can represent the number of classes that are not closed by the algebraic expression $100 - n$.

PRACTICE 5

Of s shrubs in front of a building, 3 survived the winter. How many shrubs died over the winter?

Evaluating Algebraic Expressions

In this section, we look at how to evaluate algebraic expressions. Let's begin with a simple example.

Suppose that the balance in a savings account is \$200. If d dollars is then deposited, the balance will be $(200 + d)$ dollars.

To evaluate the expression $200 + d$ for a particular value of d, we replace d with that number. If \$50 is deposited, we replace d by 50:

$$200 + d = 200 + 50 = 250$$

So the new balance will be \$250.

The following rule is helpful for evaluating expressions.

To Evaluate an Algebraic Expression

- Substitute the given value for each variable.
- Carry out the computation.

Now let's consider some more examples.

EXAMPLE 6

Evaluate each algebraic expression.

Solution

Algebraic Expression	Value
a. $n + 8$, if $n = 15$	$15 + 8 = 23$
b. $9 - z$, if $z = 7.89$	$9 - 7.89 = 1.11$
c. $\frac{2}{3}r$, if $r = 18$	$\frac{2}{3} \cdot 18 = 12$
d. $y \div 4$, if $y = 3.6$	$3.6 \div 4 = 0.9$

PRACTICE 6

Find the value of each algebraic expression.

Algebraic Expression	Value
a. $\frac{s}{4}$, if $s = 100$	
b. $0.2y$, if $y = 1.9$	
c. $x - 4.2$, if $x = 9$	
d. $25 + z$, if $z = 1.6$	

The following examples illustrate how to write and evaluate expressions to solve word problems.

EXAMPLE 7

Power consumption for a period of time is measured in watt-hours, where a watt-hour means 1 watt of power for 1 hour. How many watt-hours of energy will a 60-watt bulb consume in h hours? In 3 hours?

Solution The expression that represents the number of watt-hours used in h hours is $60h$. So for $h = 3$, the number of watt-hours is $60 \cdot 3$, or 180. Therefore, 180 watt-hours of energy will be consumed in 3 hours.

PRACTICE 7

To avoid paying private mortgage insurance, home buyers are required to make a down payment that is at least one-fifth of the purchase price of the home. What is the required down payment for a home with a purchase price of p dollars? With a purchase price of \$349,000?

EXAMPLE 8

Suppose that there are 180 days in the local school year. How many days was a student present at school if she was absent d days? 9 days?

Solution If d represents the number of days that the student was absent, the expression $180 - d$ represents the number of days that she was present. If she was absent 9 days, we substitute 9 for d in the expression:

$$180 - d = 180 - 9 = 171$$

So the student was present 171 days.

PRACTICE 8

At a coffee shop, a lunch bill came to \$15.45 plus the tip. What is the total amount of the lunch, including a tip of t dollars? A tip of \$3?

CULTURAL NOTE

Solving an equation to identify an unknown number is similar to using a balance scale to determine an unknown weight. In this picture that dates from 3,400 years ago, an Egyptian weighs gold rings against a counterbalance in the form of a bull's head.

The balance scale is an ancient measuring device. These scales were used by Sumerians for weighing precious metals and gems at least 9,000 years ago.

Source: O. A.W. Dilke, *Mathematics and Measurement* (Berkeley: University of California Press/British Museum, 1987).

4.1 Exercises

FOR EXTRA HELP 

Mathematically Speaking

Fill in each blank with the most appropriate term or phrase from the given list.

arithmetic	constant	evaluate
translate	variable	algebraic

1. A(n) _______ is a letter that represents an unknown number.

2. A(n) _______ is a known number.

3. A(n) _______ expression combines variables, constants, and arithmetic operations.

4. To _______ an algebraic expression, replace each variable with the given number, and carry out the computation.

Translate each algebraic expression to two different word phrases.

5. $t + 9$

6. $8 + r$

7. $c - 12$

8. $x - 5$

9. $c \div 3$

10. $\frac{z}{7}$

11. $10s$

12. $11t$

13. $y - 10$

14. $w - 1$

15. $7a$

16. $4x$

17. $x \div 6$

18. $\frac{y}{5}$

19. $x - \frac{1}{2}$

20. $x - \frac{1}{3}$

21. $\frac{1}{4}w$

22. $\frac{4}{5}y$

23. $2 - x$

24. $8 - y$

25. $1 + x$

26. $n + 7$

27. $3p$

28. $2x$

29. $n - 1.1$

30. $x - 6.5$

31. $y \div 0.9$

32. $\frac{n}{2.4}$

Translate each word phrase to an algebraic expression.

33. x plus 10

34. d increased by 12

35. 1 less than n

36. b decreased by 9

37. the sum of y and 5

38. 11 more than x

39. t divided by 6

40. r over 2

41. the product of 10 and y

42. 5 times p

43. w minus 5

44. the difference between n and 5

45. n increased by $\frac{4}{5}$

46. the sum of x and 0.4

47. the quotient of z and 3

48. n divided by 10

49. $\frac{2}{7}$ of x

50. the product of $\frac{2}{3}$ and y

51. 6 subtracted from k

52. 8 less than z

53. 12 more than a number

54. the sum of a number and 18

55. the difference between a number and 5.1

56. $\frac{3}{4}$ less than a number

Evaluate each algebraic expression.

57. $y + 7$, if $y = 19$

58. $3 + n$, if $n = 2.9$

59. $7 - x$, if $x = 4.5$

60. $y - 19$, if $y = 25$

61. $\frac{3}{4}p$, if $p = 20$

62. $\frac{4}{5}n$, if $n = 1\frac{1}{4}$

63. $x \div 2$, if $x = 2\frac{1}{3}$

64. $\frac{n}{3}$, if $n = 7.5$

65. $p - 7.9$, if $p = 9$

66. $20.1 + y$, if $y = 7$

67. $x \div \frac{5}{6}$, if $x = \frac{1}{6}$

68. $\frac{1}{3}y$, if $y = \frac{1}{2}$

Complete each table.

69.

x	$x + 8$
1	
2	
3	
4	

70.

x	$10 - x$
1	
2	
3	
4	

71.

n	$n - 0.2$
1	
2	
3	
4	

72.

b	$b + 2.5$
1	
2	
3	
4	

73.

x	$\frac{3}{4}x$
4	
8	
12	
16	

74.

n	$\frac{2}{3}n$
3	
6	
9	
12	

75.

z	$\frac{z}{2}$
2	
4	
6	
8	

76.

y	$\frac{y}{5}$
5	
10	
15	
20	

Mixed Practice

Solve.

77. Translate the phrase "7 less than x" to an algebraic expression.

78. Evaluate the algebraic expression $0.5t$, if $t = 8$.

79. Translate the algebraic expression $\frac{n}{2}$ to two different phrases.

80. Evaluate the algebraic expression $\frac{1}{4}y$, if $y = \frac{2}{3}$.

81. Translate the phrase "the product of 3.5 and t" to an algebraic expression.

82. Evaluate the algebraic expression $x + 1$, if $x = 4$.

83. Translate the algebraic expression $x + 6$ to two different phrases.

84. Evaluate the algebraic expression $n - 20$, if $n = 30$.

Applications

Solve.

85. A patient receives m milligrams of medication per dose. Her doctor orders her medication to be decreased by 25 milligrams. How much medication will she then receive per dose?

86. When you take out a mortgage, each monthly payment has two parts. One part goes toward the principal and the other toward the interest. If the principal payment is \$344.86 and the interest payment is i, write an algebraic expression for the total payment.

87. For the triangle shown, write an expression for the sum of the measures of the three angles.

88. Write an expression for the sum of the lengths of the sides in the trapezoid shown.

89. If a long-distance trucker drives at a speed of r miles per hour for t hours, she will travel a distance of $r \cdot t$ miles. How far will she travel at a speed of 55 miles per hour in 4 hours?

90. If a basketball player makes b baskets in a attempts, his field goal average is defined to be $\frac{b}{a}$. Find the field goal average of a player who made 12 baskets in 25 attempts.

91. A bank charges customers a fee of \$1.50 for each withdrawal made at its ATMs.

a. Write an expression for the total fee charged to a customer for w of these withdrawals

b. Find the total fee if the customer makes 9 withdrawals.

92. A computer network technician charges \$80 per hour for labor.

a. Write an expression for the cost of h hours of work.

b. Find the cost of a networking job that takes $2\frac{1}{2}$ hours.

• *Check your answers on page A-7.*

MINDSTRETCHERS

Mathematical Reasoning

1. Consider the expression $x + x$.

a. Why does this expression mean the same as the expression $2x$?

b. What does the expression $\underbrace{x + x + x + \cdots + x}_{n \text{ times}}$ mean in terms of multiplication?

Groupwork

2. Working with a partner, consider the areas of the following rectangles. For some values of x, the rectangle on the left has a larger area; for other values of x, the rectangle on the right is larger.

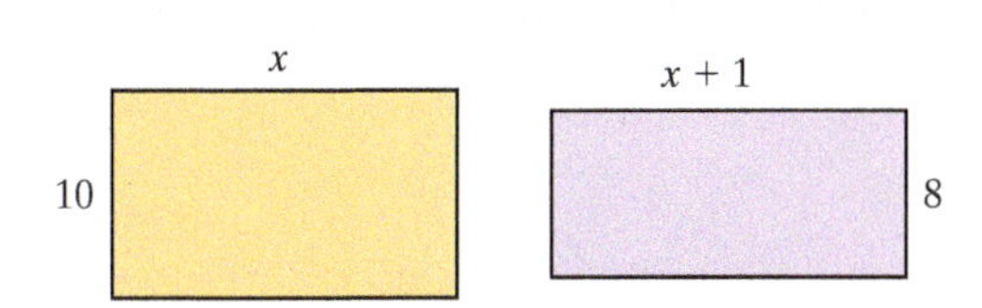

a. Find a value of x for which the rectangle on the left has the larger area.

b. Find a value of x for which the area of the rectangle on the right is larger.

Writing

3. Algebra is universal; that is, it is used in all countries of the world regardless of the language spoken. If you know how to speak a language other than English, translate each of the following algebraic expressions to that language.

a. $7x$ **b.** $x - 2$ **c.** $3 + x$ **d.** $\frac{x}{3}$

4.2 Solving Addition and Subtraction Equations

OBJECTIVES

- To translate sentences to equations involving addition or subtraction
- To solve addition and subtraction equations
- To solve word problems involving equations with addition or subtraction

What an Equation Is

An equation contains two expressions separated by an equal sign.

$$\underbrace{x + 3}_{\text{Left side}} \overset{\text{Equal sign}}{=} \underbrace{5}_{\text{Right side}}$$

Definition

An **equation** is a mathematical statement that two expressions are equal.

For example,

$$1 + 2 = 3$$
$$x - 5 = 6$$
$$2 + 7 + 3 = 12$$
$$3x = 9$$

are all equations.

Equations are used to solve a wide range of problems. A key step in solving a problem is to translate the sentences that describe the problem to an equation that models the problem. In this section, we focus on equations that involve either addition or subtraction. In the next section, we consider equations involving multiplication or division.

Translating Sentences to Equations

In translating sentences to equations, certain words and phrases mean the same as the equal sign:

- equals
- is
- is equal to
- is the same as
- yields
- results in

Let's look at some examples of translating sentences to equations that involve addition or subtraction and vice versa.

EXAMPLE 1

Translate each sentence in the table to an equation.

Solution

Sentence	Equation
a. The sum of y and 3 is equal to $7\frac{1}{2}$.	$y + 3 = 7\frac{1}{2}$
b. The difference between x and 9 is the same as 14.	$x - 9 = 14$
c. Increasing a number by 1.5 yields 3.	$n + 1.5 = 3$
d. 6 less than a number is 10.	$n - 6 = 10$

PRACTICE 1

Write an equation for each word phrase or sentence.

Sentence	Equation
a. n decreased by 5.1 is 9.	
b. y plus 2 is equal to 12.	
c. The difference between a number and 4 is the same as 11.	
d. 5 more than a number is $7\frac{3}{4}$.	

EXAMPLE 2

In a savings account, the previous balance, P, plus a deposit of \$7.50 equals the new balance of \$43.25. Write an equation that represents this situation.

Solution

The previous balance plus the deposit equals the new balance.

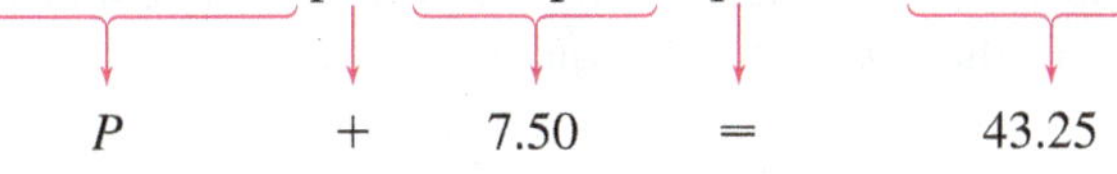

$P \quad + \quad 7.50 \quad = \quad 43.25$

So the equation is $P + 7.50 = 43.25$.

PRACTICE 2

The sale price of a jacket is \$49.95. This amount is \$6 less than the regular price p. Write an equation that represents this situation.

Equations Involving Addition and Subtraction

Suppose that you are told that five *more than some number* is equal to seven. You can find that number by solving the addition equation $x + 5 = 7$. To solve an equation means to find a number that, when substituted for the variable x, makes the equation a true statement. Such a number is called a *solution* of the equation.

To solve the equation $x + 5 = 7$, we can think of a balance scale like the one shown.

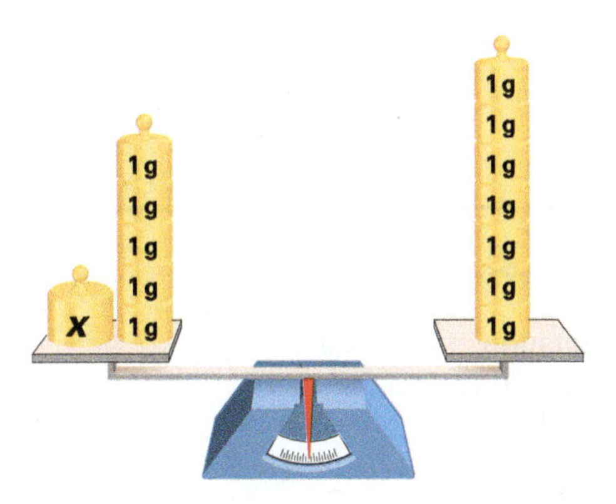

For the balance to remain level, whatever we do to one side, we must also do to the other side. In this case, if we subtract 5 g from each side of the balance, we can conclude that the unknown weight, x, must be 2 g, as shown below. So 2 is the solution of the equation $x + 5 = 7$.

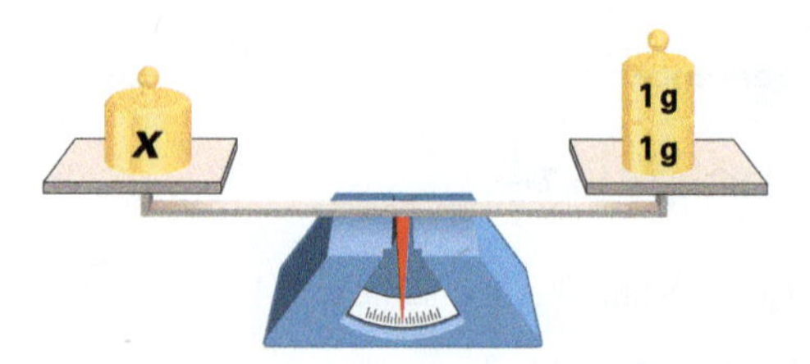

Similarly in the *subtraction equation* $x - 5 = 7$, if we add 5 to each side of the equation, we find that x equals 12.

In solving these and other equations, the key is to **isolate the variable**, that is, to get the variable alone on one side of the equation.

These examples suggest the following rule.

To Solve Addition or Subtraction Equations

- For an addition equation, *subtract* the same number from each side of the equation in order to isolate the variable on one side.
- For a subtraction equation, *add* the same number to each side of the equation in order to isolate the variable on one side.
- In either case, check the solution by substituting the value of the unknown in the original equation to verify that the resulting equation is true.

Because addition and subtraction are **opposite operations**, one operation "undoes" the other. The following examples illustrate how to perform an opposite operation to each side of an equation when you are solving for the unknown.

EXAMPLE 3

Solve and check: $y + 9 = 17$

Solution

$$y + 9 = 17$$

$$y + \underbrace{9 - 9} = \underbrace{17 - 9}$$ Subtract 9 from each side of the equation.

$$\underbrace{y + 0} = 8$$

$$y = 8$$ Any number added to 0 is the number.

Check

$$y + 9 = 17$$

$$8 + 9 \stackrel{?}{=} 17$$ Substitute 8 for y in the original equation.

$$17 \stackrel{\checkmark}{=} 17$$ The equation is true, so 8 is the solution to the equation.

Note that, because 9 was added to y in the original equation, we solved by subtracting 9 from both sides of the equation in order to isolate the variable.

PRACTICE 3

Solve and check: $x + 5 = 14$

EXAMPLE 4

Solve and check: $n - 2.5 = 0.7$

Solution

$$n - 2.5 = 0.7$$

$$\underbrace{n - 2.5 + 2.5}_{} = \underbrace{0.7 + 2.5}_{}$$ Add 2.5 to each side of the equation.

$$\underbrace{n - 0}_{} = 3.2$$

$$n = 3.2$$

Check

$$n - 2.5 = 0.7$$

$$3.2 - 2.5 \stackrel{?}{=} 0.7$$ Substitute 3.2 for n in the original equation.

$$0.7 \stackrel{\checkmark}{=} 0.7$$

Can you explain why checking an answer is important?

PRACTICE 4

Solve and check: $t - 0.9 = 1.8$

EXAMPLE 5

Solve and check: $x + \frac{1}{3} = 3\frac{1}{2}$

Solution

$$x + \frac{1}{3} = 3\frac{1}{2}$$

$$x + \frac{1}{3} - \frac{1}{3} = 3\frac{1}{2} - \frac{1}{3}$$ Subtract $\frac{1}{3}$ from each side of the equation.

$$x = 3\frac{1}{6}$$

Check

$$x + \frac{1}{3} = 3\frac{1}{2}$$

$$3\frac{1}{6} + \frac{1}{3} \stackrel{?}{=} 3\frac{1}{2}$$ Substitute $3\frac{1}{6}$ for x in the original equation.

$$3\frac{1}{2} \stackrel{\checkmark}{=} 3\frac{1}{2}$$

PRACTICE 5

Solve and check: $m + \frac{1}{4} = 5\frac{1}{2}$

Equations are often useful **mathematical models** of real-world situations, as the following examples show. To derive these models, we need to be able to translate word sentences to algebraic equations, which we then solve.

EXAMPLE 6

Write each sentence as an equation. Then solve and check.

Solution

	Sentence	Equation	Check
a.	15 is equal to y increased by 9.	$15 = 9 + y$ $15 - 9 = 9 - 9 + y$ $6 = y$, or $y = 6$	$15 = 9 + y$ $15 \stackrel{?}{=} 9 + 6$ $15 \stackrel{\checkmark}{=} 15$
b.	10 is equal to m decreased by 8.	$10 = m - 8$ $10 + 8 = m - 8 + 8$ $18 = m$, or $m = 18$	$10 = m - 8$ $10 \stackrel{?}{=} 18 - 8$ $10 \stackrel{\checkmark}{=} 10$

Note that we isolated the variable on the right side of the equation instead of the left side. The result is the same.

PRACTICE 6

Translate each sentence to an algebraic equation. Then solve and check.

a. 11 is 4 less than m.

b. The sum of 12 and n equals 21.

EXAMPLE 7

Suppose that a chemistry experiment requires students to find the mass of the water in a flask. If the mass of the flask with water is 21.49 grams and the mass of the empty flask is 9.56 grams, write an equation to find the mass of the water. Then solve and check.

Solution Recall that some problems can be solved by drawing a diagram. Let's use that strategy here.

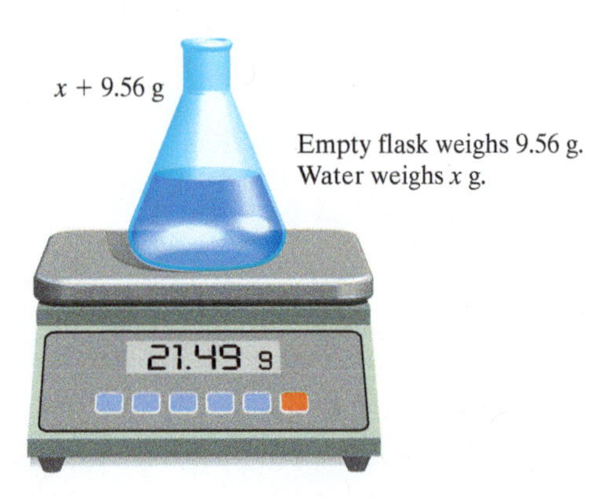

The diagram suggests the equation $21.49 = x + 9.56$, where x represents the mass of the water. Solving this equation, we get

$$21.49 = x + 9.56$$
$$21.49 - 9.56 = x + 9.56 - 9.56$$
$$11.93 = x, \text{ or } x = 11.93$$

The weight of the water is 11.93 grams.

Check
$$21.49 = x + 9.56$$
$$21.49 \stackrel{?}{=} 11.93 + 9.56$$
$$21.49 \stackrel{\checkmark}{=} 21.49$$

PRACTICE 7

An online discount book retailer charges a shipping fee of \$3.99. The total cost of a book, including the shipping fee, was \$27.18. Write an equation to determine the cost of the book without the shipping fee. Then solve and check.

EXAMPLE 8

Harvard College (in Cambridge, Massachusetts) and the College of William and Mary (in Williamsburg, Virginia) are the two oldest institutions of higher learning in the United States. Harvard, founded in 1636, is 57 years older than William and Mary. When was William and Mary founded? (***Source:*** *The Top Ten of Everything, 2006*)

Solution Let x represent the year in which William and Mary was founded. We know that 57 years earlier than the year x is 1636, the year in which Harvard was founded. This gives us the equation

$$x - 57 = 1636$$

Now we solve for the unknown:

$$x - 57 + 57 = 1636 + 57$$
$$x = 1693$$

So William and Mary was founded in 1693.

PRACTICE 8

The two U.S. states with the largest area are Alaska and Texas. The land area of Texas, 262,000 square miles, is approximately 308,000 square miles smaller than that of Alaska. Write an equation to determine Alaska's land area. Then solve and check. (***Source:*** *Time Almanac, 2006*)

4.2 Exercises

FOR EXTRA HELP      

Mathematically Speaking

Fill in each blank with the most appropriate term or phrase from the given list.

constant	subtract	equation
translates	simplifies	variable
add	sentence	

1. A(n) _______ is a mathematical statement that two expressions are equal.

2. A solution of an equation is a number that when substituted for the _______ makes the equation a true statement.

3. In the equation $x + 2 = 5$, _______ from each side of the equation in order to isolate the variable.

4. The equation $x - 1 = 6$ _______ to the sentence "The difference between x and 1 is 6."

Translate each sentence to an equation.

5. z minus 9 is 25.

6. x decreased by 7 yields 29.

7. The sum of 7 and x is 25.

8. m plus 19 equals 34.

9. t decreased by 3.1 equals 4.

10. r minus 5 is equal to 6.4.

11. $\frac{3}{2}$ increased by a number yields $\frac{9}{2}$.

12. The sum of a number and $2\frac{1}{3}$ is 8.

13. $3\frac{1}{2}$ less than a number is equal to 7.

14. The difference between a number and $1\frac{1}{2}$ is the same as $7\frac{1}{4}$.

By answering yes or no, indicate whether the value of x shown is a solution of the given equation.

15.

Equation	Value of x	Solution?
a. $x + 1 = 9$	8	
b. $x - 3 = 4$	5	
c. $x + 0.2 = 5$	4.8	
d. $x - \frac{1}{2} = 1$	$\frac{1}{2}$	

16.

Equation	Value of x	Solution?
a. $x - 39 = 5$	44	
b. $x - 2 = 6$	4	
c. $x + 2.8 = 4$	1.2	
d. $x - \frac{2}{3} = 1$	$1\frac{2}{3}$	

Identify the operation to perform on each side of the equation to isolate the variable.

17. $x + 4 = 6$

18. $x - 6 = 9$

19. $x - 11 = 4$

20. $x + 10 = 17$

21. $x - 7 = 24$

22. $x + 21 = 25$

23. $3 = x + 2$

24. $10 = x - 3$

Solve and check.

25. $a - 7 = 24$

26. $x - 9 = 13$

27. $y + 19 = 21$

28. $z + 23 = 31$

29. $x - 2 = 10$

30. $t - 4 = 19$

31. $n + 9 = 13$

32. $d + 12 = 12$

33. $5 + m = 7$

34. $17 + d = 20$

35. $39 = y - 51$

36. $44 = c - 3$

37. $z + 2.4 = 5.3$

38. $t + 2.3 = 6.7$

39. $n - 8 = 0.9$

40. $c - 0.7 = 6$

41. $y + 8.1 = 9$

42. $a + 0.7 = 2$

43. $x + \frac{1}{3} = 9$

44. $z + \frac{2}{5} = 11$

45. $m - 1\frac{1}{3} = 4$

46. $s - 4\frac{1}{2} = 8$

47. $x + 3\frac{1}{4} = 7$

48. $t + 1\frac{1}{2} = 5$

49. $c - 14\frac{1}{5} = 33$

50. $a - 9\frac{7}{10} = 27\frac{2}{3}$

51. $x - 3.4 = 9.6$

52. $m - 12.5 = 13.7$

53. $5 = y - 1\frac{1}{4}$

54. $3 = t - 1\frac{2}{3}$

55. $5\frac{3}{4} = a + 2\frac{1}{3}$

56. $4\frac{1}{3} = n + 3\frac{1}{2}$

57. $2.3 = x - 5.9$

58. $4.1 = d - 6.9$

59. $y - 7.01 = 12.9$

60. $x - 3.2 = 5.23$

61. $x + 3.443 = 8$

62. $x + 0.035 = 2.004$

63. $2.986 = y - 7.265$

64. $3.184 = y - 1.273$

Translate each sentence to an equation. Solve and check.

65. 3 more than n is 11.

66. The sum of x and 15 is the same as 33.

67. 6 less than y equals 7.

68. The difference between t and 4 yields 1.

69. If 10 is added to n, the sum is 19.

70. 25 added to a number m gives a result of 53.

71. x increased by 3.6 is equal to 9.

72. n plus $3\frac{1}{2}$ equals 7.

73. A number minus $4\frac{1}{3}$ is the same as $2\frac{2}{3}$.

74. A number decreased by 1.6 is 5.9.

Choose the equation that best describes the situation.

75. After 6 months of dieting and exercising, an athlete lost $8\frac{1}{2}$ pounds. If she now weighs 135 pounds, what was her original weight?

a. $w + 8\frac{1}{2} = 135$ **b.** $w - 126\frac{1}{2} = 8\frac{1}{2}$

c. $w - 8\frac{1}{2} = 135$ **d.** $w + 135 = 143\frac{1}{2}$

76. A teenager has d dollars. After buying an Xbox 360 for \$59.99, he has \$6.01 left. How many dollars did he have at first?

a. $d + 59.99 = 66$ **b.** $d - 59.99 = 6.01$

c. $d - 59.99 = 66$ **d.** $d + 6.01 = 59.99$

77. A gigabyte (GB) is a unit used to measure computer memory. The hard drive of a computer has 10.18 GB of used space, with the rest free space. If the total capacity of the hard drive is 69.52 GB, find the amount of free space on the hard drive.

a. $x + 10.18 = 69.52$ **b.** $x - 10.18 = 69.52$

c. $x + 69.52 = 10.18$ **d.** $x - 69.52 = 10.18$

78. At a certain college, tuition costs a student $2,000 a semester. If the student received $1,250 in financial aid, how much more money does he need for the semester's tuition?

a. $x + 1{,}250 = 2{,}000$ **b.** $x + 2{,}000 = 3{,}250$

c. $x - 2{,}000 = 1{,}250$ **d.** $x - 1{,}250 = 2{,}000$

Mixed Practice

Solve and check.

79. $10 = a - 4.5$

80. $x + \frac{1}{2} = 6$

Solve.

81. The life expectancy in the United States of a female born in the year 1990 was 78.8 years. A decade later, it was 0.9 years greater. Choose the equation to find the life expectancy of a female born in the year 2000. (*Source:* The National Center for Health Statistics)

a. $x + 78.8 = 0.9$ **b.** $x - 0.9 = 78.8$

c. $x + 0.9 = 78.8$ **d.** $x - 10 = 0.9$

82. Identify the operation to perform on each side of the equation $y - 1 = 5$ to isolate the variable.

83. Is 3 a solution to the equation $10 - x = 7$?

84. Is 6 a solution to the equation $x + 4.5 = 7.5$?

85. Identify the operation to perform on each side of the equation $n + 2 = 10$ to isolate the variable.

86. Translate the sentence "x decreased by 4 is 10" to an equation.

87. The hygienist at a dentist's office cleaned a patient's teeth. The total bill came to $125, which was partially covered by dental insurance. If the patient paid $60 out of pocket toward the bill, choose the equation to find how much of the bill was covered by insurance.

a. $x + 60 = 125$ **b.** $x - 60 = 125$

c. $x + 125 = 60$ **d.** $x - 125 = 60$

88. Translate the sentence "The sum of 4.2 and n is 8" to an equation.

Applications

Write an equation. Solve and check.

89. An article on Broadway shows reported that this week the box office receipts for a particular show were $621,000. If that amount was $13,000 less than last week's, how much money did the show take in last week?

90. The first algebra textbook was written by the Arab mathematician Muhammad ibn Musa al-Khwarazmi. The title of that book, which gave rise to the word *algebra*, was *Aljabr wa'lmuqabalah*, meaning "the art of bringing together unknowns to match a known quantity." If the book appeared in the year 825, how many years ago was this? (*Source:* R.V. Young, *Notable Mathematicians*)

91. In the triangle shown, angles A and B are complementary, that is, the sum of their measures is 90°. Find x, the number of degrees in angle B.

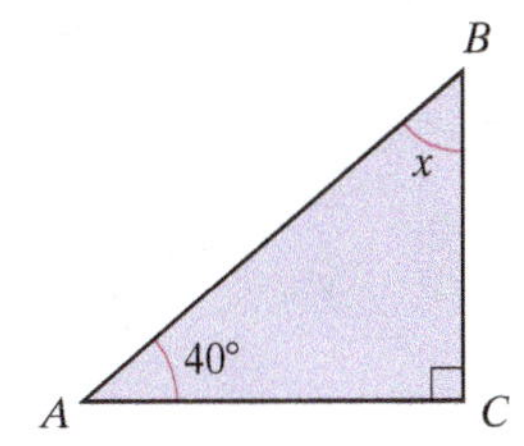

92. In the following diagram, angles ABD and CBD are supplementary, that is, the sum of their measures is 180°. If the measure of angle ABD is 109°, find y.

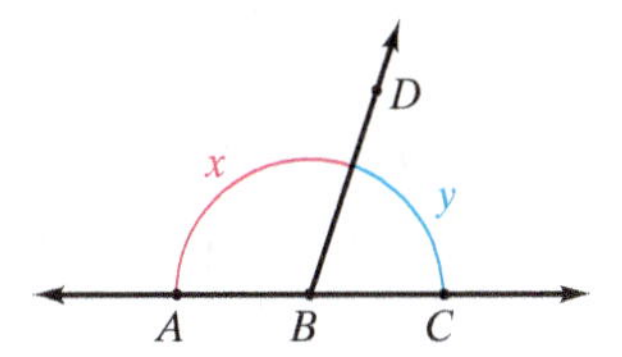

93. A local community college increased the cost of a credit hour by $12 this year. If the cost of a credit hour this year is $96, what was the cost last year?

94. Mount Kilimanjaro, the highest elevation on the continent of Africa, is 299 meters lower than Mount McKinley, the highest elevation on the continent of North America. If Mount Kilimanjaro is 5,895 meters high, how high is Mount McKinley? (**Source:** *The World Factbook, 2006*)

95. On a state freeway, the minimum speed limit is 45 miles per hour. This is 20 miles per hour lower than the maximum speed limit. What is the maximum speed limit?

96. The melting point of silver is 1,763 degrees Fahrenheit. This is 185 degrees less than the melting point of gold. What is the melting point of gold? (**Source:** *The New York Times Almanac, 2006*)

Use a calculator to solve the following problems, giving (a) the equation, (b) the exact answer, and (c) an estimate of the answer.

97. In a recent year, the U.S. charity that received the greatest private support ($1,324,089,000) was the Salvation Army. The charity that received the second greatest private support ($794,000,000) was the American Cancer Society. How much more money did the Salvation Army receive? (**Source:** *The Chronicle of Philanthropy, 2004*)

98. During a recession, an automobile company laid off 18,578 employees, reducing its workforce to 46,894. Write an equation that describes the number of employees the company had before the recession. Then solve and check.

• *Check your answers on page A-8.*

MINDSTRETCHERS

Groupwork

1. Working with a partner, compare the equations $x - 4 = 6$ and $x - a = b$.

   a. Use what you know about the first equation to solve the second equation for x.

   b. What are the similarities and the differences between the two equations?

Writing

2. Equations often serve as models for solving word problems. Write two different word problems corresponding to each of the following equations.

   a. $x + 4 = 9$
   -
   -

   b. $x - 1 = 5$
   -
   -

Critical Thinking

3. In the magic square at the right, the sum of each row, column, and diagonal is the same. Find that sum and write and solve equations to get the values of f, g, h, r, and t.

f	6	11
g	10	h
r	14	t

4.3 Solving Multiplication and Division Equations

Translating Sentences to Equations

OBJECTIVES

- To translate sentences to equations involving multiplication or division
- To solve multiplication and division equations
- To solve word problems involving equations with multiplication or division

In order to translate sentences involving multiplication or division to equations, we must recall the key words that indicate when to multiply and when to divide.

EXAMPLE 1

Translate each sentence in the table to an equation.

Solution

Sentence	Equation
a. The product of 3 and x is equal to 0.6.	$3x = 0.6$
b. The quotient of y and 4 is 15.	$\frac{y}{4} = 15$
c. Two-thirds of a number is 9.	$\frac{2}{3}n = 9$
d. One-half is equal to some number over 6.	$\frac{1}{2} = \frac{n}{6}$

PRACTICE 1

Write an equation for each sentence.

Sentence	Equation
a. Twice x is the same as 14.	
b. The quotient of a and 6 is 1.5.	
c. Some number divided by 0.3 is equal to 1.	
d. Ten is equal to one-half of some number.	

EXAMPLE 2

A house that sold for \$125,000 is twice its assessed value, x. Write an equation to represent this situation.

Solution The selling price of the house is twice its assessed value, x.

$$\underbrace{125{,}000}_{\text{The selling price of the house}} \overset{\text{is}}{=} \underbrace{2x}_{\text{twice its assessed value, } x}$$

So the equation is $125{,}000 = 2x$.

PRACTICE 2

The area of a rectangle is equal to the product of its length (3 feet) and its width (w). The rectangle's area is 15 square feet. Represent this relationship in an equation.

Equations Involving Multiplication and Division

As with addition equations, we can also solve *multiplication equations* by thinking of a balance scale like the one shown below at the left.

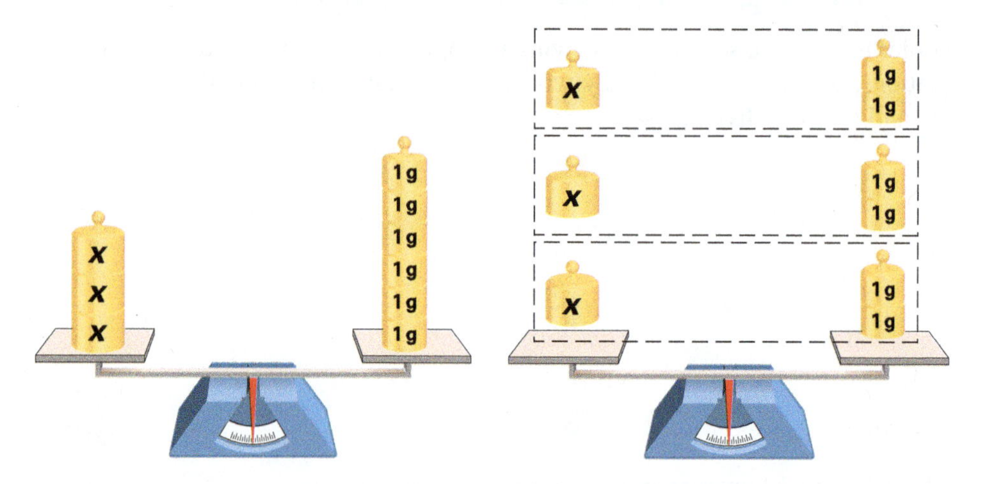

For example, consider the sentence "Three times some number x equals six," which translates to the multiplication equation $3x = 6$. We want to find the number for the variable x that, when substituted, makes this equation a true statement. To keep the balance level, whatever we do to one side we must do to the other side. In this case, dividing each side of the balance by 3 shows that in each group the unknown, x, must equal 2, as shown above at the right.

Similarly, in the division equation $\frac{x}{4} = 3$, we can multiply each side of the equation by 4 and then conclude that x equals 12.

These examples suggest the following rule.

To Solve Multiplication or Division Equations

- For a multiplication equation, divide by the same number on each side of the equation in order to isolate the variable on one side.
- For a division equation, multiply by the same number on each side of the equation in order to isolate the variable on one side.
- In either case, check the solution by substituting the value of the unknown in the original equation to verify that the resulting equation is true.

Because multiplication and division are opposite operations, one "undoes" the other. The following examples show how to perform the opposite operation on each side of an equation to solve for the unknown.

EXAMPLE 3

Solve and check: $5x = 20$

Solution $5x = 20$

$$\frac{5x}{5} = \frac{20}{5}$$ Divide each side of the equation by 5: $\frac{5x}{5} = 1x$, or x.

$$x = 4$$

Check $5x = 20$

$$5(4) \stackrel{?}{=} 20$$ Substitute 4 for x in the original equation.

$$20 \stackrel{\checkmark}{=} 20$$ The equation is true, so 4 is the solution to the original equation.

In Example 3, can you explain why $1x = x$?

PRACTICE 3

Solve and check: $6x = 30$

EXAMPLE 4

Solve and check: $5 = \frac{y}{2}$

Solution $5 = \frac{y}{2}$

Multiply each side of the equation by 2:

$$2 \cdot 5 = 2 \cdot \frac{y}{2}$$ $2 \cdot \frac{y}{2} = \frac{2}{1} \cdot \frac{y}{2} = 1y$, or y.

$$10 = y, \text{ or } y = 10$$

Check $5 = \frac{y}{2}$

$$5 \stackrel{?}{=} \frac{10}{2}$$ Substitute 10 for y in the original equation.

$$5 \stackrel{\checkmark}{=} 5$$

PRACTICE 4

Solve and check: $1 = \frac{a}{6}$

EXAMPLE 5

Solve and check: $0.2n = 4$

Solution $0.2n = 4$

$$\frac{0.2n}{0.2} = \frac{4}{0.2}$$ Divide each side by 0.2: $0.2\overline{)4.0}$ gives $20.$, or 20.

$$n = 20$$

Check $0.2n = 4$

$$0.2(20) \stackrel{?}{=} 4$$ Substitute 20 for n in the original equation.

$$4.0 \stackrel{?}{=} 4$$

$$4 \stackrel{\checkmark}{=} 4$$

PRACTICE 5

Solve and check: $1.5x = 6$

EXAMPLE 6

Solve and check: $\frac{m}{0.5} = 1.3$

Solution

$$\frac{m}{0.5} = 1.3$$

$$(0.5)\frac{m}{0.5} = (0.5)(1.3) \quad \text{Multiply each side by 0.5.}$$

$$m = 0.65$$

Check

$$\frac{m}{0.5} = 1.3$$

$$\frac{0.65}{0.5} \stackrel{?}{=} 1.3 \quad \text{Substitute 0.65 for } m \text{ in the original equation.}$$

$$1.3 \stackrel{\checkmark}{=} 1.3$$

PRACTICE 6

Solve and check: $\frac{a}{2.4} = 1.2$

EXAMPLE 7

Solve and check: $\frac{2}{3}n = 6$

Solution

$$\frac{2}{3}n = 6$$

$$\frac{2}{3}n \div \frac{2}{3} = 6 \div \frac{2}{3} \quad \text{Divide each side by } \frac{2}{3}.$$

$$\left(\frac{2}{3}n\right)\left(\frac{3}{2}\right) = 6\left(\frac{3}{2}\right)$$

$$\left(\frac{2}{3}\right)\left(\frac{3}{2}\right)n = 6\left(\frac{3}{2}\right)$$

$$n = 9$$

Check

$$\frac{2}{3}n = 6$$

$$\frac{2}{3}(9) \stackrel{?}{=} 6 \quad \text{Substitute 9 for } n \text{ in the original equation.}$$

$$\frac{2}{3}\left(\frac{9}{1}\right) \stackrel{?}{=} 6$$

$$6 \stackrel{\checkmark}{=} 6$$

PRACTICE 7

Solve and check: $\frac{3}{4}x = 12$

As in the case of addition and subtraction equations, multiplication and division equations can be useful mathematical models of real-world situations. To derive these models, we translate word sentences to algebraic equations and solve.

EXAMPLE 8

Write each sentence as an algebraic equation. Then solve and check.

Solution

Sentence	Equation	Check
a. Thirty-five is equal to the product of 5 and x.	$35 = 5x$ $\frac{35}{5} = \frac{5x}{5}$ $7 = x$, or $x = 7$	$35 = 5x$ $35 \stackrel{?}{=} 5(7)$ $35 \stackrel{\checkmark}{=} 35$
b. One equals p divided by 3.	$1 = \frac{p}{3}$ $3 \cdot 1 = 3 \cdot \frac{p}{3}$ $3 = p$, or $p = 3$	$1 = \frac{p}{3}$ $1 \stackrel{?}{=} \frac{3}{3}$ $1 \stackrel{\checkmark}{=} 1$

PRACTICE 8

Translate each sentence to an equation. Then solve and check.

Sentence	Equation	Check
a. Twelve is equal to the quotient of z and 6.		
b. Sixteen equals twice x.		

EXAMPLE 9

A baseball player runs 360 feet when hitting a home run.

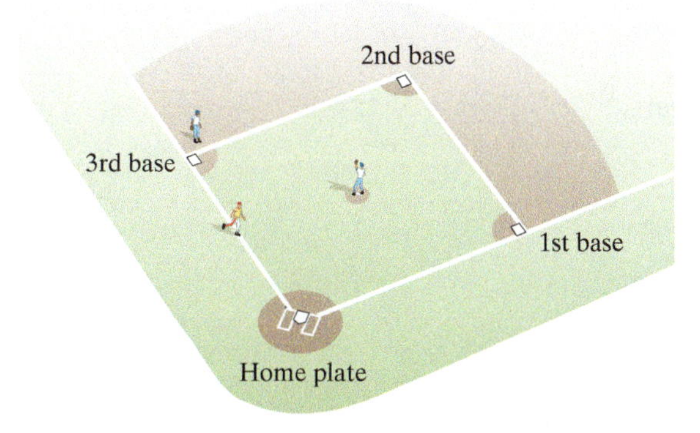

If the distances between successive bases on a baseball diamond are equal, how far is it from third base to home plate? Write an equation. Then solve and check.

Solution Let x equal the distance between successive bases. Since these distances are equal, $4x$ represents the distance around the bases.

But we know that the distance around the bases also equals 360 feet. So $4x = 360$. We solve this equation for x.

$$4x = 360$$

$$\frac{4x}{4} = \frac{360}{4} \quad \text{Divide each side by 4.}$$

$$x = 90$$

The distance from third base to home plate is 90 feet.

PRACTICE 9

The Pentagon is the headquarters of the U.S. Department of Defense.

The distance around the Pentagon is about 1.6 kilometers. If each side of the Pentagon is the same length, write an equation to find that length. Then solve and check.
(*Source:* Gene Gurney, *The Pentagon*)

EXAMPLE 10

Botswana and Andorra have the lowest and highest life expectancy, respectively, of any countries in the world. Botswana's life expectancy is only about 34 years, which is approximately $\frac{2}{5}$ of Andorra's. What is the life expectancy in Andorra? (*Source:* U.S. Bureau of the Census, International Database)

Solution Let a equal Andorra's life expectancy. Botswana's life expectancy, 34 years, is equal to $\frac{2}{5}$ of a, so we write the following equation:

$$34 = \frac{2}{5}a$$

We can solve this equation by dividing both sides by $\frac{2}{5}$.

$$34 \div \frac{2}{5} = \frac{2}{5}a \div \frac{2}{5}$$

$$(34)\left(\frac{5}{2}\right) = \left(\frac{2}{5}a\right)\left(\frac{5}{2}\right)$$

$$(34)\left(\frac{5}{2}\right) = \left(\frac{2}{5}\right)\left(\frac{5}{2}\right)a$$

$$85 = a, \text{ or } a = 85$$

So the life expectancy of Andorra is approximately 85 years.

PRACTICE 10

Six months after buying a used car, a couple sold it, taking a loss equal to $\frac{3}{8}$ of the car's original price. If their loss was $525, what was the original price? Write an equation. Then solve and check.

4.3 Exercises

FOR EXTRA HELP

MyMathLab | Math XL PRACTICE | WATCH | DOWNLOAD | READ | REVIEW

Mathematically Speaking

Fill in each blank with the most appropriate term or phrase from the given list.

divide	expression	equation
addition	division	checked
substituting	solved	evaluating
		multiply

1. In the equation $2x = 6$, _______ each side of the equation by 2 in order to isolate the variable.

2. In the equation $\frac{x}{5} = 3$, _______ each side of the equation by 5 in order to isolate the variable.

3. Check whether a number is a solution of an equation by _______ the number for the variable in the equation.

4. An equation is _______ by finding its solution.

5. The equal sign separates the two sides of a(n) _______.

6. Multiplication and _______ are opposite operations.

Translate each sentence to an equation.

7. $\frac{3}{4}$ of a number y is 12.

8. The product of $\frac{2}{3}$ and x is 20.

9. A number x divided by 7 is equal to $\frac{7}{2}$.

10. The quotient of z and 1.5 is 10.

11. $\frac{1}{3}$ of x is 2.

12. 2 times m is equal to 11.

13. The quotient of a number and 3 is equal to $\frac{1}{3}$.

14. A quantity divided by 100 is 0.36.

15. The product of 9 and an amount is the same as 27.

16. $\frac{4}{5}$ of a price is equal to 24.

By answering yes or no, indicate whether the value of x shown is a solution of the given equation.

17.

Equation	Value of x	Solution?
a. $7x = 21$	3	
b. $3x = 12$	36	
c. $\frac{x}{4} = 8$	2	
d. $\frac{x}{0.2} = 4$	8	

18.

Equation	Value of x	Solution?
a. $\frac{x}{3} = 10$	30	
b. $2.5x = 5$	2	
c. $2x = \frac{1}{3}$	$\frac{1}{6}$	
d. $\frac{x}{0.4} = 3$	12	

Identify the operation to perform on each side of the equation to isolate the variable.

19. $3x = 15$

20. $6y = 18$

21. $\frac{x}{2} = 9$

22. $\frac{y}{6} = 1$

23. $\frac{3}{4}a = 21$

24. $\frac{2}{3}m = 14$

25. $1.5b = 15$

26. $2.6x = 52$

Solve and check.

27. $5x = 30$

28. $8y = 8$

29. $\frac{x}{2} = 9$

30. $\frac{n}{9} = 3$

31. $36 = 9n$

32. $125 = 5x$

33. $\frac{x}{7} = 13$

34. $\frac{w}{10} = 21$

35. $1.7y = 6.8$

36. $0.5a = 7.5$

37. $2.1b = 42$

38. $1.5x = 45$

39. $\frac{m}{15} = 10.5$

40. $\frac{p}{10} = 12.1$

41. $\frac{t}{0.4} = 1$

42. $\frac{n}{0.5} = 6$

43. $\frac{2}{3}x = 1$

44. $\frac{1}{8}n = 3$

45. $\frac{1}{4}x = 9$

46. $\frac{3}{7}t = 15$

47. $17t = 51$

48. $100x = 400$

49. $10y = 4$

50. $100n = 50$

51. $7 = \frac{n}{100}$

52. $40 = \frac{p}{10}$

53. $2.5 = \frac{x}{5}$

54. $4.6 = \frac{z}{2}$

55. $2 = 4x$

56. $3 = 5x$

57. $\frac{14}{3} = \frac{7}{9}m$

58. $\frac{4}{9} = \frac{2}{3}a$

Solve. Round to the nearest tenth. Check.

59. $3.14x = 21.3834$

60. $2.54x = 78.25$

61. $\frac{x}{1.414} = 3.5$

62. $\frac{x}{1.732} = 1.732$

Translate each sentence to an equation. Solve and check.

63. The product of 8 and n is 56.

64. The product of 12 and m is 3.

65. $\frac{3}{4}$ of a number y is equal to 18.

66. $\frac{1}{3}$ of a number x is 16.

67. A number x divided by 5 is 11.

68. A number y divided by 100 is 10.

69. Twice x is equal to 36.

70. 3 times m is 90.

71. $\frac{1}{2}$ of an amount is 4.

72. $\frac{5}{7}$ of a number is 10.

73. A number divided by 5 is equal to $1\frac{3}{5}$.

74. An amount divided by 14 is equal to $1\frac{1}{2}$.

75. The quotient of a number and 2.5 is 10.

76. A quantity divided by 15 equals 3.6.

Choose the equation that best describes each situation.

77. Suppose that a teenager spends \$20, which is $\frac{1}{4}$ of his total savings, m. How much money did he have in the beginning?

a. $m - \frac{1}{4} = 20$ **b.** $4m = 20$

c. $m + \frac{1}{4} = 20$ **d.** $\frac{1}{4}m = 20$

78. Find the weight of a child if $\frac{1}{3}$ of her weight is 9 pounds.

a. $3x = 9$ **b.** $\frac{1}{3}x = 9$

c. $x + 3 = 9$ **d.** $x + \frac{1}{3} = 9$

79. A high school student plans to buy an MP3 player 8 weeks from now. If the MP3 player costs \$140, how much money must the student save each week in order to buy it?

a. $8c = 140$ **b.** $c + 8 = 140$

c. $\frac{c}{8} = 140$ **d.** $c - 8 = 140$

80. The student government at a college sold tickets to a play. From the ticket sales, they collected \$300, which was twice the cost of the play. How much did the play cost?

a. $\frac{n}{2} = 300$ **b.** $n - 2 = 300$

c. $2n = 300$ **d.** $n + 2 = 300$

Mixed Practice

Solve and check.

81. $11 = 2x$

82. $\frac{x}{6} = 9$

Solve.

83. The cost of dinner at a restaurant was split evenly among three friends. If each friend paid \$25.75, choose the equation to find the amount on the check.

a. $x + 3 = 25.75$ **b.** $3x = 25.75$

c. $x - 3 = 25.75$ **d.** $\frac{x}{3} = 25.75$

84. Identify the operation to perform on each side of the equation $\frac{n}{2} = 3$ to isolate the variable.

85. Translate the sentence "The quotient of y and 3 is 6" to an equation.

86. Is 2 a solution of the equation $\frac{n}{3} = 6$?

87. Translate the sentence "Twice x is 5" to an equation.

88. The marriage rate in China is approximately 4 times that of the United States. ("Marriage rate" is the number of marriages per 1,000 annually in the latest year for which data are available.) The marriage rate in China is 35.9. Choose the equation to find the approximate marriage rate in the United States. (***Source:*** *The Top Ten of Everything, 2006*)

a. $35.9 = \frac{x}{4}$ **b.** $35.9 = 4x$

c. $35.9 = x + 4$ **d.** $35.9 = x - 4$

89. Is 25 a solution of the equation $0.4x = 10$?

90. Identify the operation to perform on each side of the equation $4x = 7$ to isolate the variable.

Application

Write an equation. Solve and check.

91. In the city block shown below, the perimeter is 60 units. Find the length of one side of the square city block.

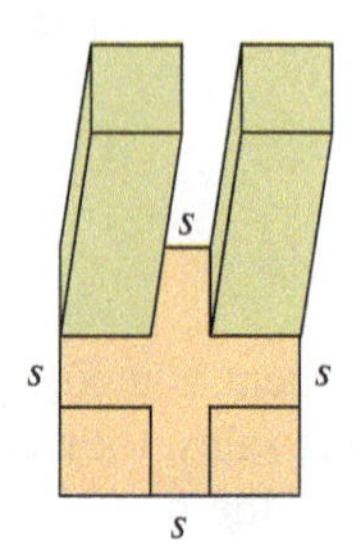

92. The area of the basketball court shown below is 4,700 square feet. Find the width of the court.

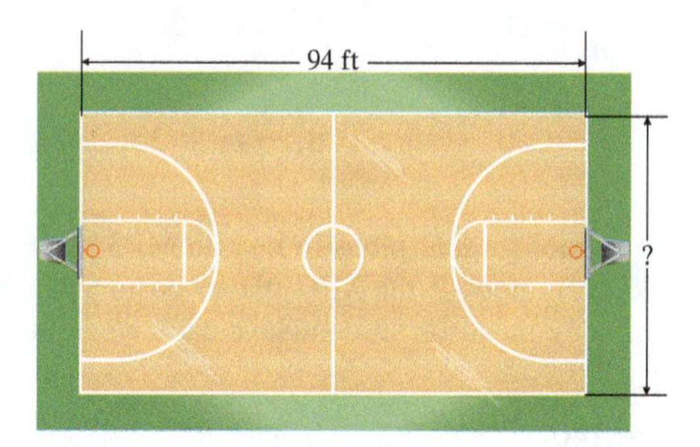

93. In an Ironman 70.3 triathlon, athletes must complete a 56-mile bike ride. This is one-half the distance of the bike ride in a regular Ironman triathlon. What distance must an athlete bike in a regular Ironman triathlon? (*Source:* World Triathlon Corporation)

94. According to the nutrition label, one packet of regular instant oatmeal has five-eighths the calories of one packet of maple and brown sugar instant oatmeal. If the regular oatmeal has 100 calories, how many calories does the maple and brown sugar oatmeal have?

95. An online DVD movie-rental service charges customers a monthly fee for unlimited DVD rentals. If a customer paid a total of $119.88 for one year of rental service, what was the monthly rental fee?

96. One plan offered by a long-distance phone service provider charges $0.07 per minute for long-distance phone calls. A customer using this plan was charged $22.26 for long-distance calls this month. How many minutes of long-distance calls did she make this month?

97. A lab technician prepared an alcohol-and-water solution that contained 60 milliliters of alcohol. This was two-fifths of the total amount of the solution.

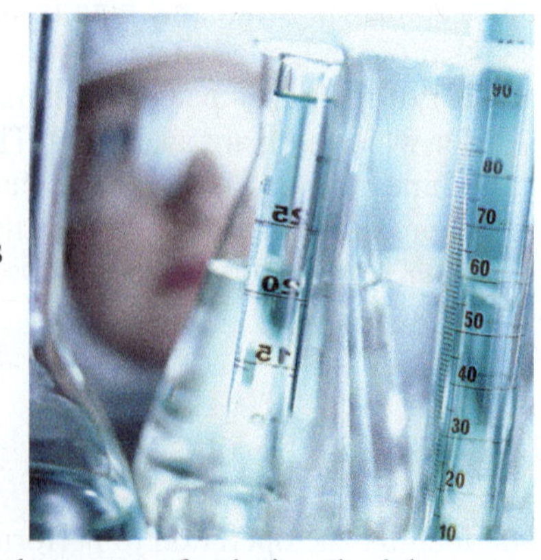

a. What was the total amount of solution the lab technician prepared?

b. How much water was in the solution?

98. A sales representative invested $5,500 of his sales bonus in the stock market. This represents one-third of his total sales bonus.

a. What was his total sales bonus?

b. How much of his sales bonus was not invested in the stock market?

Use a calculator to solve the following problems, giving (a) the equation, (b) the exact answer, and (c) an estimate of the answer.

99. The population density of a country is the quotient of the country's population and its land area (in square miles). According to the last census, the population density of the United States was approximately 79.6 persons per square mile. Use this approximation to determine the U.S. population at the time of the census, if the land area of the United States was 3,537,441 square miles. (*Source:* U.S. Bureau of the Census)

100. In a recent year, the top two U.S. airlines in terms of passenger traffic were American Airlines and United Airlines. American flew 119,987,000,000 passenger-miles, which was approximately 1.18 times the number of passenger-miles that United flew. According to this approximation, how many passenger-miles did United fly? (*Source:* International Civil Aviation Organization)

MINDSTRETCHERS

Writing

1. Write two different word problems that are applications of each equation.

a. $4x = 20$

-
-

b. $\frac{x}{2} = 5$

-
-

Groupwork

2. The equations $\frac{r}{7} = 2$ and $\frac{7}{r} = 2$ are similar in form. Working with a partner, answer the following questions.

a. How would you solve the first equation for r?

b. How can you use what you know about the first equation to solve the second equation for r?

c. What are the similarities and differences between the two equations?

Critical Thinking

3. In a magic square with four rows and four columns, the sum of the entries in each row, column, and diagonal is the same. If the entries are the consecutive whole numbers 1 through 16, what is the sum of the numbers in each diagonal?

KEY CONCEPTS AND SKILLS

CONCEPT SKILL

Concept/Skill	Description	Example
[4.1] **Variable**	A letter that represents an unknown number.	x, y, t
[4.1] **Constant**	A known number.	$2, \frac{1}{3}, 5.6$
[4.1] **Algebraic Expression**	An expression that combines variables, constants, and arithmetic operations.	$x + 3, \frac{1}{8}n$
[4.1] **To Evaluate an Algebraic Expression**	• Substitute the given value for each variable. • Carry out the computation.	Evaluate $8 - x$ for $x = 3.5$: $8 - x = 8 - 3.5$, or 4.5
[4.2] **Equation**	A mathematical statement that two expressions are equal.	$2 + 4 = 6, x + 5 = 7$
[4.2] **To Solve Addition or Subtraction Equations**	• For an addition equation, subtract the same number from each side of the equation in order to isolate the variable on one side.	$y + 9 = 15$ $y + 9 - 9 = 15 - 9$ $y = 6$ Check $y + 9 = 15$ $6 + 9 \stackrel{?}{=} 15$ $15 \stackrel{\checkmark}{=} 15$
	• For a subtraction equation, add the same number to each side of the equation in order to isolate the variable on one side. • In either case check the solution by substituting the value of the unknown in the original equation to verify that the resulting equation is true.	$w - 6\frac{1}{2} = 8$ $w - 6\frac{1}{2} + 6\frac{1}{2} = 8 + 6\frac{1}{2}$ $w = 14\frac{1}{2}$ Check $w - 6\frac{1}{2} = 8$ $14\frac{1}{2} - 6\frac{1}{2} \stackrel{?}{=} 8$ $8 \stackrel{\checkmark}{=} 8$
[4.3] **To Solve Multiplication or Division Equations**	• For a multiplication equation, divide by the same number on each side of the equation in order to isolate the variable on one side.	$1.3r = 26$ $\frac{1.3r}{1.3} = \frac{26}{1.3}$ $r = 20$ Check $1.3r = 26$ $1.3(20) \stackrel{?}{=} 26$ $26 \stackrel{\checkmark}{=} 26$
	• For a division equation, multiply by the same number on each side of the equation in order to isolate the variable on one side. • In either case check the solution by substituting the value of the unknown in the original equation to verify that the resulting equation is true.	$\frac{x}{7} = 8$ $7 \cdot \frac{x}{7} = 7 \cdot 8$ $x = 56$ Check $\frac{x}{7} = 8$ $\frac{56}{7} \stackrel{?}{=} 8$ $8 \stackrel{\checkmark}{=} 8$

Chapter 4

Review Exercises

To help you review this chapter, solve these problems.

[4.1] *Translate each algebraic expression to words.*

1. $x + 1$

2. $y + 4$

3. $w - 1$

4. $s - 3$

5. $\frac{c}{7}$

6. $\frac{a}{10}$

7. $2x$

8. $6y$

9. $y \div 0.1$

10. $n \div 1.6$

11. $\frac{1}{3}x$

12. $\frac{1}{10}w$

Translate each word phrase to an algebraic expression.

13. Nine more than m

14. The sum of b and $\frac{1}{2}$

15. y decreased by 1.4

16. Three less than z

17. The quotient of 3 and x

18. n divided by 2.5

19. The product of an amount and 3

20. Twelve times some number

Evaluate each algebraic expression.

21. $b + 8$, for $b = 4$

22. $d + 12$, for $d = 7$

23. $a - 5$, for $a = 5$

24. $c - 9$, for $c = 15$

25. $1.5x$, for $x = 0.2$

26. $1.3t$, for $t = 5$

27. $\frac{1}{2}n$, for $n = 3$

28. $\frac{1}{6}a$, for $a = 2\frac{1}{2}$

29. $w - 9.6$, for $w = 10$

30. $v - 3\frac{1}{2}$, for $v = 8$

31. $\frac{m}{1.5}$, for $m = 2.4$

32. $\frac{x}{0.2}$, for $x = 1.8$

[4.2] *Solve and check.*

33. $x + 11 = 20$

34. $y + 15 = 24$

35. $n - 19 = 7$

36. $b - 12 = 8$

37. $a + 2.5 = 6$

38. $c + 1.6 = 9.1$

39. $x - 1.8 = 9.2$

40. $y - 1.4 = 0.6$

41. $w + 1\frac{1}{2} = 3$

42. $s + \frac{2}{3} = 1$

43. $c - 1\frac{1}{4} = 5\frac{1}{2}$

44. $p - 6 = 5\frac{2}{3}$

45. $7 = m + 2$

46. $10 = n + 10$

47. $39 = c - 39$

48. $72 = y - 18$

49. $38 + n = 49$

50. $37 + x = 62$

51. $4.0875 + x = 35.136$

52. $24.625 = m - 1.9975$

[4.2–4.3] *Translate each sentence to an equation.*

53. n decreased by 19 is 35.

54. 37 less than an amount equals 234.

55. 9 increased by a number is equal to $15\frac{1}{2}$.

56. 26 more than s is $30\frac{1}{3}$.

57. Twice y is 16.

58. The product of t and 25 is 175.

59. 34 is equal to n divided by 19.

60. 17 is the quotient of z and 13.

61. $\frac{1}{3}$ of a number equals 27.

62. $\frac{2}{5}$ of a number equals 4.

By answering yes or no, indicate whether the value of x shown is a solution to the given equation.

63.

Equation	Value of x	Solution?
a. $0.3x = 6$	2	
b. $x - \frac{1}{2} = 1\frac{2}{3}$	$2\frac{1}{6}$	
c. $\frac{x}{0.5} = 7$	3.5	
d. $x + 0.1 = 3$	3.1	

64.

Equation	Value of x	Solution?
a. $0.2x = 6$	30	
b. $x + \frac{1}{2} = 1\frac{2}{3}$	$\frac{5}{6}$	
c. $\frac{x}{0.2} = 4.1$	8.2	
d. $x + 0.5 = 7.4$	6.9	

[4.3] *Solve and check.*

65. $2x = 10$

66. $8t = 16$

67. $\frac{a}{7} = 15$

68. $\frac{n}{6} = 9$

69. $9y = 81$

70. $10r = 100$

71. $\frac{w}{10} = 9$

72. $\frac{x}{100} = 1$

73. $1.5y = 30$

74. $1.2a = 144$

75. $\frac{1}{8}n = 4$

76. $\frac{1}{2}b = 16$

77. $\frac{m}{1.5} = 2.1$

78. $\frac{z}{0.3} = 1.9$

79. $100x = 40$

80. $10t = 5$

81. $0.3 = \frac{m}{4}$

82. $1.4 = \frac{b}{7}$

83. $0.866x = 10.825$

84. $\frac{x}{0.707} = 2.1$

Mixed Applications

Write an algebraic expression for each problem. Then evaluate the expression for the given amount.

85. The temperature increases 2 degrees an hour. By how many degrees will the temperature increase in h hours? In 3 hours?

86. During the fall term, a math tutor works 20 hours per week. What is the tutor's hourly wage if she earns d dollars per week? $191 per week?

87. The local supermarket sells a certain fruit for 89¢ per pound. How much will p pounds cost? 3 pounds?

88. After having borrowed $3,000 from a bank, a customer must pay the amount borrowed plus a finance charge. How much will he pay the bank if the finance charge is d dollars? $225?

Write an equation. Then solve and check.

89. After depositing $238 in a checking account, the balance will be $517. What was the balance before the deposit?

90. Hurricane Gilbert was one of the strongest storms to hit the Western Hemisphere in the twentieth century. A newspaper reported that the hurricane left 500,000 people, or about one-fourth of the population of Jamaica, homeless. Approximately how many people lived in Jamaica? (*Source:* J. B. Elsner and A. B. Kara, *Hurricanes of the North Atlantic*)

91. Drinking bottled water is more popular in some countries than in others. In a recent year, the per capita consumption of bottled water for Italians was about 177 liters, or approximately $2\frac{1}{2}$ times as much as it was for Americans. Using this approximation, find the per capita consumption for Americans, to the nearest liter. (*Source:* Euromonitor)

92. A bowler's final score is the sum of her handicap and scratch score (actual score). If a bowler has a final score of 225 and a handicap of 50, what was her scratch score?

93. On the Moon, a person weighs about one-sixth of his or her weight on Earth. What is the weight on Earth of an astronaut who weighs 30 pounds on the Moon?

94. The Nile River is about 1.8 times as long as the Missouri River. If the Nile is about 6,696 kilometers long, approximately how long is the Missouri? (*Source:* Eliot Elisofon, *The Nile*)

95. The normal body temperature is 98.6°F. An ill patient had a temperature of 101°F. This temperature is how many degrees above normal?

96. This year, a community college received 8,957 applications for admission, which amounts to 256 fewer than were received last year. How many applications did the community college receive last year?

• *Check your answers on page A-8.*

Chapter 4 POSTTEST

FOR EXTRA HELP

Test solutions are found on the enclosed CD.

To see if you have mastered the topics in this chapter, take this test.

Write each algebraic expression in words.

1. $x + \frac{1}{2}$

2. $\frac{a}{3}$

Translate each word phrase to an algebraic expression.

3. 10 less than a number

4. The quotient of 8 and p

Evaluate each algebraic expression.

5. $a - 1.5$, for $a = 1.5$

6. $\frac{b}{9}$, for $b = 2\frac{1}{4}$

Translate each sentence to an equation.

7. The difference between x and 6 is $4\frac{1}{2}$.

8. The quotient of y and 8 is 3.2.

Solve and check.

9. $x + 10 = 10$

10. $y - 6 = 6$

11. $81 = 3n$

12. $82 = \frac{a}{9}$

13. $m - 1.8 = 6$

14. $1.5n = 75$

15. $10x = 5\frac{1}{2}$

16. $\frac{n}{100} = 7.6$

Write an equation. Then solve and check.

17. A recipe for seafood stew requires $2\frac{1}{4}$ pounds of fish. After buying $1\frac{3}{4}$ pounds of bluefish, a chef decides to fill out the recipe with codfish. How many pounds of codfish should he buy?

18. A newspaper reported that this year, 30,000 elephants—$\frac{1}{3}$ of all the elephants in a certain country—had been hunted down and killed for their ivory tusks. How many elephants were there at the beginning of the year?

19. The population of the world in the year 2000 is expected to be about two-thirds of the projected world population in 2050. If the population in 2000 was about 6 billion people, what is the projected world population 50 years later? (***Source:*** U.S. Bureau of the Census, International Database, 2006)

20. In chemistry, an endothermic reaction is one that absorbs heat. As a result of an endothermic reaction, the temperature of a solution dropped by 19.8 degrees Celsius to 7.6 degrees Celsius. What was the temperature of the solution before the reaction took place? (***Source:*** Timberlake, *Chemistry: An Introduction to General, Organic, and Biological Chemistry*)

Cumulative Review Exercises

To help you review, solve the following:

1. Subtract: $8\frac{1}{4} - 2\frac{7}{8}$

2. Find the quotient: $7.5 \div 1{,}000$

3. Decide whether 2 is a solution to the equation $w + 3 = 5$

4. Multiply: 804×29

5. Round 3.14159 to the nearest hundredth.

6. Solve and check: $n - 3.8 = 4$

7. Solve and check: $\frac{x}{2} = 16$

8. In animating a cartoon, artists had to draw 24 images to appear during 1 second of screen time. How many images did they have to draw to produce a 5-minute cartoon?

9. Farmers depend on bees to pollinate many crop plants, such as apples and cherries. In the American Midwest, the acreage of crops is large as compared with the number of bees, so farmers are especially concerned if the number of beehives declines. When the number of beehives in the state of Illinois dropped from 101,000 to 46,000, how big a drop was this? (***Source:*** http://www.ag.uiuc.edu)

10. Dental insurance reimbursed a patient \$200 on a bill of \$700. Did the patient get less or more than $\frac{1}{3}$ of his money back? Explain.

• *Check your answers on page A-8.*

CHAPTER 5

Ratio and Proportion

Ratio and Proportion and Pharmacology

Many of the medicines that pharmacists dispense come in solutions. An example is aminophylline, a medicine that people with asthma take to ease their breathing.

To prepare a solution of aminophylline, pharmacists dissolve 250 milligrams of aminophylline for every 10 cubic centimeters (cc or cm^3) of sterile water.

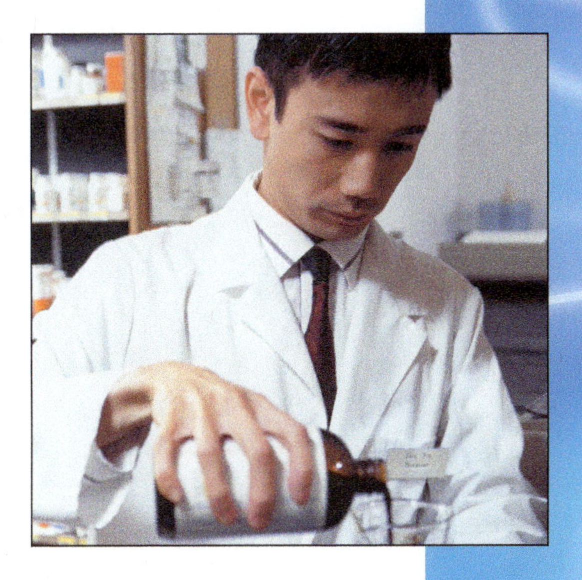

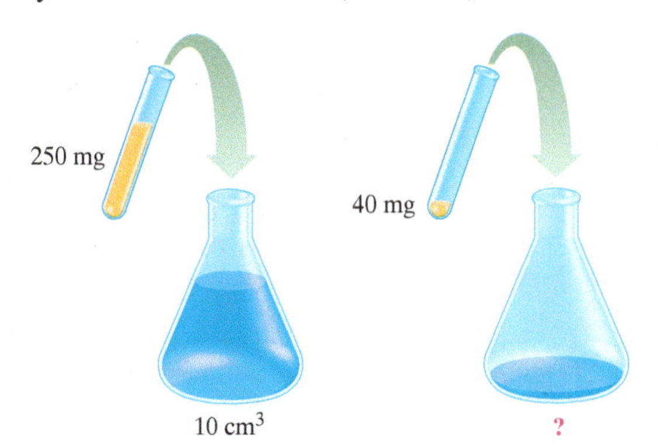

In filling a prescription for, say, 40 milligrams of aminophylline, we must determine the amount of sterile water needed. Pharmacists use the concepts of ratio and proportion to establish how much of the sterile water to dispense. (*Source:* U.S. Army Medical Department Center and School, *Pharmacology Math for the Practical Nurse*)

Chapter 5 PRETEST

To see if you have already mastered the topics in this chapter, take this test.

Write each ratio or rate in simplest form.

1. 6 to 8

2. 40 to 100

3. \$30 to \$18

4. 19 feet to 51 feet

5. 48 gallons of water in 15 minutes

6. 10 milligrams every 6 hours

Find the unit rate.

7. 12 dental assistants for every 6 dentists

8. 35 calculators for 35 students

Determine the unit price.

9. \$690 for 3 boxes of ceramic tiles

10. 12 bottles of lemon iced tea for \$6.00

Determine whether each proportion is true or false.

11. $\frac{2}{3} = \frac{16}{24}$

12. $\frac{32}{20} = \frac{8}{3}$

Solve and check.

13. $\frac{6}{8} = \frac{x}{12}$

14. $\frac{21}{x} = \frac{2}{3}$

15. $\frac{\frac{1}{2}}{4} = \frac{2}{x}$

16. $\frac{x}{6} = \frac{8}{0.3}$

Solve.

17. A contractor combines 80 pounds of sand with 100 pounds of gravel. In this mixture, what is the ratio of sand to gravel?

18. A machine at a potato chip factory can peel 12,000 pounds of potatoes in 60 minutes. At this rate, how many pounds of potatoes can it peel per minute?

19. In the first quarter of a year, a company paid \$66,000 for security. On the basis of this expense, project how much money the company will pay for security in one year.

20. The scale on a map is 3 inches to 31 miles. If two cities are 8.4 inches apart on the map, what is the actual distance, to the nearest mile, between the two cities?

• *Check your answers on page A-8.*

5.1 Introduction to Ratios

OBJECTIVES

- To write ratios of like quantities in simplest form
- To write ratios of unlike quantities in simplest form
- To solve word problems involving ratios

What Ratios Are and Why They Are Important

We frequently need to compare quantities. Sports, medicine, and business are just a few areas where we use **ratios** to make comparisons. Consider the ratios in the following examples.

- The volleyball team won 4 games for every 3 they lost.
- A physician assistant prepared a 1-to-25 boric acid solution.
- The stock's price-to-earnings ratio was 13 to 1.

Can you think of other examples of ratios in your daily life?

The preceding examples illustrate the following definition of a ratio.

Definition

A **ratio** is a comparison of two quantities expressed as a quotient.

There are, in general, three basic ways to write a ratio. For instance, we can write the ratio 1 to 25 as

$$1 \text{ to } 25 \qquad 1:25 \qquad \frac{1}{25}$$

No matter which notation we use for this ratio, it is read "1 to 25."

Simplifying Ratios

Because a ratio can be written as a fraction, we can say that, as with any fraction, a ratio is in simplest form (or reduced to lowest terms) when 1 is the only common factor of the numerator and denominator.

Let's consider some examples of writing ratios in simplest form.

EXAMPLE 1

Write the ratio 10 to 5 in simplest form.

Solution The ratio 10 to 5 expressed as a fraction is $\frac{10}{5}$.

$$\frac{10}{5} = \frac{10 \div 5}{5 \div 5} = \frac{2}{1}$$

So the ratio 10 to 5 is the same as the ratio 2 to 1. Note that the ratio 2 to 1 means that the first number is twice as large as the second number.

PRACTICE 1

Write the ratio 8:12 in simplest form.

Frequently, we deal with quantities that have units, such as months or feet. When both quantities in a ratio have the same unit, they are called **like quantities**. In a ratio of like quantities, the units drop out.

EXAMPLE 2

Express the ratio 5 months to 3 months in simplest form.

Solution The ratio 5 months to 3 months expressed as a fraction is $\frac{5 \text{ months}}{3 \text{ months}}$. Simplifying, we get $\frac{5}{3}$, which is already in lowest terms. Note that with ratios we do not rewrite improper fractions as mixed numbers because our answer must be a comparison of *two* numbers.

PRACTICE 2

Express in simplest form the ratio 9 feet to 5 feet.

EXAMPLE 3

A young couple put \$58,000 down and financed \$232,000 when buying a new home. What is the ratio of the amount put down to the purchase price of the home?

Solution This is a two-step problem. First we must find the purchase price of the home.

$$\$58{,}000 + \$232{,}000 = \$290{,}000$$

Then we write the ratio of the amount put down to the purchase price of the home.

$$\frac{\text{Amount put down}}{\text{Purchase price}} = \frac{\$58{,}000}{\$290{,}000} = \frac{1}{5}$$

The ratio is 1 to 5, which means the couple put down \$1 of every \$5 of the purchase price.

PRACTICE 3

A sales representative invests \$1,500 of his \$6,000 bonus in a high-risk fund and the remainder of the money in a low-risk fund. What is the ratio of the amount he placed in the high-risk fund to that in the low-risk fund?

Now let's compare **unlike quantities**, that is, quantities that have different units or are different kinds of measurement. Such a comparison is called a **rate**.

Definition

A **rate** is a ratio of unlike quantities.

For instance, suppose that your rate of pay is \$52 for each 8 hours of work. Simplifying this rate, we get

$$\frac{\$52}{8 \textbf{ hours}} = \frac{\$13}{2 \textbf{ hours}}$$

So you are paid \$13 for every 2 hours that you worked. Note that the units are expressed as part of the answer.

EXAMPLE 4

Simplify each rate.

a. 350 miles to 18 gallons of gas

b. 18 trees to produce 2,000 pounds of paper

Solution

a. 350 miles to 18 gallons $= \dfrac{350 \text{ miles}}{18 \text{ gallons}} = \dfrac{175 \text{ miles}}{9 \text{ gallons}}$

b. 18 trees to 2,000 pounds $= \dfrac{18 \text{ trees}}{2{,}000 \text{ pounds}} = \dfrac{9 \text{ trees}}{1{,}000 \text{ pounds}}$

PRACTICE 4

Express each rate in simplest form.

a. 150 milliliters of medication infused every 60 minutes

b. 18 pounds lost in 12 weeks

Examples 1, 2, 3, and 4 illustrate the following rule for simplifying a ratio or rate.

To Simplify a Ratio or Rate

- Write the ratio or rate as a fraction.
- Express the fraction in simplest form.
- If the quantities are alike, drop the units. If the quantities are unlike, keep the units.

Frequently, we want to find a particular kind of rate called a *unit rate*. In the rate $\dfrac{\$13}{2 \text{ hours}}$, for instance, it would be useful to know what is earned for each hour (that is, the hourly wage). We need to rewrite $\dfrac{\$13}{2 \text{ hours}}$ so that the denominator is 1 hour.

$$\frac{\$13}{2 \text{ hours}} = \frac{\$13 \div 2}{2 \text{ hours} \div 2} = \frac{\$6.50}{1 \text{ hour}} = \$6.50 \text{ per hour, or } \$6.50/\text{hr}$$

Note that "per" means "divided by."

Here, we divided the numbers in both the numerator and denominator by the number in the denominator.

Definition

A **unit rate** is a rate in which the number in the denominator is 1.

EXAMPLE 5

Write as a unit rate.

a. 275 miles in 5 hours

b. $3,453 for 6 weeks

Solution First, we write each rate as a fraction. Then we divide numbers in the numerator and denominator by the number in the denominator, getting 1 in the denominator.

PRACTICE 5

Express as a unit rate.

a. a fall of 192 feet in 4 seconds

b. 15 hits in 40 times at bat

a. 275 miles in 5 hours $= \frac{275 \text{ miles}}{5 \text{ hours}} = \frac{275 \text{ miles} \div 5}{5 \text{ hours} \div 5} = \frac{55 \text{ miles}}{1 \text{ hour}}$, or 55 mph

b. \$3,453 for 6 weeks $= \frac{\$3{,}453}{6 \text{ weeks}} = \frac{\$3{,}453 \div 6}{6 \text{ weeks} \div 6} = \frac{\$575.50}{1 \text{ week}}$, or \$575.50 per week

EXAMPLE 6

In the United States, there are approximately 1,000 public two-year colleges, with a total enrollment of about 6,000,000 students. What is the enrollment per college? (***Source:*** *The Chronicle of Higher Education*, as reported in *Time Almanac 2006*)

Solution $\frac{6{,}000{,}000 \text{ students}}{1{,}000 \text{ colleges}} = \frac{6{,}000 \text{ students}}{1 \text{ college}}$

So the enrollment is 6,000 students per public two-year college.

PRACTICE 6

A hummingbird beats its wings 2,500 times in 5 minutes. What is the number of times it beats its wings per minute? (***Source:*** The National Zoo)

In order to get the better buy, we sometimes compare prices by computing the price of a single item. This **unit price** is a type of unit rate.

Definition

A **unit price** is the price of one item, or one unit.

To find a unit price, we write the ratio of the total price of the units to the number of units and then simplify.

$$\text{Unit price} = \frac{\text{Total price}}{\text{Number of units}}$$

Let's consider some examples of unit pricing.

EXAMPLE 7

Find the unit price.

a. \$300 for 12 months of membership

b. 6 credits for \$234

c. 10-ounce box of wheat flakes for \$2.76

Solution

a. $\frac{\$300}{12 \text{ months}} = \$25/\text{month}$

b. $\frac{\$234}{6 \text{ credits}} = \$39/\text{credit}$

c. $\frac{\$2.76}{10 \text{ ounces}} = \$0.276/\text{ounce}$

$\approx \$0.28/\text{ounce}$ rounded to the nearest cent

PRACTICE 7

Determine the unit price.

a. 4 supersaver flights for \$696

b. \$22 for 8 hours of parking

c. \$19.80 for 20 song downloads

EXAMPLE 8

For the following two bottles of aspirin, which is the better buy?

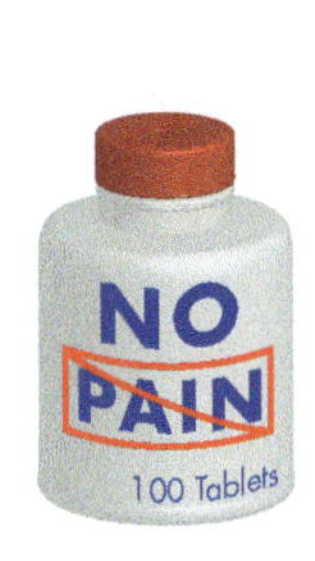

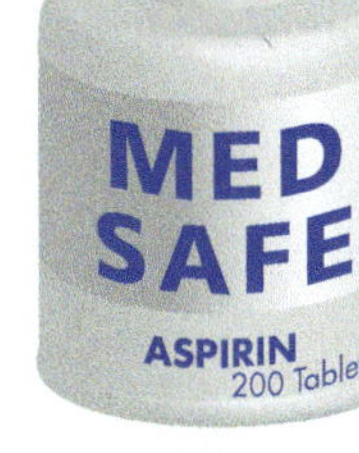

\$6.49 for 100 tablets **\$9.99 for 200 tablets**

Solution First, we find the unit price for each bottle of aspirin.

$$\text{Unit price} = \frac{\text{Total price}}{\text{Number of units}} = \frac{\$6.49}{100} = \$0.0649 \approx \$0.06 \text{ per tablet}$$

$$\text{Unit price} = \frac{\text{Total price}}{\text{Number of units}} = \frac{\$9.99}{200} = \$0.04995 \approx \$0.05 \text{ per tablet}$$

Since $\$0.05 < \$0.06$, the better buy is the 200-tablet bottle of aspirin.

PRACTICE 8

Which can of coffee has the lower unit price?

39-oz can for \$10.39 **13-oz can for \$3.69**

CULTURAL NOTE

The shape of a grand piano is dictated by the length of its strings. When a stretched string vibrates, it produces a particular pitch, say C. A second string of comparable tension will produce another pitch, which depends on the ratio of the string lengths. For instance, if the ratio of the second string to the first string is 18 to 16, then plucking the second string will produce the pitch B.

Around 500 B.C., the followers of the mathematician Pythagoras learned to adjust string lengths in various ratios so as to produce an entire scale. Thus the concept of ratio is central to the construction of pianos, violins, and many other musical instruments.

Sources:

John R. Pierce, *The Science of Musical Sound* (New York: Scientific American Library, 1983)

David Bergamini, *Mathematics* (New York: Time-Life Books, 1971)

5.1 Exercises

FOR EXTRA HELP       

PRACTICE WATCH DOWNLOAD READ REVIEW

Mathematically Speaking

Fill in each blank with the most appropriate term or phrase from the given list.

weight of a unit	numerator	unlike
like	difference	quotient
simplest form	fractional form	number of units
		denominator

1. A ratio is a comparison of two quantities expressed as a(n) _______.
2. A rate is a ratio of _______ quantities.
3. A ratio is said to be in _______ when 1 is the only common factor of the numerator and denominator.
4. Quantities that have the same units are called _______ quantities.
5. A unit rate is a rate in which the number in the _______ is 1.
6. To find the unit price, divide the total price of the units by the _______.

Write each ratio in simplest form.

7. 6 to 9
8. 9 to 12
9. 10 to 15
10. 21 to 27
11. 55 to 35
12. 8 to 10
13. 12 to 8
14. 25 to $1\frac{1}{4}$
15. 2.5 to 10
16. 1.25 to 100
17. 60 minutes to 45 minutes
18. \$40 to \$25
19. 10 feet to 10 feet
20. 75 tons to 75 tons
21. 30¢ to 18¢
22. 66 years to 32 years
23. 7 miles per hour to 24 miles per hour
24. 21 gallons to 20 gallons
25. 1,000 acres to 50 acres
26. 2,000 miles to 25 miles
27. 8 grams to 7 grams
28. 19 ounces to 51 ounces
29. 24 seconds to 30 seconds
30. 28 milliliters to 42 milliliters

Write each rate in simplest form.

31. 25 telephone calls in 10 days
32. 42 gallons in 4 minutes
33. 288 calories burned in 40 minutes
34. 190 e-mails in 25 days
35. 2 million hits on a website in 6 months
36. 50 million troy ounces of gold produced in 12 months
37. 68 baskets in 120 attempts
38. 18 boxes of cookies for \$45
39. 296 points in 16 games

40. 12 knockouts in 16 fights

41. 500 square feet of carpeting for $1,645

42. 300 full-time students to 200 part-time students

43. 48 males for every 9 females

44. 3 case workers for every 80 clients

45. 40 Democrats for every 35 Republicans

46. $12,500 in 6 months

47. 2 pounds of zucchini for 16 servings

48. 57 hours of work in 9 days

49. 1,535 flights in 15 days

50. 25 pounds of plaster for 2,500 square feet of wall

51. 3 pounds of grass seeds for 600 square feet of lawn

52. 684 parts manufactured in 24 hours

Determine the unit rate.

53. 3,375 revolutions in 15 minutes

54. 3,000 houses to 1,500 acres of land

55. 120 gallons of heating oil for 15 days

56. 48 yards in 8 carries

57. 3 tanks of gas to cut 10 acres of lawn

58. 192 meters in 6 seconds

59. 8 yards of material for 5 dresses

60. 648 heartbeats in 9 minutes

61. 20 hours of homework in 10 days

62. $200 for 8 hours of work

63. A run of 5 kilometers in 20 minutes

64. 56 calories in 4 ounces of orange juice

65. 140 fat calories in 2 tablespoons of peanut butter

66. 60 children for every 5 adults

Find the unit price.

67. 12 bars of soap for $5.40

68. 4 credit hours for $200

69. 6 rolls of film that cost $17.70

70. 2 notebooks that cost $6.90

71. 3 plants for $200

72. $240,000 for a 30-second primetime television commercial spot

73. 5 nights in a hotel for $495

74. 60 minutes of Internet access for $3

Complete each table. Determine which is the better buy.

75. Security envelopes

Number of Units	Total Price	Unit Price
125	$6.69	
500	$15.49	

76. Huggies® diapers

Number of Units	Total Price	Unit Price
56	$8.46	
112	$17.47	

77. Centrum® multivitamins tablets

Number of Units	Total Price	Unit Price
180	$12.99	
250	$17.49	

78. Honey jars

Number of Units (Ounces)	Total Price	Unit Price
16	$4.00	
32	$7.50	

Fill in the table. Which is the best buy?

79. Memorex® DVD-R discs

Number of Units	Total Price	Unit Price
25	$14.99	
50	$26.55	
100	$54.99	

80. Duracell® AA batteries

Number of Units	Total Price	Unit Price
4	$5.99	
8	$6.99	
16	$10.99	

Mixed Practice

Solve.

81. To the nearest cent, find the unit price of an 18-ounce jar of creamy peanut butter that costs $2.89.

82. Complete the table. Then find the best buy.

Dove® white soap bars

Number of Units	Total Price	Unit Price
2	$3.19	
4	$5.21	
8	$10.49	

83. Simplify the rate: 4 tutors for every 30 students.

84. Write as a unit rate: 50 lots to 0.2 square mile.

85. Write the ratio 20 to 4 in simplest form.

86. Express $\frac{30 \text{ centimeters}}{45 \text{ centimeters}}$ in simplest form.

Applications

Solve. Simplify if possible.

87. The number line shown is marked off in equal units. Find the ratio of the length of the distance x to the distance y.

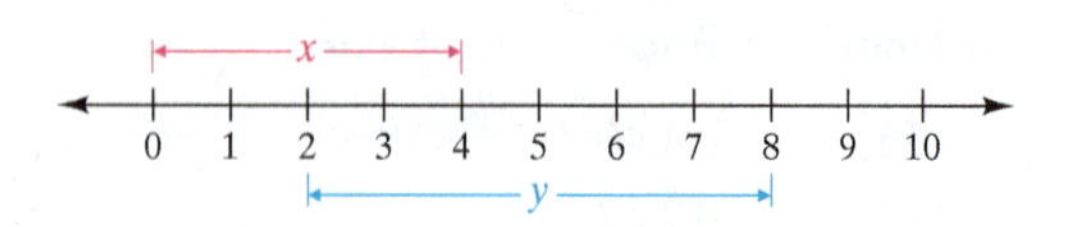

88. In the following rectangle, what is the ratio of the width to the length?

Length = 6 ft

Width = 3 ft

89. In 10 ounces of cashew nuts, there are 1,700 calories. How many calories are there per ounce?

90. For a building valued at $200,000 the property tax is $4,000. Find the ratio of the tax to the building's value.

91. On average, a person blinks 100 times in 4 minutes. How many times does a person blink in 1 minute? (*Source: Neurology*, May 1984, 677–8)

92. A bathtub contains 20 gallons of water. If the tub empties in 4 minutes, what is the rate of flow of the water per minute?

93. In a student government election, 1,000 students cast a vote for the incumbent, 900 voted for the opponent, and 100 cast a protest vote. What was the ratio of the incumbent's vote to the total number of votes?

94. At a college, 4,500 of the 7,500 students are female. What is the ratio of females to males at the college?

95. Russia has a population of approximately 143,000,000 people and a land area of about 17,000,000 square kilometers. The ratio of the number of people to the area is called the population density. What is the population density of Russia, rounded to the nearest tenth? (*Source: The World Factbook*)

96. Eighteen thousand people can ride El Toro, a roller coaster at Six Flags Great Adventure in New Jersey, in 12 hours. How many people per hour can ride El Toro? (*Source:* Six Flags Great Adventure)

97. In a recent vote, the U.S. House of Representatives voted on House Resolution 3132 as shown.

	Yea	Nay	Present Nonvoting
Republican	195	29	7
Democrat	175	23	4
Independent	1		
Totals	371	52	11

Is the ratio of the representatives who voted against the resolution (Nay) to those who voted for it (Yea) higher or lower for Democrats than for Republicans? (*Source:* http://chocola.house.gov)

98. The table below shows the breakdown of the number of patients in two hospital units at a local city hospital. Is the ratio of nurses to patients in the intensive care unit higher or lower than the ratio of nurses to patients in the medical unit?

	Intensive Care Unit	Medical Unit
Patients	25	65
Nurses	8	11

99. The following table deals with five of the longest-reigning monarchs in history.

Monarch	Country	Reign	Length of Reign (in years)
King Louis XIV	France	1643–1715	72
King John II	Liechtenstein	1858–1929	71
Emperor Franz-Josef	Austria-Hungary	1848–1916	67
Queen Victoria	United Kingdom	1837–1901	63
Emperor Hirohito	Japan	1926–1989	62

(*Source: The Top 10 of Everything 2006*)

a. What is the ratio of Emperor Hirohito's length of reign to that of Emperor Franz-Josef?

b. What is the ratio of Queen Victoria's length of reign to that of King Louis XIV?

100. The following bar graph deals with music groups that were popular and the number of chart hits that they had in the United States.

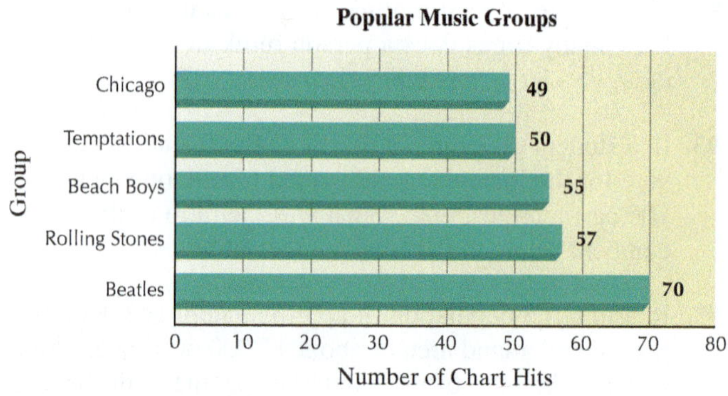

(*Source: The Top 10 of Everything 2006*)

a. Find the ratio of the number of chart hits for the Temptations as compared to the Rolling Stones.

b. What is the ratio of the number of chart hits that the Beach Boys had to that of the Beatles?

Using a calculator, solve the following problems, giving (a) the operation(s) carried out in the solution, (b) the exact answer, and (c) an estimate of the answer.

101. In the insurance industry, a **loss ratio** is the ratio of total losses paid out by an insurance company to total premiums collected for a given time period.

$$\text{Loss ratio} = \frac{\text{Losses paid}}{\text{Premiums collected}}$$

In 2 months, a certain insurance company paid losses of $6,400,000 and collected premiums of $12,472,000. What is the loss ratio?

102. Analysts for a brokerage firm prepare research reports on companies with stocks traded in various stock markets. One statistic that an analyst uses is the **price-to-earnings (P.E.) ratio**.

$$\text{P.E. ratio} = \frac{\text{Market price per share}}{\text{Earnings per share}}$$

Find the P.E. ratio for a stock that had a per-share market price of $70.75 and earnings of $5.37 per share.

• *Check your answers on page A-9.*

MINDSTRETCHERS

History

1. For a **golden rectangle**, the ratio of its length to its width is approximately 1.618 to 1 (the **golden ratio**).

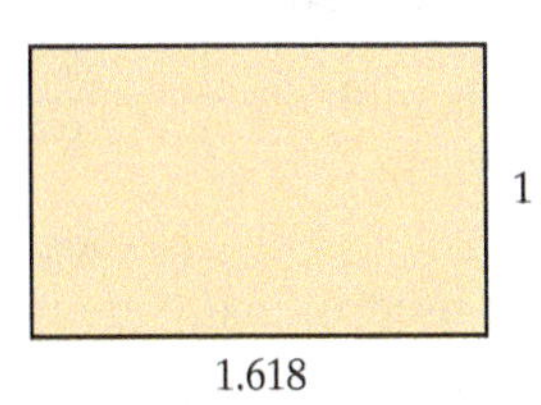

To the ancient Egyptians and Greeks, a golden rectangle was considered to be the ratio most pleasing to the eye. Show that index cards in either of the two standard sizes (3×5 and 5×8) are close approximations to the golden rectangle.

Investigation

2. The distance around a circle is called its **circumference** (C). The distance across the circle through its center is called its **diameter** (d).

   a. Use a string and ruler to measure C and d for both circles shown.

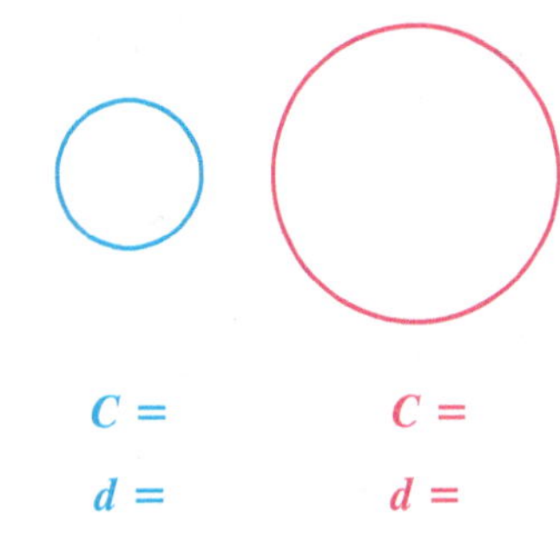

   b. Compute the ratio of C to d for each circle. Are the ratios approximately equal?

$\frac{C}{d} =$ $\frac{C}{d} =$

Writing

3. Sometimes we use *differences* rather than *quotients* to compare two quantities. Give an example of each kind of comparison and any advantages and disadvantages of each approach.

5.2 Solving Proportions

Writing Proportions

OBJECTIVES

- To write and solve proportions
- To solve word problems involving proportions

When two ratios—for instance, 1 to 2 and 4 to 8—are equal, they are said to be *in proportion*. We can write "1 is to 2 as 4 is to 8" as $\frac{1}{2} = \frac{4}{8}$. Such an equation is called a **proportion**.

Proportions are common in daily life and are used in many areas, such as finding the distance between two cities from a map with a given scale.

Definition
A **proportion** is a statement that two ratios are equal.

One way to see if a proportion is true is to determine whether the *cross products* of the ratios are equal. For example, we see that the proportion

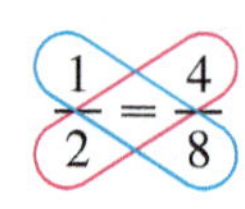

is true, because $2 \cdot 4 = 1 \cdot 8$, or $8 = 8$. However, the proportion $\frac{3}{5} = \frac{9}{10}$ is not true, since $5 \cdot 9 \neq 3 \cdot 10$.

EXAMPLE 1

Determine whether the proportion 4 is to 3 as 16 is to 12 is true.

Solution First, we write the ratios in fractional form: $\frac{4}{3} = \frac{16}{12}$.

$$\frac{4}{3} = \frac{16}{12}$$

$3 \cdot 16 \stackrel{?}{=} 4 \cdot 12$ Set the cross products equal.

$48 \stackrel{\checkmark}{=} 48$

So the proportion 4 is to 3 as 16 is to 12 is true.

PRACTICE 1

Are the ratios 10 to 4 and 15 to 6 in proportion?

EXAMPLE 2

Is $\frac{15}{9} = \frac{8}{5}$ a true proportion?

Solution $\frac{15}{9} \stackrel{?}{=} \frac{8}{5}$

$9 \cdot 8 \stackrel{?}{=} 15 \cdot 5$ Set the cross products equal.

$72 \neq 75$

The cross products are not equal. So the proportion is not true.

PRACTICE 2

Determine whether $\frac{15}{6} = \frac{8}{3}$ is a true proportion.

EXAMPLE 3

A college claims that the student-to-faculty ratio is 13 to 1. If there are 96 faculty for 1,248 students, is the college's claim true?

Solution The college claims a student-to-faculty ratio of $\frac{13}{1}$, and the actual ratio of students to faculty is $\frac{1,248}{96}$. We want to know if these two ratios are equal.

$$\begin{array}{l} \text{Students} \rightarrow \\ \text{Faculty} \rightarrow \end{array} \frac{13}{1} \stackrel{?}{=} \frac{1,248}{96} \begin{array}{l} \leftarrow \text{Students} \\ \leftarrow \text{Faculty} \end{array}$$

$$1 \cdot 1,248 \stackrel{?}{=} 13 \cdot 96 \quad \text{Set the cross products equal.}$$

$$1,248 \stackrel{\checkmark}{=} 1,248$$

Since the cross products are equal, the college's claim is true.

PRACTICE 3

A company has a policy making the compensation of its CEO proportional to the dividends that are paid to shareholders. If the dividends increase from \$72 to \$80 and the CEO's compensation is increased from \$360,000 to \$420,000, was the company's policy followed?

Solving Proportions

Suppose that you make \$840 for working 4 weeks in a book shop. At this rate of pay, how much money will you make in 10 weeks? To solve this problem, we can write a proportion in which the rates compare the amount of pay to the time worked. We want to find the amount of pay corresponding to 10 weeks, which we call x.

$$\begin{array}{l} \text{Pay} \rightarrow \\ \text{Time} \rightarrow \end{array} \frac{840}{4} = \frac{x}{10} \begin{array}{l} \leftarrow \text{Pay} \\ \leftarrow \text{Time} \end{array}$$

After setting the cross products equal, we find the missing value.

$$\frac{840}{4} = \frac{x}{10}$$

$$4 \cdot x = 840 \cdot 10$$

$$4x = 8,400$$

$$\frac{4x}{4} = \frac{8,400}{4} \quad \text{Divide each side of the equation by 4.}$$

$$x = 2,100$$

So you will make \$2,100 in 10 weeks.

We can check our solution by substituting 2,100 for x in the original proportion.

$$\frac{840}{4} = \frac{x}{10}$$

$$\frac{840}{4} \stackrel{?}{=} \frac{2,100}{10}$$

$$4 \cdot 2,100 \stackrel{?}{=} 840 \cdot 10 \quad \text{Set the cross products equal.}$$

$$8,400 \stackrel{\checkmark}{=} 8,400$$

Our solution checks.

To Solve a Proportion

- Find the cross products, and set them equal.
- Solve the resulting equation.
- Check the solution by substituting the value of the unknown in the original equation to be sure that the resulting proportion is true.

EXAMPLE 4

Solve and check: $\frac{2}{3} = \frac{x}{15}$

Solution

$\frac{2}{3} = \frac{x}{15}$

$3 \cdot x = 2 \cdot 15$ Set the cross products equal.

$3x = 30$

$\frac{3x}{3} = \frac{30}{3}$ Divide each side by 3.

$x = 10$

Check

$\frac{2}{3} = \frac{x}{15}$

$\frac{2}{3} \stackrel{?}{=} \frac{10}{15}$ Substitute 10 for x.

$2 \cdot 15 \stackrel{?}{=} 3 \cdot 10$ Set the cross products equal.

$30 \stackrel{\checkmark}{=} 30$

PRACTICE 4

Solve and check: $\frac{x}{6} = \frac{12}{9}$

EXAMPLE 5

Solve and check: $\frac{\frac{1}{4}}{12} = \frac{x}{96}$

Solution

$\frac{\frac{1}{4}}{12} = \frac{x}{96}$

$12 \cdot x = \frac{1}{4} \cdot 96$ Set the cross products equal.

$12x = 24$

$\frac{12x}{12} = \frac{24}{12}$ Divide each side by 12.

$x = 2$

So $x = 2$.

Check

$\frac{\frac{1}{4}}{12} = \frac{x}{96}$

$\frac{\frac{1}{4}}{12} \stackrel{?}{=} \frac{2}{96}$ Substitute 2 for x.

$\frac{1}{4} \cdot (96) \stackrel{?}{=} 12(2)$ Set the cross products equal.

$24 \stackrel{\checkmark}{=} 24$

PRACTICE 5

Solve and check: $\frac{\frac{1}{2}}{2} = \frac{3}{x}$

EXAMPLE 6

Forty pounds of sodium hydroxide are needed to neutralize 49 pounds of sulfuric acid. At this rate, how many pounds of sodium hydroxide are needed to neutralize 98 pounds of sulfuric acid? (*Source:* Peter Atkins and Loretta Jones, *Chemistry*)

PRACTICE 6

Saffron is a powder made from crocus flowers and is used in the manufacture of perfume. Some 8,000 crocus flowers are required to make 2 ounces of saffron. How many flowers are needed to make 16 ounces of saffron? (*Source: The World Book Encyclopedia*)

Solution Let n represent the number of pounds of sodium hydroxide needed. We set up a proportion to compare the amount of sodium hydroxide to the amount of sulfuric acid.

$$\frac{40}{49} = \frac{n}{98} \quad \begin{array}{l}\leftarrow \text{Sodium hydroxide}\\ \leftarrow \text{Sulfuric acid}\end{array}$$

$$49n = 40 \cdot 98 \quad \text{Set the cross products equal.}$$

$$49n = 3{,}920$$

$$\frac{49n}{49} = \frac{3{,}920}{49} \quad \text{Divide each side by 49.}$$

$$n = 80$$

Check

$$\frac{40}{49} = \frac{n}{98}$$

$$\frac{40}{49} \stackrel{?}{=} \frac{80}{98} \quad \text{Substitute 80 for } n.$$

$$49 \cdot 80 \stackrel{?}{=} 40 \cdot 98 \quad \text{Set the cross products equal.}$$

$$3{,}920 \stackrel{\checkmark}{=} 3{,}920$$

So 80 pounds of sodium hydroxide are needed to neutralize 98 pounds of sulfuric acid.

Tip A good way to set up a proportion is to write quantities of the same kind in the numerators and their corresponding quantities of the other kind in the denominators.

EXAMPLE 7

The scale of the following map of Nevada indicates that $\frac{1}{2}$ inch represents 100 miles. If the two cities highlighted on the map are 1.6 inches apart, what is the actual distance between them?

PRACTICE 7

A jet files 135 miles in $\frac{1}{4}$ hour. At this rate, how far can it fly in 1.5 hours?

Solution We know that $\frac{1}{2}$ inch corresponds to 100 miles. Let's set up a proportion that compares inches to miles, letting m represent the unknown number of miles.

$$\frac{\frac{1}{2}\text{ inch}}{100\text{ miles}} = \frac{1.6\text{ inches}}{m\text{ miles}}$$

$$\frac{\frac{1}{2}}{100} = \frac{1.6}{m}$$

$$\frac{1}{2}m = (100)(1.6) \quad \text{Set the cross products equal.}$$

$$\frac{1}{2}m = 160$$

$$\frac{1}{2}m \div \frac{1}{2} = 160 \div \frac{1}{2} \quad \text{Divide each side by } \frac{1}{2}.$$

$$\frac{1}{2}m \times \frac{2}{1} = 160 \times \frac{2}{1}$$

$$m = 320$$

Check

$$\frac{\frac{1}{2}}{100} = \frac{1.6}{m}$$

$$\frac{\frac{1}{2}}{100} \stackrel{?}{=} \frac{1.6}{320}$$

$$100(1.6) \stackrel{?}{=} \frac{1}{2} \cdot (320)$$

$$160 \stackrel{\checkmark}{=} 160$$

So the cities are 320 miles apart.

EXAMPLE 8

In the following diagram, the heights and shadow lengths of the two objects shown are in proportion. Find the height of the tree, h.

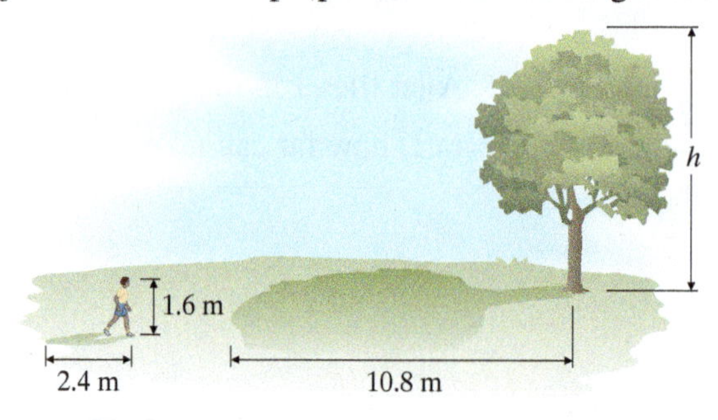

Solution The heights and shadow lengths are in proportion, so we write the following:

$$\frac{h\text{ meters (Height)}}{10.8\text{ meters (Shadow)}} = \frac{1.6\text{ meters (Height)}}{2.4\text{ meters (Shadow)}}$$

$$\frac{h}{10.8} = \frac{1.6}{2.4}$$

$$2.4h = (10.8)(1.6)$$

$$\frac{2.4h}{2.4} = \frac{17.28}{2.4}$$

$$h = 7.2$$

Check

$$\frac{h}{10.8} = \frac{1.6}{2.4}$$

$$\frac{7.2}{10.8} \stackrel{?}{=} \frac{1.6}{2.4}$$

$$(10.8)(1.6) \stackrel{?}{=} (7.2)(2.4)$$

$$17.28 \stackrel{\checkmark}{=} 17.28$$

So the height of the tree is 7.2 meters.

PRACTICE 8

The wingspans of the Boeing 777 (pictured) and 767 passenger jets are approximately in the ratio 5 to 4. Find the wingspan of the Boeing 767 to the nearest 10 feet. (*Source:* http://boeing.com/commercial/777family/pf/pf-pf_exterior_general.html)

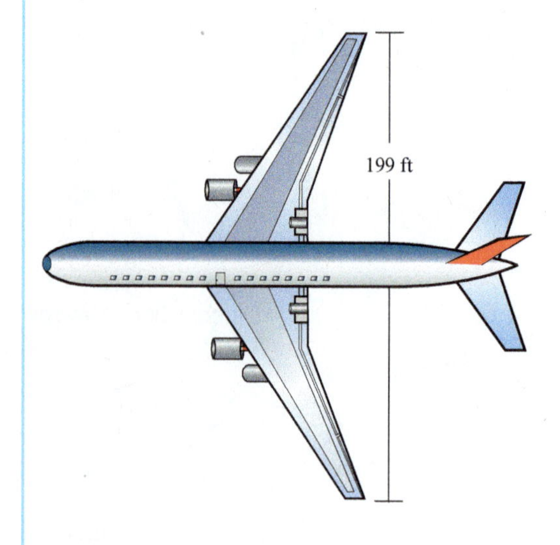

5.2 Exercises

FOR EXTRA HELP

PRACTICE WATCH DOWNLOAD READ REVIEW

Mathematically Speaking

Fill in each blank with the most appropriate term or phrase from the given list.

equation	check	like
products	solve	cross products
as	proportion	

1. A(n) _______ is a statement that two ratios are equal.

2. To determine if a proportion is true, check whether the _______ of the ratios are equal.

3. The proportion $\frac{4}{5} = \frac{8}{10}$ can be read "4 is to 5 _______ 8 is to 10."

4. To _______ the proportion $\frac{x}{2} = \frac{4}{6}$, find the value of x that makes the proportion true.

Indicate whether each statement is true or false.

5. Thirty is to 9 as 40 is to 12.

6. Nine is to 12 as 12 is to 16.

7. Two is to 3 as 7 is to 16.

8. Three is to 8 as 10 is to 27.

9. One and one-tenth is to 0.3 as 44 is to 12.

10. One and one-half is to 2 as 0.6 is to 0.8.

11. $\frac{3}{6} = \frac{2}{5}$

12. $\frac{4}{7} = \frac{5}{8}$

13. $\frac{12}{28} = \frac{18}{42}$

14. $\frac{28}{24} = \frac{7}{6}$

15. $\frac{6}{1} = \frac{3}{\frac{1}{2}}$

16. $\frac{5}{30} = \frac{\frac{1}{3}}{2}$

Solve and check.

17. $\frac{4}{8} = \frac{10}{x}$

18. $\frac{2}{3} = \frac{x}{42}$

19. $\frac{x}{19} = \frac{10}{5}$

20. $\frac{1}{6} = \frac{x}{78}$

21. $\frac{5}{x} = \frac{15}{12}$

22. $\frac{15}{x} = \frac{6}{10}$

23. $\frac{4}{1} = \frac{52}{x}$

24. $\frac{1}{17} = \frac{x}{51}$

25. $\frac{7}{4} = \frac{14}{x}$

26. $\frac{x}{6} = \frac{15}{18}$

27. $\frac{x}{8} = \frac{3}{6}$

28. $\frac{7}{5} = \frac{35}{x}$

29. $\frac{6}{21} = \frac{x}{70}$

30. $\frac{4}{x} = \frac{92}{23}$

31. $\frac{x}{12} = \frac{25}{20}$

32. $\frac{20}{25} = \frac{x}{45}$

33. $\frac{28}{x} = \frac{36}{27}$

34. $\frac{27}{63} = \frac{24}{x}$

35. $\frac{x}{10} = \frac{4}{3}$

36. $\frac{5}{6} = \frac{2}{x}$

37. $\frac{4}{x} = \frac{\frac{2}{5}}{10}$

38. $\frac{\frac{3}{4}}{6} = \frac{3}{x}$

39. $\frac{x}{27} = \frac{1.6}{24}$

40. $\frac{24}{28} = \frac{1.8}{x}$

41. $\frac{10.5}{x} = \frac{5}{10}$

42. $\frac{32}{7.2} = \frac{x}{9}$

43. $\frac{7}{0.9} = \frac{x}{36}$

44. $\frac{18}{x} = \frac{4.8}{56}$

45. $\frac{600}{x} = \frac{3}{1\frac{1}{2}}$

46. $\frac{2\frac{1}{3}}{5} = \frac{x}{12}$

47. $\frac{15}{2} = \frac{x}{2\frac{2}{3}}$

48. $\frac{x}{11} = \frac{6}{5\frac{1}{2}}$

49. $\dfrac{\frac{1}{2}}{\frac{1}{5}} = \dfrac{x}{4}$

50. $\dfrac{2}{\frac{4}{5}} = \dfrac{\frac{2}{3}}{x}$

51. $\dfrac{\frac{1}{3}}{x} = \dfrac{2}{1.2}$

52. $\dfrac{2.5}{x} = \dfrac{\frac{1}{4}}{50}$

53. $\dfrac{x}{0.16} = \dfrac{0.15}{4.8}$

54. $\dfrac{1.5}{1.25} = \dfrac{x}{0.5}$

Mixed Practice

55. Solve and check: $\dfrac{\frac{3}{4}}{15} = \dfrac{x}{8}$

56. Solve and check: $\dfrac{1.6}{x} = \dfrac{2.4}{27}$

57. Solve and check: $\dfrac{3}{2} = \dfrac{2\frac{2}{5}}{x}$

58. Determine whether the proportion 8 is to 1 as 2 is to $\dfrac{1}{4}$ is true.

59. Is $\dfrac{4}{9} = \dfrac{3}{8}$ a true or false statement?

60. Solve and check: $\dfrac{x}{9} = \dfrac{5}{6}$

Applications

Solve and check.

61. An average adult's heart beats 8 times every 6 seconds, whereas a newborn baby's heart beats 7 times every 3 seconds. Determine whether these rates are the same. (*Source: Mosby's Medical, Nursing, and Allied Health Dictionary*)

62. A full-time student at a community college pays tuition of \$1,296 for 12 credits, and a part-time student pays \$1,008 for 9 credits. Are the tuition rates the same?

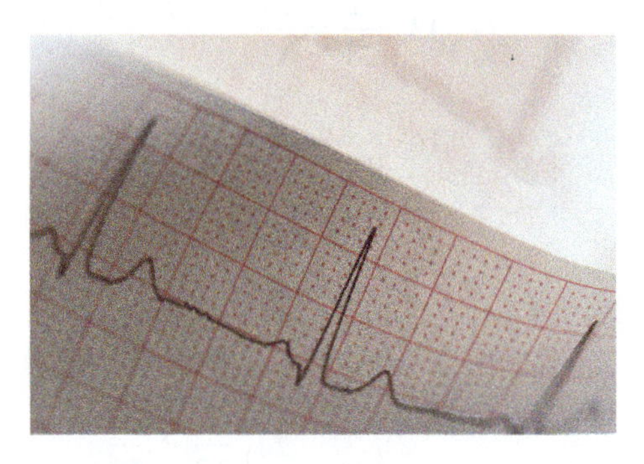

63. A dripping faucet wastes about 15 gallons of water daily. About how much water is wasted in 3 hours? (*Hint:* 1 day = 24 hours)

64. An intravenous fluid is infused at a rate of 2.5 milliliters per minute. How many milliliters are infused per hour?

65. The recommended daily allowance of protein for adults is 0.8 grams for every 2.2 pounds of body weight. If you weigh 150 pounds, how many grams of protein to the nearest tenth should you consume each day? (*Source: The Nutrition Desk Reference*)

66. A homeowner is preparing a solution of insecticide and water to spray her house plants. The directions on the insecticide bottle instruct her to mix 1 part of insecticide with 50 parts of water. How much water must she mix with 2 tablespoons of insecticide?

67. In water molecules, for every 2 hydrogen atoms there is 1 oxygen atom. How many hydrogen atoms combine with 50 oxygen atoms to form water molecules?

68. The scale on a map is $\frac{1}{4}$ centimeter to 50 kilometers. Find the actual distance between two towns represented by 10 centimeters on the map.

69. The following rectangular photo is to be enlarged so that the width of the enlargement is 25 inches. If the dimensions of the photo are to remain in proportion, what should the length of the enlargement be?

70. Architects now use computers to render their designs. If the actual length of the kitchen shown on the computer-generated floor plan is 10 feet, what is the actual length of the dining area?

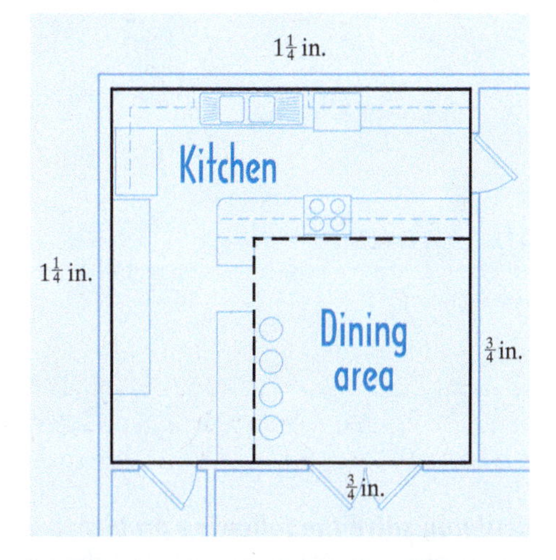

71. A popular scale for building model railroads is the N scale, where model trains are $\frac{1}{160}$ the size of actual trains. At this scale, what is the model size of a boxcar that is actually 40 feet long?

72. Thirty gallons of oil flow through a pipe in 4 hours. At this rate, how long will it take 280 gallons to flow through this same pipe?

73. The ratio of your federal income tax to state income tax is 10 to 3. How much is your state income tax if your federal income tax is $2,000?

74. A computer can download a 1,558-kilobyte file in 38 seconds. At this rate, how long will it take to download a 2,009-kilobyte file?

75. To determine the number of fish in a lake, researchers tagged 150 of them. In a later sample, they found that 6 of 480 fish were tagged. About how many fish were in the lake?

76. On a particular day, 115 Japanese yen were worth the same as 1 U.S. dollar. If a shirt cost 2,300 yen, what was its value in U.S. dollars?

77. A 5-speed bicycle has a chain linking the pedal sprocket and the gears on the rear wheel. The ratio of pedal turns to rear-wheel turns in first gear is 9 to 14. How many times in first gear does the rear wheel turn if the pedals turn 180 times?

78. The tallest land animal is the giraffe. How tall is a giraffe that casts a shadow 320 centimeters long, if a man nearby who is 180 centimeters tall casts a shadow 100 centimeters long? (*Source: Encyclopedia of Mammals*)

79. A tablet of medication consists of two substances in the ratio of 9 to 5. If the tablet contains 140 milligrams of medication, how much of each substance is in the tablet?

80. A certain metal is 5 parts tin and 2 parts lead. How many kilograms of each are there in 28 kilograms of the metal?

81. The nutrition label from a box of General Mills Total cereal indicates that a $\frac{3}{4}$-cup serving contains 23 grams of carbohydrates and 2 grams of protein.

a. How many grams of carbohydrates are there in 3 cups of cereal?

b. What is the amount of protein in $1\frac{1}{2}$ cups of cereal?

82. The following recipe is for raspberry muffins.

Raspberry Muffins *Serves 12*

1 spray of cooking spray
1 1/2 cup whole wheat self-rising flour
4 Tbsp reduced-calorie margarine, softened
2 oz ready-to-eat crisp rice cereal, divided (about 2 cups)
1 1/2 cups raspberries, divided
2/3 cup unpacked brown sugar, divided
2 large eggs, lightly beaten
2/3 cup buttermilk

(*Source:* http://weightwatchers.com)

a. How many cups of whole wheat self-rising flour are needed for a serving of 18?

b. What is the number of servings if $4\frac{1}{2}$ cups of flour are used, with other ingredients increased proportionately?

Using a calculator, solve the following problems, giving (a) the operation(s) carried out in the solution, (b) the exact answer, and (c) an estimate of the answer.

83. A senator reported that 640 metric tons of spent nuclear fuel had produced 660,000 gallons of nuclear waste. At this rate, how much nuclear waste would be produced by 810 metric tons of fuel?

84. A car uses 0.16 gallon of gas to travel through a tunnel 3.6 miles long. At this rate, how many gallons of gas are needed to travel 2,885 miles across country?

• *Check your answers on page A-9.*

MINDSTRETCHERS

Mathematical Reasoning

1. Pictorial comparisons (called *analogies*) are used on many standardized tests. Fill in the blank.

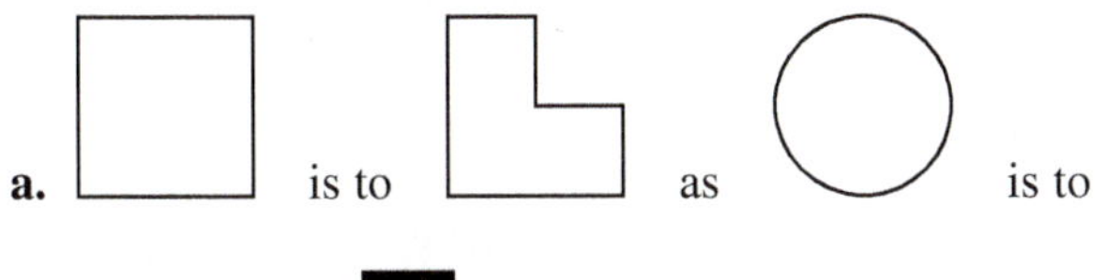

a. is to as is to ______.

b. is to as is to ______.

Groupwork

2. Work with a partner on the following.

a. Complete the following table.

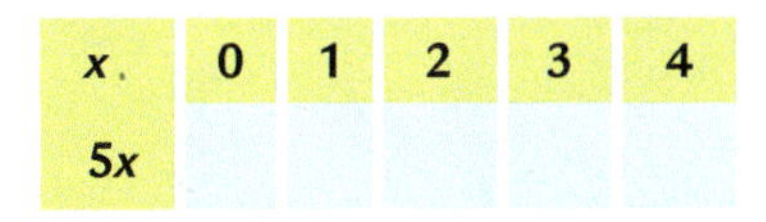

x	0	1	2	3	4
$5x$					

b. Write as many true proportions as you can, based on the values in the table.

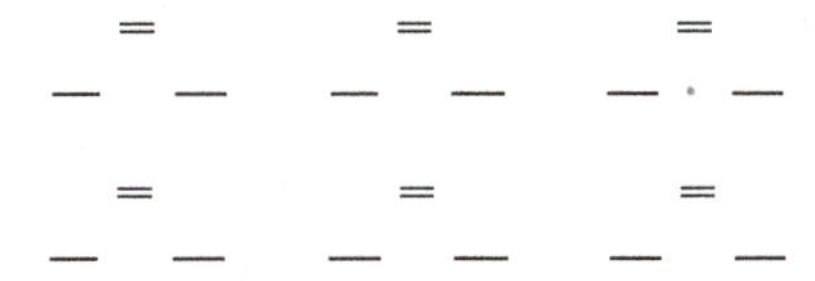

Technology

3. On the Web, there are many currency calculators that convert a given amount of a first currency into the equivalent amount of a second currency. Locate one such calculator. Use your knowledge of proportions to confirm that the currency calculator is working correctly.

KEY CONCEPTS AND SKILLS

CONCEPT SKILL

Concept/Skill	Description	Example
[5.1] Ratio	A comparison of two quantities expressed as a quotient.	3 to 4, $\frac{3}{4}$, or 3:4
[5.1] Rate	A ratio of unlike quantities.	$\frac{10 \text{ students}}{3 \text{ tutors}}$
[5.1] To simplify a ratio	• Write the ratio as a fraction. • Express the fraction in simplest form. • If the quantities are alike, drop the units. If the quantities are unlike, keep the units.	9:27 is the same as 1:3, because $\frac{9}{27} = \frac{1}{3}$ 21 hours to 56 hours $= \frac{21 \text{ hours}}{56 \text{ hours}} = \frac{21}{56} = \frac{3}{8}$ 175 miles per 7 gallons $= \frac{175 \text{ miles}}{7 \text{ gallons}} = \frac{25 \text{ miles}}{1 \text{ gallon}}$, or 25 mpg
[5.1] Unit rate	A rate in which the number in the denominator is 1.	$\frac{180 \text{ calories}}{1 \text{ ounce}}$, or 180 calories per ounce, or 180 cal/oz
[5.1] Unit price	The price of one item, or one unit.	\$0.69 per can, or \$0.69/can
[5.2] Proportion	A statement that two ratios are equal.	$\frac{5}{8} = \frac{15}{24}$
[5.2] To solve a proportion	• Find the cross products, and set them equal. • Solve the resulting equation. • Check the solution by substituting the value of the unknown in the original equation to verify that the resulting proportion is true.	$\frac{6}{9} = \frac{2}{x}$ $6x = 18$ $x = 3$ **Check** $\frac{6}{9} = \frac{2}{x}$ $\frac{6}{9} \stackrel{?}{=} \frac{2}{3}$ $6 \cdot 3 \stackrel{?}{=} 9 \cdot 2$ $18 \stackrel{\checkmark}{=} 18$

Chapter 5

Review Exercises

To help you review this chapter, solve these problems.

[5.1] *Write each ratio or rate in simplest form.*

1. 10 to 15
2. 28 to 56
3. 3 to 4
4. 50 to 16
5. 10,400 votes to 6,500 votes
6. 9 cups to 12 cups
7. 88 feet in 10 seconds
8. 45 applicants for 10 positions

Write each ratio as a unit rate.

9. 4 pounds of grass seed to plant in 1,600 square feet of lawn
10. 75 billion telephone calls in 150 days
11. 48 yards in 6 downs
12. 3,200 square feet covered by 8 gallons of paint
13. 21,000,000 vehicles produced in 2 years
14. 532,000 commuters traveled in 7 days

Find the unit price for each item.

15. $475 for 4 nights
16. $19.45 for 5 DVD movie rentals
17. $80,000 for 64 computer stations
18. $9,364 for 100 shares of stock

Fill in each table. Which is the better buy?

19. *The New Yorker* magazine issues

Number of Units	Total Price	Unit Price
47	$11.95	
92	$29.90	

20. Custom laser checks

Number of Units	Total Price	Unit Price
300	$59.99	
525	$74.99	

Complete each table. Determine the best buy.

21. GNC green tea extract capsules

Number of Units	Total Price	Unit Price
90	$7.19	
180	$7.43	
360	$17.91	

22. Johnson's baby oil bottles

Number of Units (Fluid Ounces)	Total Price	Unit Price
4	$1.89	
14	$3.59	
20	$4.69	

[5.2] *Indicate whether each proportion is true or false.*

23. $\frac{15}{25} = \frac{3}{5}$
24. $\frac{3}{1} = \frac{1}{3}$
25. $\frac{50}{45} = \frac{10}{8}$
26. $\frac{15}{6} = \frac{5}{2}$

Solve and check.

27. $\frac{1}{2} = \frac{x}{12}$
28. $\frac{9}{12} = \frac{x}{4}$
29. $\frac{12}{x} = \frac{3}{8}$
30. $\frac{x}{72} = \frac{5}{12}$
31. $\frac{1.6}{7.2} = \frac{x}{9}$
32. $\frac{x}{12} = \frac{1.2}{1.8}$
33. $\frac{5}{\frac{1}{2}} = \frac{7}{x}$
34. $\frac{3}{5} = \frac{x}{\frac{2}{3}}$

35. $\dfrac{2\frac{1}{4}}{x} = \dfrac{1}{30}$

36. $\dfrac{3}{1\frac{3}{5}} = \dfrac{x}{24}$

37. $\dfrac{\frac{5}{6}}{x} = \dfrac{2}{1.8}$

38. $\dfrac{\frac{2}{3}}{4} = \dfrac{x}{0.9}$

39. $\dfrac{0.36}{4.2} = \dfrac{2.4}{x}$

40. $\dfrac{x}{0.21} = \dfrac{0.12}{0.18}$

Mixed Applications

Solve and check.

41. An airplane has 12 first-class seats and 180 seats in coach. What is the ratio of first-class seats to coach seats?

42. A computer store sells $23,000 worth of desktop computers and $45,000 worth of laptop computers in a given month. What is the ratio of desktop to laptop computer sales?

43. If a personal care attendant earns $540 for a 6-day workweek, how much does she earn per day?

44. A glacier in Alaska moves about 2 inches in 16 months. How far does the glacier move per month?

45. In a recent year, approximately 200,000,000 of the 300,000,000 people in the United States were Internet users. What is the ratio of Internet users to the total population? (***Source:*** Internet World Stats, 2006)

46. A city's public libraries spend about $9.50 in operating expenses for every book they circulate. If their operating expenses amount to $475,000, how many books circulate?

47. In a college's day-care center, the required staff-to-child ratio is 2 to 5. If there are 60 children and 12 staff in the day-care center, is the center in compliance with the requirement?

48. Despite the director's protests, the 1924 silent film *Greed* was edited down from about 42 reels of film to 10 reels. If the original version was about 9 hours long, about how long was the edited version? (***Source:*** *The Film Encyclopedia*)

49. A sports car engine has an 8-to-1 compression ratio. Before compression, the fuel mixture in a cylinder takes up 440 cubic centimeters of space. How much space does the fuel mixture occupy when fully compressed?

50. On an architectural drawing of a planned community, a measurement of 25 feet is represented by 0.5 inches. If two houses are actually 62.5 feet apart, what is the distance between them on the drawing?

51. The density of a substance is the ratio of its mass to its volume. To the nearest hundredth, find the density of gasoline if a volume of 317.45 cubic centimeters has a mass of 216.21 grams.

52. The state of New Jersey has an area of 7,417 square miles and a population of 8,717,925. Of all the U.S. states, it has the highest population density—more than a dozen times that of the nation. Compute New Jersey's population density, the ratio of the number of people to the area, rounded to the nearest tenth. (***Source:*** U.S. Bureau of the Census)

• *Check your answers on page A-9.*

Chapter 5 POSTTEST

FOR EXTRA HELP  

Test solutions are found on the enclosed CD.

To see if you have mastered the topics in this chapter, take this test.

Write each ratio or rate in simplest form.

1. 8 to 12

2. 15 to 42

3. 55 ounces to 31 ounces

4. 180 miles to 15 miles

5. 65 revolutions in 60 seconds

6. 3 centimeters for every 75 kilometers

Find the unit rate.

7. 340 miles in 5 hours

8. 200-meter dash in 25 seconds

Determine the unit price.

9. $4,080 for 30 days

10. 25 greeting cards for $20

Determine whether each proportion is true or false.

11. $\frac{8}{21} \stackrel{?}{=} \frac{16}{40}$

12. $\frac{7}{3} \stackrel{?}{=} \frac{63}{27}$

Solve and check.

13. $\frac{15}{x} = \frac{6}{10}$

14. $\frac{102}{17} = \frac{36}{x}$

15. $\frac{0.9}{36} = \frac{0.7}{x}$

16. $\frac{\frac{1}{3}}{4} = \frac{x}{12}$

Solve.

17. To advertise his business, an owner can purchase 3 million e-mail addresses for $120 or 5 million e-mail addresses for $175. Which is the better buy?

18. A house was originally worth $95,000 but increased in value to $110,000 after 5 years. What is the ratio of the increase to the original value?

19. A man $6\frac{1}{4}$ feet tall casts a 5-foot shadow. A nearby tree casts a 20-foot shadow. If the heights and shadow lengths of the man and tree are proportional, how tall is the tree?

20. A nurse takes his patient's pulse. What is the patient's pulse per minute if it beats 12 times in 15 seconds?

• *Check your answers on page A-10.*

Cumulative Review Exercises

To help you review, solve the following:

1. Find the difference: $3\frac{1}{10} - 2\frac{7}{10}$

2. Multiply: $8.2 \times 1{,}000$

3. Solve and check: $x + 6.5 = 9$

4. Simplify the ratio: 2.5 to 10

5. Find the unit price: 3 yards for $12

6. Estimate: $12\frac{1}{7} \div 3\frac{9}{10}$

7. Solve and check: $\frac{\frac{1}{2}}{4} = \frac{x}{6}$

8. What is the area of the singles tennis court shaded in the diagram?

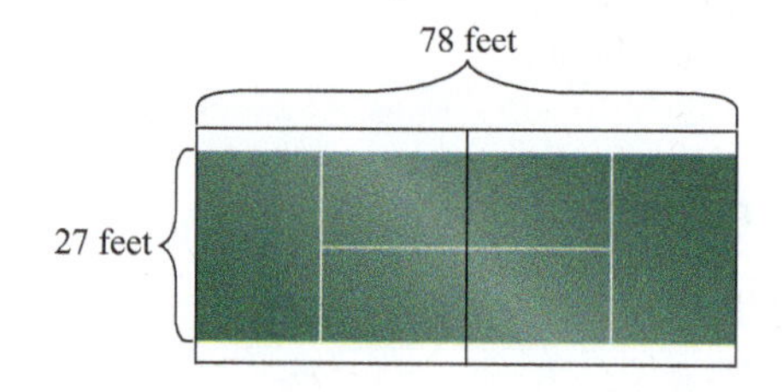

9. A rule of thumb for growing lily bulbs is to plant them 3 times as deep as they are wide. How deep should a gardener plant a lily bulb that is 2.5 inches wide?

10. At a legal firm, a part-time employee works 10 hours a week and makes $120. At this rate of pay, how much would the employee make for working 15 hours a week?

• *Check your answers on page A-10.*

CHAPTER 6

Percents

Percents and Surveys

In a recent primary election, three candidates, including Deval Patrick, were running for one political party's nomination for governor of Massachusetts. Several weeks before the primary election, a newspaper reported the results of a poll of 501 likely primary voters: 31% of those surveyed supported Patrick, with 30% in favor of a second candidate and 27% a third candidate. But every poll and survey has a *margin of error*. Allowing for this margin of error, the poll really indicated that the percent of voters in the primary supporting Patrick would probably be 26.6% to 35.4%. Because of this wide margin of error, the newspaper didn't use the headline "Patrick leads!". Instead, it correctly reported that the three candidates were in a virtual dead heat.

(*Source: The Boston Globe*)

Chapter 6 PRETEST

To see if you have already mastered the topics in this chapter, take this test.

Rewrite.

1. 5% as a fraction

2. $37\frac{1}{2}\%$ as a fraction

3. 250% as a decimal

4. 3% as a decimal

5. 0.007 as a percent

6. 8 as a percent

7. $\frac{2}{3}$ as a percent, rounded to the nearest whole percent

8. $1\frac{1}{10}$ as a percent

Solve.

9. What is 75% of 50 feet?

10. Find 110% of 50.

11. Estimate 84% of $61.77.

12. 2% of what number is 5?

13. What percent of 10 is 4?

14. What percent of 4 is 10?

15. In a municipal savings account, a city employee earned 4% interest on $350. How much money did the employee earn in interest?

16. The number of students enrolled at a community college rose from 2,475 last year to 2,673 this year. What was the percent increase in the college's enrollment?

17. In the depths of the Great Depression, 24% of the U.S. civilian labor force was unemployed. Write this percent as a simplified fraction. (*Source:* U.S. Bureau of the Census)

18. In a chemistry lab, a student dissolved 10 milliliters of acid in 30 milliliters of water. What percent of the solution was acid?

19. For parties of 8 or more, a restaurant automatically adds an 18% tip to the restaurant check. What tip would be added to a dinner check for a party of 10 if the total bill was $339.50?

20. A patient's health insurance covered 80% of the cost of her operation. She paid the remainder, which came to $2,000. Find the total cost of the operation.

• *Check your answers on page A-10.*

6.1 Introduction to Percents

OBJECTIVES

- To read and write percents
- To find the fraction and the decimal equivalent of a given percent
- To find the percent equivalent of a given fraction or decimal
- To solve word problems involving percent conversions

What Percents Are and Why They Are Important

Percent means divided by 100. So 50% (read "fifty percent") means 50 divided by 100 (or 50 out of 100).

A percent can also be thought of as a ratio or a fraction with denominator 100. For example, we can look at 50% either as the ratio of 50 parts to 100 parts or as the fraction $\frac{50}{100}$, or $\frac{1}{2}$. We can also think of 50% as 0.50, or 0.5, since a fraction can be written as a decimal.

In the diagram at the right, 50 of the 100 squares are shaded. This shaded portion represents 50%.

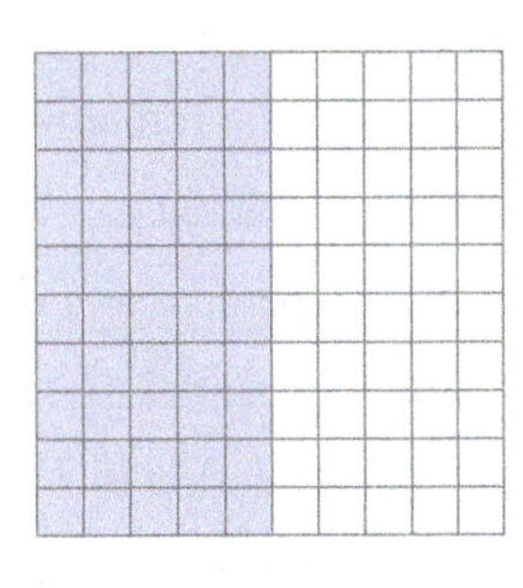

We can use diagrams to represent other percents.

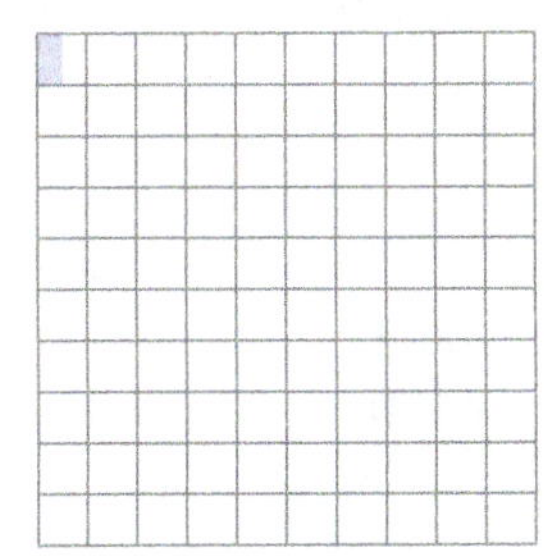

In the diagram to the left, $\frac{1}{2}$% is equivalent to the shaded portion,

$$\frac{\frac{1}{2}}{100}, \text{ or } \frac{1}{200}$$

The entire diagram at the right is shaded, so 100% means $\frac{100}{100}$, or 1.

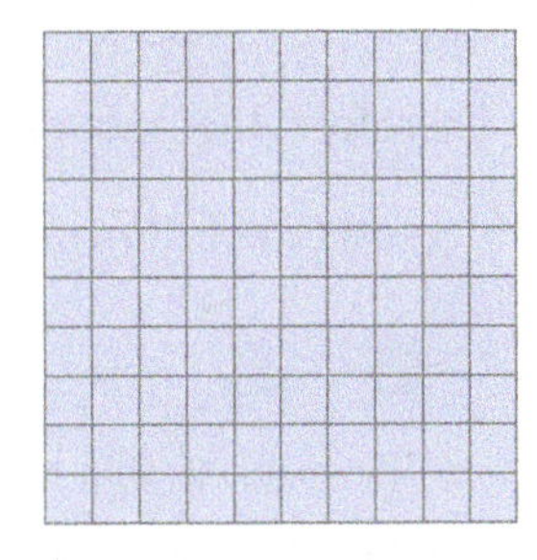

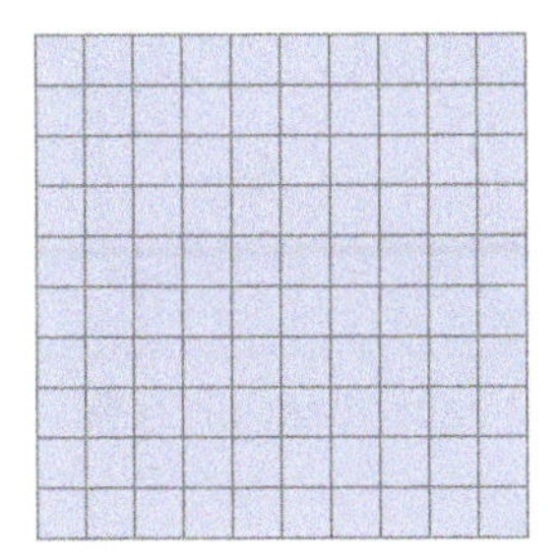

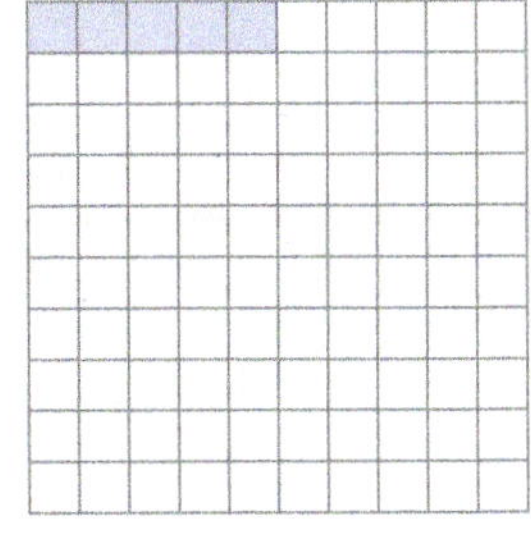

We can express 105% as $\frac{105}{100}$, or $1\frac{1}{20}$, as shown by the shaded portions of the diagrams.

Percents are commonly used, as the following statements taken from a single page of a newspaper illustrate.

- About 10% of the city's budget goes to sanitation.
- Blanket Sale—30% to 40% off!
- The number of victims of the epidemic increased by 125% in just 6 months.

A key reason for using percents so frequently is that they are easy to compare. For instance, we can tell right away that a discount of 30% is larger than a discount of 22%, simply by comparing the whole numbers 30 and 22.

To see how percents relate to fractions and decimals, let's consider finding equivalent fractions, decimals, and percents. In Chapter 3, we discussed two of the six types of conversions:

- changing a decimal to a fraction, and
- changing a fraction to a decimal.

Here, we consider the remaining four types of conversion:

- changing a percent to a fraction,
- changing a percent to a decimal,
- changing a decimal to a percent, and
- changing a fraction to a percent.

Note that each type of conversion changes the way the number is written—but not the number itself.

Changing a Percent to a Fraction

Suppose that we want to rewrite a percent—say, 30%—as a fraction. Because percent means divided by 100, we simply drop the % sign, place 30 over 100, and simplify.

$$30\% = \frac{30}{100} = \frac{3}{10}$$

Therefore, the fraction $\frac{3}{10}$ is just another way of writing the percent 30%. This result suggests the following rule.

To Change a Percent to the Equivalent Fraction

- Drop the % sign from the given percent and place the number over 100.
- Simplify the resulting fraction, if possible.

EXAMPLE 1

Write 7% as a fraction.

Solution To change this percent to a fraction, we drop the percent sign and write the 7 over 100. The fraction is already reduced to lowest term.

$$7\% = \frac{7}{100}$$

PRACTICE 1

Find the fractional equivalent of 21%.

EXAMPLE 2

Express 150% as a fraction.

Solution $150\% = \frac{150}{100} = \frac{3}{2}$, or $1\frac{1}{2}$

Note that the answer is larger than 1 because the original percent was more than 100%.

PRACTICE 2

What is the fractional equivalent of 225%?

EXAMPLE 3

Express $33\frac{1}{3}\%$ as a fraction.

Solution To find the equivalent fraction, we first drop the % sign and then put the number over 100.

$$\frac{33\frac{1}{3}}{100} = 33\frac{1}{3} \div 100 = 33\frac{1}{3} \div \frac{100}{1} = \frac{100}{3} \div \frac{100}{1} = \frac{\overset{1}{\cancel{100}}}{3} \times \frac{1}{\underset{1}{\cancel{100}}} = \frac{1}{3}$$

So $33\frac{1}{3}\%$ expressed as a fraction is $\frac{1}{3}$.

PRACTICE 3

Change $12\frac{1}{2}\%$ to a fraction.

EXAMPLE 4

The Ring of Fire contains 75% of the volcanoes on Earth. Express this percent as a fraction. (*Source:* *National Geographic*)

Ring of Fire

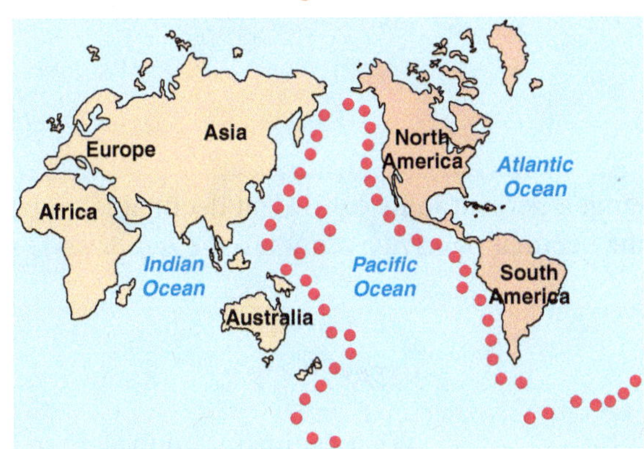

Solution $75\% = \frac{75}{100} = \frac{3}{4}$

So $\frac{3}{4}$ of the volcanoes on Earth are located in the Ring of Fire.

PRACTICE 4

About 86% of California's coastline is eroding. Express this percent as a fraction. (*Source:* Surfrider Foundation)

Changing a Percent to a Decimal

Now let's consider rewriting a percent as a decimal. For instance, take 75%. We begin by writing this percent as a fraction.

$$75\% = \frac{75}{100}$$

Converting this fraction to a decimal, we divide:

$$100\overline{)75.00}\quad 0.75$$

Note that we could have gotten this answer simply by moving the decimal point two places to the left and dropping the % sign.

$$75\% = 75.\% = .75, \text{ or } 0.75$$

This example suggests the following rule.

To Change a Percent to the Equivalent Decimal

- Drop the % sign from the given percent and divide the number by 100.

EXAMPLE 5

Change 42% to a decimal.

Solution Drop the % sign from 42% and divide by 100. Recall that to divide a decimal by 100, we can simply move the decimal point two places to the left.

$$42\% = 42.\% = .42$$

So the decimal equivalent of 42% is 0.42.

PRACTICE 5

Express 31% as a decimal.

Tip A shortcut for changing a percent to its equivalent decimal is dropping the percent sign and moving the decimal point *two places* to the *left*.

EXAMPLE 6

Find the decimal equivalent of 1%.

Solution The unwritten decimal point lies to the right of the 1. Moving the decimal point two places to the left and dropping the % sign, we get:

$$1\% = 01.\% = .01, \text{ or } 0.01$$

Note that we inserted a 0 as a placeholder, because there was only a single digit to the left of the 1.

PRACTICE 6

What decimal is equivalent to 5%?

EXAMPLE 7

Convert 37.5% to a decimal.

Solution 37.5% = .375, or 0.375

In this problem, the given number is a percent even though it involves a decimal point.

PRACTICE 7

Rewrite 48.2% as a decimal.

EXAMPLE 8

Change $12\frac{1}{2}\%$ to a decimal.

Solution To find the decimal equivalent of $12\frac{1}{2}\%$, we begin by converting the fraction $\frac{1}{2}$ in the mixed number to its decimal equivalent 0.5. Then we move the decimal point to the left two places, dropping the % sign.

$$\frac{1}{2} = 2\overline{)1.0} \quad 0.5, \text{ or } .5$$

$$12\frac{1}{2}\% = 12.5\% = .125, \text{ or } 0.125$$

PRACTICE 8

Express the following percent as a decimal: $62\frac{1}{4}\%$.

EXAMPLE 9

In 1945 at the end of World War II, the public debt of the United States was 602% of what it had been in 1940. Write this percent as a decimal.

Solution 602% = 602.%, or 6.02

The 1945 U.S. debt was 6.02 times what it had been 5 years earlier.

PRACTICE 9

In 2014, the enrollment in American schools and colleges is expected to be 112% of the enrollment in 2005. Express this percent as a decimal.

(***Source:*** National Center for Education Statistics)

Changing a Decimal to a Percent

Suppose that we want to change 0.75 to a percent. Because 100% is the same as 1, we can multiply this number by 100% without changing its value.

$$0.75 \times 100\% = 75\%$$

Note that we could have gotten this answer simply by moving the decimal point to the right two places and adding the % sign.

$$0.75 = 075.\% = 75\%$$

Note that we dropped the decimal point in the answer because it was to the right of the units digit.

To Change a Decimal to the Equivalent Percent

- Multiply the number by 100 and insert a % sign.

EXAMPLE 10

Write 0.425 as a percent.

Solution We multiply 0.425 by 100 and add a % sign.

$$0.425 = 0.425 \times 100\% = 42.5\%, \text{ or } 42\frac{1}{2}\%$$

PRACTICE 10

What percent is equivalent to the decimal 0.025?

Tip A shortcut for changing a decimal to its equivalent percent is inserting a % sign and moving the decimal point *two places* to the *right*.

EXAMPLE 11

Convert 0.03 to a percent.

Solution $0.03 = 003.\% = 3\%$

PRACTICE 11

Change 0.09 to a percent.

EXAMPLE 12

Express 0.1 as a percent.

Solution In the given number, only a single digit is to the right of the decimal point. So to move the decimal point two places to the right, we need to insert a 0 as a placeholder.

$$0.1 = 0.10 = 10.\% = 10\%$$

PRACTICE 12

What percent is equivalent to 0.7?

EXAMPLE 13

What percent is equivalent to 2?

Solution Recall that a whole number such as 2 has a decimal point understood to its right. We move the decimal point two places to the right.

$$2 = 2. = 2.00 = 200.\% = 200\%$$

So the answer is 200%, which makes sense: 200% is double 100%, just as 2 is double 1.

PRACTICE 13

Rewrite 3 as a percent.

EXAMPLE 14

Express 0.2483 as a percent, rounded to the nearest whole percent.

Solution First, we obtain the exact percent equivalent.

$$0.2483 = 24.83\%$$

To round this number to the nearest whole percent, we underline the digit 4. Then we check the critical digit immediately to its right. This digit is 8, so we round up.

$$2\underline{4}.83\% \approx 25.\% = 25\%$$

PRACTICE 14

Convert 0.714 to a percent, rounded to the nearest whole percent.

EXAMPLE 15

Red blood cells make up about 0.4 of the total blood volume in the human body, whereas 55% of the total blood volume is plasma. Which makes up more of the blood volume—red blood cells or plasma? Explain. (*Source:* Mayo Clinic; Merck)

Solution We want to compare the decimal 0.4 and the percent 55%. One way is to change the decimal to a percent.

$$0.4 = 0.40 = 40.\% = 40\%$$

Since 40% is less than 55%, we conclude that plasma makes up more of the blood volume.

PRACTICE 15

Air is a mixture of many gases. For example, 0.78 of air is nitrogen, and 0.93% is argon. Is there more nitrogen or argon in air? Explain.

Changing a Fraction to a Percent

Now let's change a fraction to a percent. Consider, for instance, the fraction $\frac{1}{5}$.

To convert this fraction to a percent, multiply $\frac{1}{5}$ by 100%, which is equal to 1.

$$\frac{1}{5} = \frac{1}{5} \times 100\% = \frac{1}{\cancel{5}_1} \times \frac{\cancel{100}^{20}}{1}\% = 20\%$$

To Change a Fraction to the Equivalent Percent

- Multiply the fraction by 100 and insert a % sign.

EXAMPLE 16

Rewrite $\frac{7}{20}$ as a percent.

Solution To change the given fraction to a percent, we multiply by 100 and insert a % sign.

$$\frac{7}{20} = \frac{7}{20} \times 100\% = \frac{7}{\cancel{20}_1} \times \frac{\cancel{100}^{5}}{1}\% = 35\%$$

PRACTICE 16

Convert $\frac{4}{25}$ to a percent.

EXAMPLE 17

Which is larger: 130% or $1\frac{3}{8}$?

Solution To compare, let's express $1\frac{3}{8}$ as a percent.

$$1\frac{3}{8} = 1\frac{3}{8} \times 100\% = \frac{11}{8} \times \frac{100}{1}\%$$

$$= \frac{11}{\underset{2}{\cancel{8}}} \times \frac{\overset{25}{\cancel{100}}}{1}\% = \frac{275}{2}\% = 137\frac{1}{2}\%$$

Because $137\frac{1}{2}\%$ is larger than 130%, so is $1\frac{3}{8}$.

PRACTICE 17

True or false: $\frac{2}{3} > 60\%$. Justify your answer.

EXAMPLE 18

A student got 28 of 30 questions correct on a test. If all the questions were equal in value, what was the student's grade, rounded to the nearest whole percent?

Solution The student answered $\frac{28}{30}$ of the questions right. To find the student's grade, we change this fraction to a percent.

$$\frac{28}{30} = \frac{28}{30} \times 100\% = \frac{28}{\underset{3}{\cancel{30}}} \times \frac{\overset{10}{\cancel{100}}}{1}\%$$

$$= \frac{280}{3}\% = 93\frac{1}{3}\% = 93.3\ldots\% \approx 93\%$$

Note that the critical digit is 3, so we round down. The rounded grade was therefore 93%.

PRACTICE 18

An administrative assistant spends \$490 out of her monthly salary of \$1,834 on rent. What percent of her monthly salary is spent on rent, rounded to the nearest whole percent?

Before going any further, study the table of common fraction, decimal, and percent equivalents in the Appendix at the back of this book. These numbers come up frequently and are useful reference points for estimating the answer to percent word problems, as we demonstrate in Section 6.2.

6.1 Exercises

FOR EXTRA HELP

Mathematically Speaking

Fill in each blank with the most appropriate term or phrase from the given list.

right	fraction	percent	divide
decimal	left	whole number	multiply

1. A(n) _______ is a ratio or fraction with denominator 100.

2. To change a percent to the equivalent _______, drop the % sign from the given percent, and place the number over 100.

3. To change a percent to the equivalent decimal, move the decimal point two places to the _______ and drop the % sign.

4. To change a fraction to the equivalent percent, _______ the fraction by 100 and insert a % sign.

Change each percent to a fraction or mixed number. Simplify.

5. 8%

6. 3%

7. 250%

8. 110%

9. 33%

10. 41%

11. 18%

12. 6%

13. 14%

14. 45%

15. 65%

16. 92%

17. $\frac{3}{4}\%$

18. $\frac{1}{10}\%$

19. $\frac{3}{10}\%$

20. $\frac{1}{5}\%$

21. $7\frac{1}{2}\%$

22. $2\frac{1}{2}\%$

23. $14\frac{2}{7}\%$

24. $28\frac{4}{7}\%$

Convert each percent to a decimal.

25. 6%

26. 9%

27. 72%

28. 25%

29. 0.1%

30. 0.2%

31. 102%

32. 113%

33. 42.5%

34. 10.5%

35. 500%

36. 400%

37. $106\frac{9}{10}\%$

38. $201\frac{1}{10}\%$

39. $3\frac{1}{2}\%$

40. $2\frac{4}{5}\%$

41. $\frac{9}{10}\%$

42. $\frac{7}{10}\%$

43. $\frac{3}{4}\%$

44. $\frac{1}{4}\%$

Express each decimal as a percent.

45. 0.31

46. 0.05

47. 0.17

48. 0.18

49. 0.3

50. 0.4

51. 0.04

52. 0.875

53. 0.125

54. 0.27

55. 1.29

56. 1.07

57. 2.9

58. 3.5

59. 2.87

60. 12.91

61. 1.016

62. 1.003

63. 9

64. 7

Change each fraction to a percent.

65. $\frac{3}{10}$ **66.** $\frac{1}{2}$ **67.** $\frac{1}{10}$ **68.** $\frac{3}{20}$

69. $\frac{4}{25}$ **70.** $\frac{6}{25}$ **71.** $\frac{9}{10}$ **72.** $\frac{7}{10}$

73. $\frac{3}{50}$ **74.** $\frac{1}{50}$ **75.** $\frac{5}{9}$ **76.** $\frac{2}{9}$

77. $\frac{1}{9}$ **78.** $\frac{4}{7}$ **79.** 6 **80.** 8

81. $1\frac{1}{2}$ **82.** $2\frac{3}{5}$ **83.** $2\frac{1}{6}$ **84.** $1\frac{1}{3}$

Replace ▢ with < or >.

85. $2\frac{1}{4}$ ▢ 240% **86.** $\frac{5}{6}$ ▢ 80% **87.** $\frac{1}{2}\%$ ▢ 50% **88.** $\frac{1}{40}$ ▢ $\frac{1}{4}\%$

Express as a percent, rounded to the nearest whole percent.

89. $\frac{4}{9}$ **90.** $\frac{3}{7}$ **91.** 2.2469 **92.** 1.1633

Complete each table.

93.

Fraction	Decimal	Percent
		$33\frac{1}{3}\%$
	0.666 ...	
	0.25	
		75%
		20%
$\frac{2}{5}$		
	0.6	

94.

Fraction	Decimal	Percent
	0.8	
$\frac{1}{6}$		
$\frac{5}{6}$		
		12.5%
	0.375	
		$62\frac{1}{2}\%$
	0.875	

Mixed Practice

Solve.

95. Change 104% to a mixed number.

96. What percent is equivalent to $\frac{2}{5}$?

97. Express $3\frac{1}{6}$ as a percent.

98. Express $62\frac{1}{2}\%$ as a fraction.

99. Convert 27.5% to a decimal.

100. Find the decimal equivalent to $\frac{3}{8}\%$

101. What percent is equivalent to 3.1?

102. Change 0.003 to a percent.

103. Which is smaller, $2\frac{5}{9}$ or 254%?

104. Express 1.2753 to the nearest whole percent.

Applications

Solve.

105. It is estimated that by the year 2010, 79% of all e-mail sent worldwide each day will be spam e-mail. Express this percent as a decimal. (***Source:*** Radicati Group)

106. According to a recent study, 65% of children have had an imaginary companion by age 7. Express this percent as a fraction. (***Source:*** University of Washington, http://uwnews.org)

107. The following graph shows the percent of people in various age groups who say that they get at least some of their news from cell phones, PDAs, or podcasts.

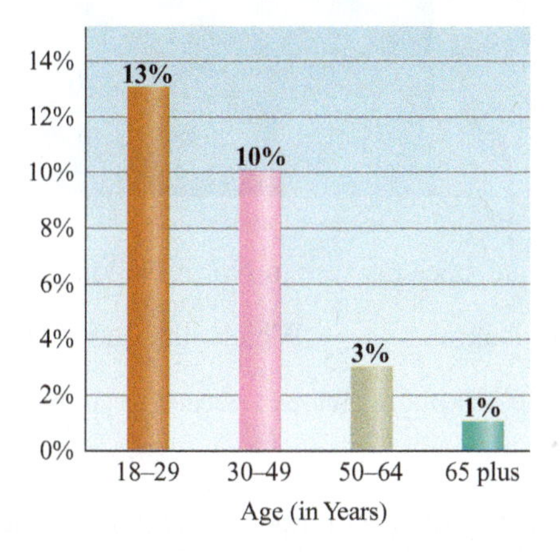

What fraction of people in the 30–49 age group do *not* get their news in this way? (***Source:*** *USA Today*, August 14, 2006)

108. The following graph shows the distribution of investments for a retiree. Express as a decimal the percent of investments that are in equities.

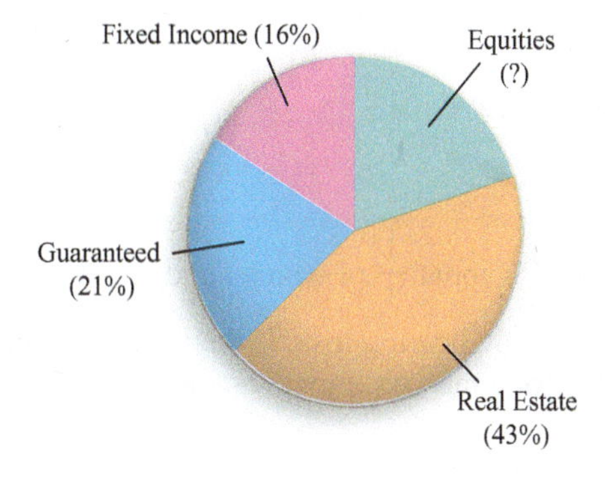

109. According to the nutrition label, one large egg contains 6 grams of protein. This is 10% of the daily value (DV) for protein. Express this percent as a fraction.

110. A bank offers a Visa credit card with a fixed annual percentage rate (APR) of 16.99%. Express the APR as a decimal.

111. New York City has the largest population of any U.S. city. But, Juneau, Alaska, has the greatest area—9.0 times that of New York City. Write this decimal as a percent. (***Source:*** U.S. Bureau of the Census)

112. A medical school accepted $\frac{2}{5}$ of its applicants. What percent of the applicants did the school accept?

113. When the recession ended, the factory's output grew by 135%. Write this percent as a simplified mixed number.

114. According to a survey, 78% of the arguments that couples have are about money. Express this percent as a decimal.

115. In Nevada, the federal government controls about 84.5% of the land. Convert this percent to a decimal. (*Source:* Republican Study Committee)

116. After an oil spill, 15% of the wildlife survived. Express this percent as a fraction.

117. By age 75, about $\frac{1}{3}$ of women and about 40% of men have chronic hearing loss. Is this condition more common among men or among women? Explain.

118. The state sales tax rate in North Dakota is 5%, and in South Dakota it is $\frac{1}{25}$. Which state has a lower sales tax rate? Explain. (*Source:* Federation of Tax Administrators, 2006)

119. A quality control inspector found 2 defective machine parts out of 500 manufactured.

a. What percent of the machine parts manufactured were defective?

b. What percent of the machine parts manufactured were not defective?

120. In a survey of several hundred children, 3 out of every 25 children indicated that they wanted to become professional athletes when they grow up.

a. What percent of the children wanted to become professional athletes?

b. What percent of the children did not want to become professional athletes? (*Source: National Geographic Kids*)

Use a calculator to solve the following problems, giving (a) the operation(s) you used, (b) the exact answer, and (c) an estimate of the answer.

121. In the 2004 U.S. presidential election, 122,294,978 voters turned out. At the time, the voting-age population numbered 221,256,931. What percent of the voting-age population turned out? (*Source: Time Almanac 2006*)

122. The first Social Security retirement benefits were paid in 1940 to Ida May Fuller of Vermont. She had paid in a total of $24.85 and got back $20,897 before her death in 1975. Express the ratio of what she got back to what she put in as a percent. (*Source:* James Trager, *The People's Chronology*)

• *Check your answers on page A-10.*

MINDSTRETCHERS

Mathematical Reasoning

1. By mistake, you move the decimal point to the right instead of to the left when changing a percent to a decimal. Your answer is how many times as large as the correct answer?

Writing

2. A study of the salt content of seawater showed that the average salt content varies from 33‰ to 37‰, where the symbol ‰ (read "per mil") means "for every thousand." Explain why you think the scientist who wrote this study did not use the % symbol.

Critical Thinking

3. What percent of the region shown is shaded in?

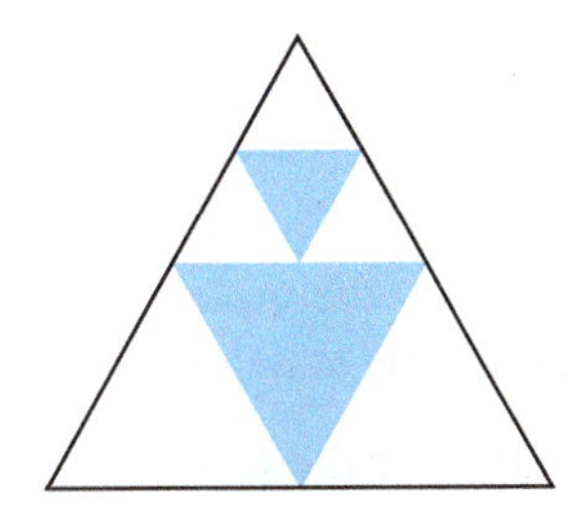

CULTURAL NOTE

Sources: Frank J. Swetz, *Capitalism and Arithmetic: The New Math of the 15th Century,* Open Court, 1987; Carolyn Webber and Aaron Wildavsky, *A History of Taxation and Expenditure in the Western World,* Simon and Schuster, 1986.

Throughout history, the concepts of percent and taxation have been interrelated. At the peak of the Roman Empire, the Emperor Augustus instituted an inheritance tax of 5% to provide retirement funds for the military. Another emperor, Julius Caesar, imposed a 1% sales tax on the population. And in Roman Asia, tax collectors exacted a tithe of 10% on crops. If landowners could not pay, the collectors offered to loan them funds at interest rates that ranged from 12% up to 48%.

Roman taxation served as a model for modern countries when these countries developed their own systems of taxation many centuries later.

6.2 Solving Percent Problems

The Three Basic Types of Percent Problems

OBJECTIVES

- To identify the amount, the base, and the percent in a percent problem
- To find the amount, the base, or the percent in a percent problem
- To estimate the amount in a percent problem
- To solve word problems involving a percent

Frequently, we think of a percent not in isolation but rather in connection with another number. In other words, we take *a percent of a number*.

Consider, for example, the problem of taking 50% of 8. This problem is equivalent to finding $\frac{1}{2}$ of 8, which gives us 4.

Note that this percent problem, like all others, involves three numbers.

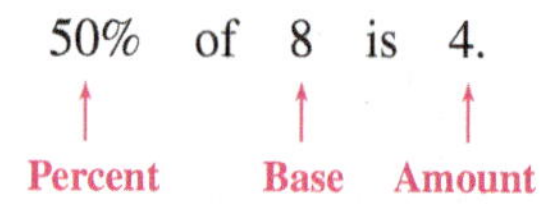

- The 50% is called the **percent** (or the **rate**). The percent always contains the % sign.
- The 8 is called the **base**. The base of a percent—the number that we are taking the percent of—always follows the word *of* in the statement of the problem.
- The remaining number 4 is called the **amount** (or the **part**).

Percent problems involve finding one of the three numbers. For example, if we omit the 4 in "50% of 8 is 4," we ask the question: What is 50% of 8? Omitting the 8, we ask: 50% of what number is 4? And omitting the 50%, we ask: What percent of 8 is 4?

There are several ways to solve these three basic percent questions. In this section, we discuss two ways—the translation method and the proportion method.

The Translation Method

In the translation method, a percent problem has the form

The percent of the base is the amount.

The percent problem gives only two of the three quantities. To find the missing quantity using the translation method, we translate to a simple equation that we then solve.

This method depends on translating the words in the given problem to the appropriate mathematical symbols.

Word(s)	Math Symbol
What, what number, what percent	x (or some other letter)
is	$=$
of	$\times$ or $\cdot$
percent, %	Percent value expressed as a decimal or fraction

Let's translate several percent problems to equations.

EXAMPLE 1

Translate each question to an equation, using the translation method.

a. What is 10% of 2? **b.** 20% of what number is 5?

c. What percent of 8 is 4?

Solution

a. In this problem, we are looking for the amount.

What is 10% of 2?
$$\downarrow \quad \downarrow \quad \downarrow \quad \downarrow \quad \downarrow$$
$$x \quad = \quad 0.1 \quad \cdot \quad 2$$

b. Here, we are looking for the base, that is, the number after the word *of*.

20% of what number is 5?
$$\downarrow \quad \downarrow \quad \downarrow \quad \downarrow \quad \downarrow$$
$$0.2 \quad \cdot \quad x \quad = \quad 5$$

c. This problem asks "what percent?" So we are looking for the percent.

What percent of 8 is 4?
$$\downarrow \quad \downarrow \quad \downarrow \quad \downarrow \quad \downarrow$$
$$x \quad \cdot \quad 8 \quad = \quad 4$$

PRACTICE 1

Use the translation method to set up an equation.

a. What is 70% of 80?

b. 50% of what number is 10?

c. What percent of 40 is 20?

Finding an Amount

Now let's apply the translation method to solve the type of percent problem in which we are given both the percent and the base and are looking for the amount.

EXAMPLE 2

What is 25% of 8?

Solution First, we translate the question to an equation.

What is 25% of 8?
$$\downarrow \quad \downarrow \quad \downarrow \quad \downarrow \quad \downarrow$$
$$x \quad = \quad \frac{1}{4} \quad \cdot \quad 8$$

Then we solve this equation:

$$x = \frac{1}{4} \cdot 8 = \frac{1}{\cancel{4}_1} \cdot \frac{\cancel{8}^2}{1} = \frac{2}{1}$$
$$= 2$$

So 2 is 25% of 8. Would we have gotten the same answer if we had translated 25% to 0.25?

PRACTICE 2

What is 20% of 40?

EXAMPLE 3

Find 200% of 30.

Solution We can reword the problem as a question.

What is 200% of 30?
$$\downarrow \quad \downarrow \quad \downarrow \quad \downarrow \quad \downarrow$$
$$x \quad = \quad 2 \quad \cdot \quad 30$$

Solving this equation, we get $x = 60$. So 200% of 30 is 60.

PRACTICE 3

150% of 8 is what number?

Tip When the percent is less than 100%, the amount is *less* than the base. When the percent is more than 100%, the amount is *more* than the base.

In a percent problem, we frequently want to *estimate* the amount. Sometimes an approximate answer is good enough, and other times we use the estimate to check an exact answer we have already computed.

EXAMPLE 4

Approximately how much is 67% of 14.8?

Solution Here is one way to estimate the answer. We note that 67% is close to $66\frac{2}{3}\%$, which is equivalent to the fraction $\frac{2}{3}$. Also, we see that 14.8 rounds to 15. So the answer to the given question is close to the answer to the following question.

$$\text{What is } \frac{2}{3} \text{ of } 15?$$

$$\downarrow \quad \downarrow \quad \downarrow \quad \downarrow \quad \downarrow$$

$$x = \frac{2}{3} \cdot 15$$

We multiply mentally. $x = \frac{2}{\cancel{3}_1} \cdot \frac{\cancel{15}^5}{1} = \frac{10}{1} = 10$

So 67% of 14.8 is approximately 10. (The exact answer is 9.916, which is reasonably close to our estimate.)

PRACTICE 4

Estimate 49.3% of 401.6.

EXAMPLE 5

A marketing account manager has $3\frac{1}{2}\%$ of her monthly salary put into a 401(k) plan. How much did she put into the 401(k) plan if her monthly salary is $3,200?

Solution We are looking for the monthly amount placed into the 401(k) plan, which is $3\frac{1}{2}\%$ of $3,200.

$$\text{What is } 3\frac{1}{2}\% \text{ of } \$3{,}200?$$

$$\downarrow \quad \downarrow \quad \downarrow \quad \downarrow \quad \downarrow$$

$$x = (0.035) \cdot (3{,}200)$$

$$= 112$$

So she has $112 per month put into the 401(k) plan. Note that this amount has the same unit (dollars) as the base.

PRACTICE 5

Of the 600 workers at a factory, $8\frac{1}{2}\%$ belong to a union. How many workers are in the union?

Finding a Base

Now let's consider some examples of using the translation method to find the base when we know the percent and the amount.

EXAMPLE 6

4% of what number is 8?

Solution We begin by writing the appropriate equation.

$$\begin{array}{ccccc} 4\% & \text{of} & \text{what number} & \text{is} & 8? \\ \downarrow & \downarrow & \downarrow & \downarrow & \downarrow \\ 0.04 & \cdot & x & = & 8 \end{array}$$

Next, we solve this equation.

$$0.04x = 8$$

$$\frac{0.04}{0.04}x = \frac{8}{0.04} \quad \text{Divide each side by 0.04.}$$

$$x = \frac{8}{0.04} = 200 \quad 0.04\overline{)8.00} = 4\overline{)800.}\ \ 200.$$

So 4% of 200 is 8.

PRACTICE 6

6 is 12% of what number?

EXAMPLE 7

108 is 120% of what number?

Solution We consider the following question:

$$\begin{array}{ccccc} 120\% & \text{of} & \text{what number} & \text{is} & 108? \\ \downarrow & \downarrow & \downarrow & \downarrow & \downarrow \\ 1.2 & \cdot & x & = & 108 \end{array}$$

Solving, we get:

$$1.2x = 108$$

$$\frac{1.2}{1.2}x = \frac{108}{1.2}$$

$$x = 90$$

So 120% of 90 is 108.

PRACTICE 7

250% of what number is 18?

EXAMPLE 8

A college awarded financial aid to 3,843 students, which was 45% of the total number of students enrolled at the college. What was the student enrollment at the college?

Solution We must answer the following question:

$$\begin{array}{ccccc} 45\% & \text{of} & \text{what number} & \text{is} & 3{,}843? \\ \downarrow & \downarrow & \downarrow & \downarrow & \downarrow \\ 0.45 & \cdot & x & = & 3{,}843 \end{array}$$

Next we solve the equation.

$$0.45x = 3{,}843$$

$$\frac{0.45x}{0.45} = \frac{3{,}843}{0.45}$$

$$x = 8{,}540$$

So 8,540 students were enrolled at the college.

PRACTICE 8

There was a glut of office space in a city, with 400,000 square feet, or 16% of the total office space, vacant. How much office space did the city have?

Finding a Percent

Finally, let's look at the third type of percent problem, where we are given the base and the amount and are looking for the percent.

EXAMPLE 9

What percent of 80 is 60?

Solution We begin by writing the appropriate equation.

What percent of 80 is 60?

$$x \cdot 80 = 60$$

$$80x = 60 \quad \text{Write the equation in standard form.}$$

$$\frac{80}{80}x = \frac{60}{80} \quad \text{Divide each side by 80.}$$

$$x = \frac{\overset{3}{\cancel{60}}}{\underset{4}{\cancel{80}}} = \frac{3}{4} \quad \text{Simplify.}$$

Since we are looking for a percent, we change $\frac{3}{4}$ to a percent. So 75% of 80 is 60.

$$x = \frac{3}{4} = \frac{3}{\underset{1}{\cancel{4}}} \cdot \frac{\overset{25}{\cancel{100}}}{1}\% = 75\%$$

PRACTICE 9

What percent of 6 is 5?

EXAMPLE 10

What percent of 60 is 80?

Solution We begin by writing the appropriate equation, as shown to the right.

What percent of 60 is 80?

$$x \cdot 60 = 80$$

$$60x = 80$$

$$\frac{60}{60}x = \frac{80}{60}$$

$$x = \frac{\overset{4}{\cancel{80}}}{\underset{3}{\cancel{60}}} = \frac{4}{3}$$

Finally, we want to change $\frac{4}{3}$ to a percent.

$$x = \frac{4}{3} = \frac{4}{3} \cdot \frac{100}{1}\% = \frac{400}{3}\% = 133\frac{1}{3}\%$$

So $133\frac{1}{3}\%$ of 60 is 80.

PRACTICE 10

What percent of 8 is 9?

EXAMPLE 11

A young couple buys a house for $125,000, making a down payment of $25,000 and paying the difference over time with a mortgage. What percent of the cost of the house was the down payment?

Solution We write the question as shown to the right.

What percent of $125,000 is $25,000?

$$x \cdot 125{,}000 = 25{,}000$$

$$125{,}000x = 25{,}000$$

$$\frac{\cancel{125{,}000}}{\cancel{125{,}000}}x = \frac{25{,}\cancel{000}}{125{,}\cancel{000}}$$

$$x = \frac{25}{125} = \frac{1}{5}$$

Next, we change $\frac{1}{5}$ to a percent. $x = \frac{1}{5} = \frac{1}{\cancel{5}_1} \cdot \frac{\cancel{100}^{20}}{1}\% = 20\%$

So the down payment was 20% of the total cost of the house.

PRACTICE 11

Of the 400 acres on a farm, 120 were used to grow corn. What percent of the total acreage was used to grow corn?

The Proportion Method

So far, we have used the translation method to solve percent problems. Now let's consider an alternative approach, the proportion method.

Using the proportion method, we view a percent relationship in the following way.

$$\frac{\textbf{Amount}}{\textbf{Base}} = \frac{\textbf{Percent}}{\textbf{100}}$$

If we are given two of the three quantities, we set up this proportion and then solve it to find the third quantity.

EXAMPLE 12

What is 60% of 35?

Solution The base (the number after the word *of*) is 35. The percent (the number followed by the % sign) is 60. The amount is unknown. We set up the proportion, substitute into it, and solve.

$$\frac{\text{Amount}}{\text{Base}} = \frac{\text{Percent}}{100}$$

$$\frac{x}{35} = \frac{60}{100}$$

$$100x = 60 \cdot 35 \quad \text{Set cross products equal.}$$

$$\frac{\cancel{100}}{\cancel{100}}x = \frac{2{,}100}{100} \quad \text{Divide each side by 100.}$$

$$x = 21$$

So 60% of 35 is 21.

PRACTICE 12

Find 108% of 250.

EXAMPLE 13

15% of what number is 21?

Solution Here, the number after the word *of* is missing, so we are looking for the base. The amount is 21, and the percent is 15. We set up the proportion, substitute into it, and solve.

$$\frac{\text{Amount}}{\text{Base}} = \frac{\text{Percent}}{100}$$

$$\frac{21}{x} = \frac{15}{100}$$

$$15x = 2{,}100 \quad \text{Set cross products equal.}$$

$$\frac{\not{15}}{\not{15}}x = \frac{2{,}100}{15} \quad \text{Divide each side by 15.}$$

$$x = 140$$

So 15% of 140 is 21.

PRACTICE 13

2% of what number is 21.6?

EXAMPLE 14

What percent of \$45 is \$30?

Solution We know that the base is 45 and that the amount is 30 and are looking for the percent.

$$\frac{30}{45} = \frac{x}{100}$$

$$45x = 3{,}000$$

$$\frac{\not{45}}{\not{45}}x = \frac{3{,}000}{45}$$

$$x = 66\frac{2}{3}$$

So we conclude that $66\frac{2}{3}$% of \$45 is \$30.

PRACTICE 14

What percent of 63 is 21?

EXAMPLE 15

A car depreciated, that is, dropped in value, by 20% during its first year. By how much did the value of the car decline if it cost \$30,500 new?

Solution The question here is: What is 20% of \$30,500? So the percent is 20, the base is \$30,500 and we are looking for the amount. We set up the proportion and solve.

$$\frac{x}{30{,}500} = \frac{20}{100}$$

$$100x = 610{,}000$$

$$\frac{\not{100}}{\not{100}}x = \frac{610{,}000}{100}$$

$$x = 6{,}100$$

So the value of the car depreciated by \$6,100.

PRACTICE 15

A credit card company requires a minimum payment of 4% of the balance. What is the minimum payment if the credit card balance is \$2,450?

EXAMPLE 16

Each day, an adult takes tablets containing 24 milligrams of zinc. If this amount is 160% of the recommended daily allowance, how many milligrams are recommended? (*Source: Podiatry Today*)

Solution Here, we are looking for the base. The question is: 160% of what amount is 24 milligrams? We set up the proportion and solve.

$$\frac{24}{x} = \frac{160}{100}$$
$$160x = 2{,}400$$
$$\frac{160}{160}x = \frac{2{,}400}{160}$$
$$x = 15$$

Therefore, the recommended daily allowance of zinc is 15 milligrams. Note that this base is less than the amount (24 milligrams). Why must that be true?

PRACTICE 16

A Nobel Prize winner had to pay the Internal Revenue Service $129,200—or 38% of his prize—in taxes. How much was his Nobel Prize worth?

EXAMPLE 17

A college accepted 1,620 of the 4,500 applicants for admission. What was the acceptance rate, expressed as a percent?

Solution The question is: What percent of 4,500 is 1,620?

$$\frac{1{,}620}{4{,}500} = \frac{x}{100}$$
$$4{,}500x = 162{,}000$$
$$\frac{4{,}500}{4{,}500}x = \frac{162{,}000}{4{,}500}$$
$$x = 36$$

So the college's acceptance rate was 36%.

PRACTICE 17

A bookkeeper's annual salary was raised from $38,000 to $39,900. What percent of her original annual salary is her new annual salary?

Percents on a Calculator

Most calculators have a percent key (%), sometimes used with the 2nd function (2nd). However, the percent key functions differently on different models. Check to see if the following approach works on your machine. If it does not, experiment to find an approach that does.

EXAMPLE 18

Use a calculator to find 50% of 8.

Solution

Press	Display
50 [2nd] [%] [×] 8 [ENTER]	50% * 8 4.

PRACTICE 18

What is 8.25% of $72.37, to the nearest cent?

6.2 Exercises

FOR EXTRA HELP   PRACTICE  WATCH  DOWNLOAD  READ  REVIEW

Mathematically Speaking

Fill in each blank with the most appropriate term or phrase from the given list.

amount	of	base
is	what	percent

1. The _______ is the number that we are taking the percent of.

2. The _______ is the result of taking the percent of the base.

3. The _______ of the base is the amount.

4. In the translation method of solving a percent problem, _______ is replaced by a multiplication symbol.

Find the amount. Check by estimating.

5. What is 75% of 8?

6. Find 50% of 48.

7. Compute 100% of 23.

8. What is 200% of 6?

9. Find 41% of 7.

10. Calculate 6% of 9.

11. What is 35% of $400?

12. 40% of 10 miles is what?

13. What is 3.1% of 20?

14. Find 0.5% of 7.

15. Compute $\frac{1}{2}$% of 20.

16. $\frac{1}{10}$% of 35 is what number?

17. What is $12\frac{1}{2}$% of 32?

18. Compute $66\frac{2}{3}$% of 33.

19. What is $7\frac{1}{8}$% of $257.13, rounded to the nearest cent?

20. Calculate 8.9% of 7,325 miles, rounded to the nearest mile.

Find the base.

21. 25% of what number is 8?

22. 30% of what number is 120?

23. $12 is 10% of how much money?

24. 1% of what salary is $195?

25. 5 is 200% of what number?

26. 70% of what amount is 14?

27. 2% of what amount of money is $5?

28. 8 meters is 20% of what length?

29. 15 is $33\frac{1}{3}$% of what number?

30. $8\frac{1}{2}$% of what number is 85?

31. 3.5 is 200% of what number?

32. 150% of what number is 8.1?

33. 0.5% of what number is 23?

34. 0.75% of what is 24?

35. 6.5% of how much money is \$3,200, rounded to the nearest cent?

36. 4,718 is $2\frac{1}{8}$% of what number?

Find the percent.

37. 50 is what percent of 100?

38. What percent of 13 is 13?

39. What percent of 8 is 6?

40. What percent of 50 is 20?

41. What percent of 12 is 10?

42. 5 is what percent of 15?

43. 2 miles is what percent of 8 miles?

44. \$16 is what percent of \$20?

45. \$30 is what percent of \$20?

46. 10 is what percent of 8?

47. 9 feet is what percent of 8 feet?

48. 35¢ is what percent of 21¢?

49. 2.5 is what percent of 4?

50. 0.1 is what percent of 8?

51. What percent of 251,749 is 76,801, rounded to the nearest percent?

52. 8,422 is what percent of 11,630, to the nearest percent?

Mixed Practice

Solve.

53. Compute $37\frac{1}{2}$% of 160

54. Calculate 0.01% of 55, rounded to the nearest hundredth.

55. What percent of 15 is 10?

56. What percent of 20 is 30?

57. 20% of what length is 35 miles?

58. 2.5% of what is 32?

59. 3 feet is what percent of 60 feet?

60. Find 7.2% of \$300.

61. 4% of what amount of money is \$20?

62. \$24 is what percent of \$300?

63. What is 40% of 25?

64. $\frac{3}{4}$% of what number is 60?

Applications

Solve.

65. During a tournament, a golfer made par on 12 of 18 holes. On what percent of the holes on the course did she make par?

66. In a dormitory, 40% of the rooms are especially equipped for disabled students. How many rooms are so equipped if the dorm has 80 rooms?

67. Flexible-fuel vehicles run on E85, an alternative fuel that is a blend of ethanol and gasoline containing 85% ethanol. How much ethanol is in 12 gallons of E85?

68. A property management company sold 80% of the condominium units in a new building with 90 units. How many units were sold?

69. Payroll deductions comprise 40% of the gross income of a student working part-time. If his deductions total $240, what is his gross income?

70. In 1862, the U.S. Congress enacted the nation's first income tax, at the rate of 3%. How much in income tax would you have paid if you made $2,500? (***Source:*** U.S. Bureau of the Census)

71. A 224-pound man joined a weight-loss program and lost 56 pounds. What percent of his original weight did he lose?

72. According to the report on a country's economic conditions, 1.5 million people, or 8% of the workforce, were unemployed. How large was the workforce?

73. In a restaurant, 60% of the tables are in the no-smoking section. If the restaurant has 90 tables, how many tables are in the no-smoking section?

74. Investors buy a studio apartment for $150,000. Of this amount, they are able to put down $30,000. The down payment is what percent of the purchase price?

75. A lab technician mixed 36 milliliters of alcohol with 84 milliliters of water to make a solution. What percent of the solution was alcohol?

76. A shopper lives in a town where the sales tax is 5%. Across the river, the tax is 4%. If it costs her $6 to make the round trip across the river, should she cross the river to buy a $250 television set?

77. A student answered 90% of the questions on a math exam correctly. If she answered 36 questions correctly, how many questions were on the exam?

78. In the first quarter of last year, a steel mill produced 300 tons of steel. If this was 20% of the year's output, find that output.

79. The following graph shows the breakdown of the projected U.S population by gender in the year 2020. If the population is expected to be 340 million people, how many more women than men will there be in 2020? (***Source:*** U.S. Bureau of the Census)

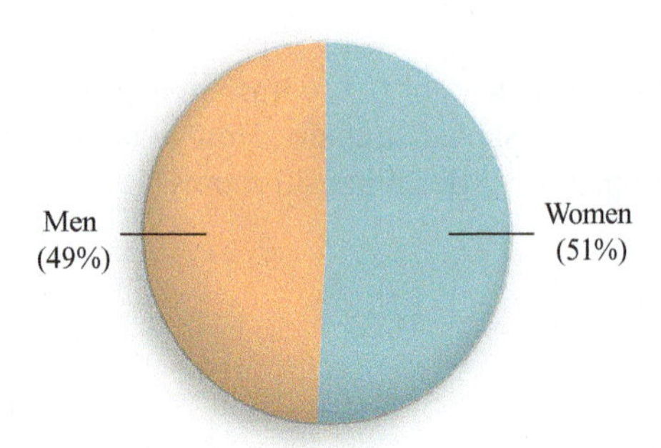

80. The percent of various kinds of degrees projected to be conferred in the year 2010 is shown in the following graph. If a total of 2,860,000 degrees are conferred, how many more Bachelor's degrees than Associate's degrees will be conferred in that year? (***Source:*** National Center for Educational Statistics)

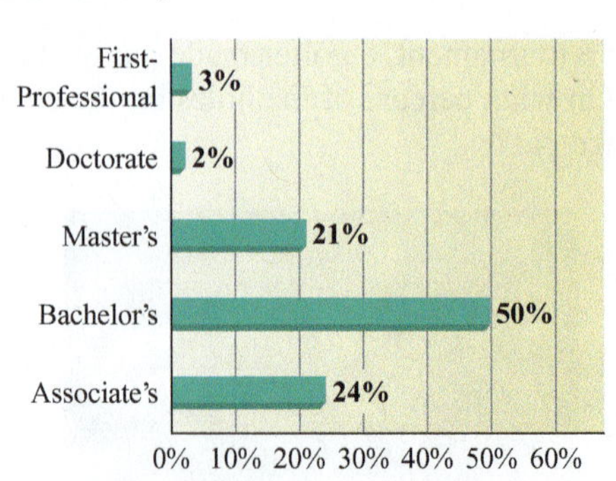

81. According to the latest census, the state of Oregon has an urban population of about 3 million people and a rural population of approximately 1 million people.

a. What is the combined population of Oregon?

b. What percent of the combined population is the urban population?
(***Source:*** U.S. 2006 Census)

82. A homeowner builds a family room addition on his 1,650-square-foot house, increasing the area of the house by 495 square feet.

a. Calculate the total area of the house with the addition.

b. What percent of the original area is the total area?

83. A company's profits amounted to 10% of its sales. If the profits were \$3 million, compute the company's sales.

84. The most commonly studied language in American colleges is Spanish, with some 750,000 enrollments. By contrast, the number of enrollments in French is approximately 200,000. The number of French enrollments is what percent of the number of Spanish enrollments, to the nearest 10%? (***Source:*** *Time Almanac 2006*)

85. In a company, 85% of the employees are female. If 765 males work for the company, what is the total number of employees?

86. A quarterback completed 15 passes or 20% of his attempted passes. How many of his attempted passes did he *not* complete?

87. A math lab coordinator is willing to spend up to 25% of her income on housing. What is the most she can spend if her annual income is \$36,000?

88. An office supply warehouse shipped 648 cases of copy paper. If this represents 72% of the total inventory, how many cases of paper did the warehouse have in its inventory?

• *Check your answers on page A-10.*

MINDSTRETCHERS

Writing

1. Do you prefer solving percent problems using the translation method or the proportion method? In a few sentences, explain why.

Critical Thinking

2. At a college, 20% of the women commute, in contrast to 30% of the men. Yet more women than men commute. Explain how this result is possible.

Technology

3. On the Web, go to the U.S. Bureau of the Census home page (http://www.census.gov). Pose a percent problem of interest to you involving data from the site, and solve the problem.

6.3 More on Percents

OBJECTIVES

- To solve percent increase and decrease problems
- To solve percent problems involving taxes, commissions, markups, and discounts
- To solve simple and compound interest problems
- To solve word problems of these types

Finding a Percent Increase or Decrease

Next, let's consider a type of "what percent" problem that deals with a *changing quantity*. If the quantity is increasing, we speak of a *percent increase*; if it is decreasing, of a *percent decrease*.

Here is an example: Last year, a family paid \$2,000 in health insurance, and this year, their health insurance bill was \$2,500. By what percent did this expense increase?

Note that this problem states the value of a quantity at two points in time. We are asked to find the percent increase between these two values.

To solve, we first compute the difference between the values.

$$2{,}500 - 2{,}000 = 500$$

Change in value

The question posed is expressed as follows:

$$\begin{array}{ccccc} \text{What percent} & \text{of} & 2{,}000 & \text{is} & 500? \\ \downarrow & \downarrow & \downarrow & \downarrow & \downarrow \\ x & \cdot & 2{,}000 & = & 500 \end{array}$$

It is important to note that the *base* here—as in all percent change problems—is the original value of the quantity.

Next, we solve the equation.

$$2{,}000x = 500$$

$$\frac{\cancel{2{,}000}}{\cancel{2{,}000}}x = \frac{500}{2{,}000}$$

$$x = \frac{1}{4} = 0.25, \text{ or } 25\%$$

So we conclude that the family's health insurance expense *increased* by 25%.

To Find a Percent Increase or Decrease

- Compute the difference between the two given values.
- Compute what percent this difference is of the *original value*.

EXAMPLE 1

The cost of a marriage license had been \$10. Later it rose to \$15. What percent increase was this?

PRACTICE 1

To accommodate a flood of tourists, businesses in town boosted the number of hotel beds from 25 to 100. What percent increase is this?

Solution The earlier value of the cost of the license was \$10, and the later value was \$15. The *change in value* is, therefore, \$15 − \$10, or \$5. So the question is as follows:

$$\begin{array}{ccccc} \text{What percent} & \text{of} & \$10 & \text{is} & \$5? \\ \downarrow & \downarrow & \downarrow & \downarrow & \downarrow \\ x & \cdot & 10 & = & 5 \end{array}$$

$$10x = 5$$

$$\frac{\cancel{10}}{\cancel{10}}x = \frac{\overset{1}{\cancel{5}}}{\underset{2}{\cancel{10}}}$$

$$x = \frac{1}{2}$$

Next, we change $\frac{1}{2}$ to a percent. $\frac{1}{2} = 0.5$, or 50%

So the cost of the license increased by 50%.

EXAMPLE 2

Suppose that an animal species is considered to be endangered if its population drops by more than 60%. If a species' population fell from 40 to 18, should we consider the animal endangered?

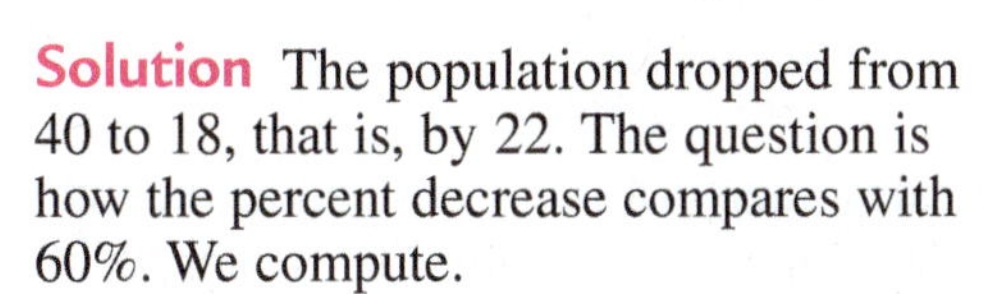

Solution The population dropped from 40 to 18, that is, by 22. The question is how the percent decrease compares with 60%. We compute.

$$\begin{array}{ccccc} \text{What percent} & \text{of} & 40 & \text{is} & 22? \\ \downarrow & \downarrow & \downarrow & \downarrow & \downarrow \\ x & \cdot & 40 & = & 22 \end{array}$$

$$40x = 22$$

$$x = \frac{22}{40}, \text{ or } \frac{11}{20}$$

We convert this fraction to a percent. $\frac{11}{20} = 0.55$, or 55%

Since the population decreased by less than 60%, the species is not considered to be endangered.

PRACTICE 2

Major financial crashes took place on both Tuesday, October 29, 1929, and Monday, October 19, 1987. On the earlier date, the stock index dropped from 300 to 230. On the latter date, it dropped from 2,250 to 1,750. As a percent, did the stock index drop more in 1929 or in 1987? (***Source:*** *The Wall Street Journal*)

Business Applications of Percent

The idea of percent is fundamental to business and finance. Percent applications are part of our lives whenever we buy or sell merchandise, pay taxes, and borrow or invest money.

Taxes

Governments levy taxes to pay for a variety of services, from supporting schools to paving roads. There are many kinds of taxes, including sales, income, property, and import taxes.

In general, the amount of a tax that we pay is a percent of a related value. For instance, sales tax is usually computed as a percent of the price of merchandise sold. Thus, in a town where the sales tax rate is 7%, we could compute the tax on any item sold by computing 7% of the price of that item.

Similarly, property tax is commonly computed by taking a given percent (the tax rate) of the property's assessed value. And an import tax is calculated by taking a specified percent of the market value of the imported item.

EXAMPLE 3

The sales tax on a \$950 digital camcorder is \$71.25. What is the sales tax rate, expressed as a percent?

Solution We must consider the following question:

71.25 is what percent of 950?

$$71.25 = x \cdot 950$$

$$950x = 71.25$$

$$\frac{950x}{950} = \frac{71.25}{950}$$

$$x = 0.075, \text{ or } 7.5\%$$

So the rate of the sales tax is 7.5%, or $7\frac{1}{2}\%$.

PRACTICE 3

When registering a new car, the owner paid a 3% excise tax on the purchase price of \$18,500. How much excise tax did he pay?

Commission

To encourage salespeople to make more sales, many of them, instead of receiving a fixed salary, are paid on **commission**. Working on commission means that the amount of money that they earn is a specified percent—say, 10%—of the total sales for which they are responsible.

On some jobs, salespeople make a flat fee in addition to a commission based on sales. On other jobs, a salesperson may earn a higher rate of commission on sales over an amount previously agreed upon, as an extra incentive.

EXAMPLE 4

A real estate agent sold a condo in San Diego for \$222,000. On this amount, she received a commission of 6%.

a. Find the amount of the commission.

b. How much money will the seller make from the sale after paying the agent's fee?

Solution

a. The commission is 6% of \$222,000.

What is 6% of \$222,000?

$$x = 0.06 \cdot 222{,}000 = 13{,}320$$

So the commission amounted to \$13,320.

b. The seller made \$222,000 – \$13,320, or \$208,680.

PRACTICE 4

A sales associate at a furniture store is paid a base monthly salary of \$1,500. In addition, she earns a 9% commission on her monthly sales. If her total sales this month is \$12,500, calculate

a. her commission, and

b. her total monthly income.

Discount

In buying or selling merchandise, the term **discount** refers to a reduction on the merchandise's original price. The rate of discount is usually expressed as a percent of the selling price.

EXAMPLE 5

A drugstore gives senior citizens a 10% discount. If some pills normally sell for $16 a bottle, how much will a senior citizen pay?

Solution Note that, because senior citizens get a discount of 10%, they pay 100% − 10%, or 90%, of the normal price.

The question then becomes: What is 90% of $16?

We multiply. $x = 0.9 \cdot 16 = 14.4$

So a senior citizen will pay $14.40 for a bottle of the pills.

Note that another way to solve this problem is first to compute the amount of the discount (10% of $16) and then to subtract this discount from the original price. With this approach, do we get the same answer?

PRACTICE 5

Find the sale price.

Markup

A retail firm must sell goods at a higher price (the selling price) than it pays for the merchandise (the cost) to stay in business. The **markup** on an item is the difference between the selling price and the cost. Often the markup rate on merchandise is expressed as a fixed percent of the selling price.

EXAMPLE 6

An online bookstore sells a best seller for $35 at a markup rate of 55% based on the selling price. How much money is the markup on the best seller?

Solution We write the question shown at the right.

What is 55% of $35?

$x = 0.55 \cdot 35 = 19.25$

So the markup on the best seller is $19.25.

PRACTICE 6

A department store buyer purchases wallets at $480 per dozen and sells them for $80 each. What percent markup, based on the selling price, is the store making?

Simple Interest

Anyone who has been late in paying a credit card bill or who has deposited money in a savings account knows about **interest**. When we loan or deposit money, we make interest. When we borrow money, we pay interest.

Interest depends on the amount of money borrowed (the **principal**), the annual rate of interest (usually expressed as a percent), and the length of time the money is borrowed (usually expressed in years). We can compute the amount of interest by multiplying the principal by the rate of interest and the number of years. This type of interest is called *simple interest* to distinguish it from *compound interest* (which we discuss later).

EXAMPLE 7

How much simple interest is earned in 1 year on a principal of \$900 at an annual interest rate of 6.5%?

Solution To compute the interest, we multiply the principal by the rate of interest and the number of years.

$$\text{Interest} = \underset{\text{Principal}}{900} \times \underset{\text{Rate of Interest}}{0.065} \times \underset{\text{Number of Years}}{1}$$
$$= 58.5$$

So \$58.50 in interest is earned.

PRACTICE 7

What is the simple interest on an investment of \$20,000 for 1 year at an annual interest rate of 7.25%?

EXAMPLE 8

A customer deposited \$825 in a savings account that each year pays 5% in simple interest, which is credited to his account. What is the account balance after 2 years?

Solution To solve this problem, let's break it into two questions:

- How much interest did the customer make after 2 years?
- What is the sum of the original deposit and that interest?

First, let's find the interest. To do this, we multiply the principal by the rate of interest and the number of years.

$$\text{Interest} = \underset{\text{Principal}}{(825)}\ \underset{\text{Rate of Interest}}{(0.05)}\ \underset{\text{Number of Years}}{(2)}$$
$$= 82.50$$

The customer made \$82.50 in interest.

Now, let's find the account balance by adding the amount of the original deposit to the interest made.

$$\text{Account Balance} = \underset{\text{Original Deposit}}{825} + \underset{\text{Interest}}{82.50}$$
$$= 907.50$$

So the account balance after 2 years is \$907.50

PRACTICE 8

A bank account pays 6% simple interest on \$1,600 for 2 years. Compute the account balance after 2 years.

Compound Interest

As we have seen, simple interest is paid on the principal. Most banks, however, pay their customers *compound interest*, which is paid on both the principal and the previous interest generated.

For instance, suppose that a bank customer has \$1,000 deposited in a savings account that pays 5% interest compounded annually. There were no withdrawals or other deposits. Let's compute the balance in the account at the end of the third year.

The following table shows the account balance after the customer has left the money in the account for 3 years. After 1 year, the account will contain \$1,050 (that is, 100% of the original \$1,000 added to 5% of \$1,000, giving us 105% of \$1,000).

Year	Balance at the End of the Year
0	\$1,000
1	\$1,000 + 0.05 × \$1,000 = \$1,050.00
2	\$1,050 + 0.05 × \$1,050 = \$1,102.50
3	\$1,102.50 + 0.05 × \$1,102.50 = \$1,157.63

The balance in the account after the third year is \$1,157.63, rounded to the nearest cent.

In the following table, for each year we multiply the account balance by 1.05 to compute the balance at the end at the next year.

Year	Balance at the End of the Year
0	\$1,000
1	1.05 × \$1,000 = \$1,050.00
2	$(1.05)^2$ × \$1,000 = \$1,102.50
3	$(1.05)^3$ × \$1,000 = \$1,157.63

So the balance at the end of the third year is $(1.05)^3$ × \$1,000, or \$1,157.63, in agreement with our previous computation. What would the balance be at the end of the fourth year? What is the relationship between the number of years the money has been invested and the power of 1.05?

In computing the preceding answer, we needed to raise the number 1.05 to a power. Before scientific calculators became available, compound interest problems were commonly solved by use of a compound interest table that contained information such as the following:

Number of Years	4%	5%	6%	7%
1	1.04000	1.05000	1.06000	1.07000
2	1.08160	1.10250	1.12360	1.14490
3	1.12486	1.15763	1.19102	1.22504

When using such a table to calculate a balance, we simply multiply the principal by the number in the table corresponding to the rate of interest and the number of years for which the principal is invested. For instance, after 3 years a principal of \$1,000 compounded at 5% per year results in a balance of 1.15763 × 1,000, or \$1,157.63, as we previously noted.

Today, problems of this type are generally solved on a calculator.

EXAMPLE 9

A couple deposited $7,000 in a bank account and did not make any withdrawals or deposits in the account for 3 years. The interest is compounded annually at a rate of 3.5%. What will be the amount in their account at the end of this period?

Solution Each year, the amount in the account is 100% + 3.5%, or 1.035 times the previous year's balance. So at the end of 3 years, the number of dollars in the account is calculated as follows:

Principal	First Year	Second Year	Third Year
↓	↓	↓	↓
7,000 ×	1.035 ×	1.035 ×	1.035

It makes sense to use a calculator to carry out this computation. One way to key in this computation on a calculator is as follows.

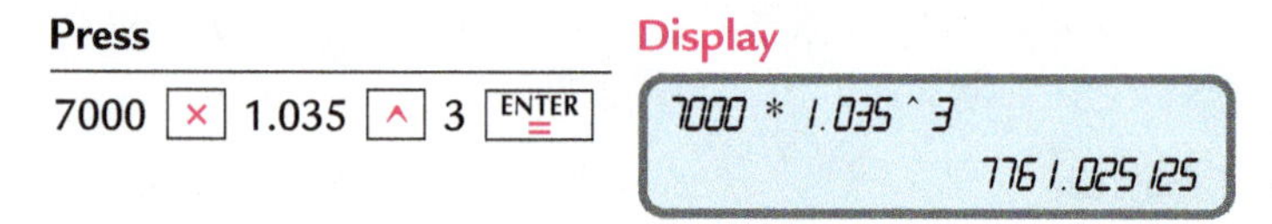

So at the end of 3 years, they have $7,761.03 in the account, rounded to the nearest cent.

PRACTICE 9

Find the balance after 4 years on a principal amount of $2,000 invested at a rate of 6% compounded annually.

6.3 Exercises

FOR EXTRA HELP  

 PRACTICE  WATCH  DOWNLOAD  READ  REVIEW

Mathematically Speaking

Fill in each blank with the most appropriate term or phrase from the given list.

discount	on salary	on commission
markup	simple	compound
final	original	

1. When computing a percent increase or decrease, the _______ value is used as the base of the percent.

2. Sellers who are paid a fixed percent of the sales for which they are responsible are said to work _______.

3. A reduction on the price of merchandise is called a(n) _______.

4. When interest is paid on both the principal and the previous interest generated, it is called _______ interest.

Find the percent increase or decrease.

5.

Original Value	New Value	Percent Increase or Decrease
\$10	\$12	
\$10	\$8	
\$6	\$18	
\$35	\$70	
\$14	\$21	
\$10	\$1	
\$8	\$6.50	
\$6	\$5.25	

6.

Original Value	New Value	Percent Increase or Decrease
\$5	\$6	
\$12	\$10	
\$4	\$9	
\$25	\$45	
\$10	\$36	
\$100	\$20	
4 ft	3 ft	
8 lb	4.5 lb	

Compute the sales tax. Round to the nearest cent.

7.

Selling Price	Rate of Sales Tax	Sales Tax
\$30.00	5%	
\$24.88	3%	
\$51.00	$7\frac{1}{2}\%$	
\$196.23	4.5%	

8.

Selling Price	Rate of Sales Tax	Sales Tax
\$40.00	6%	
\$16.98	4%	
\$85.00	$5\frac{1}{2}\%$	
\$286.38	5%	

Compute the commission. Round to the nearest cent.

9.

Sales	Rate of Commission	Commission
$700	10%	
$450	2%	
$870	$4\frac{1}{2}\%$	
$922	7.5%	

10.

Sales	Rate of Commission	Commission
$400	1%	
$670	3%	
$610	$6\frac{1}{2}\%$	
$2,500	8.25%	

Compute the discount and sale price. Round to the nearest cent.

11.

Original Price	Rate of Discount	Discount	Sale Price
$700.00	25%		
$18.00	10%		
$43.50	20%		
$16.99	5%		

12.

Original Price	Rate of Discount	Discount	Sale Price
$200.00	30%		
$21.00	50%		
$88.88	10%		
$72.50	40%		

Compute the markup and the cost. The rate of markup is based on selling price. Round to the nearest cent.

13.

Selling Price	Rate of Markup	Markup	Cost
$10.00	50%		
$23.00	70%		
$18.40	10%		
$13.55	60%		

14.

Selling Price	Rate of Markup	Markup	Cost
$20.00	40%		
$81.00	25%		
$74.20	30%		
$300.00	8.5%		

Calculate the simple interest and the final balance. Round to the nearest cent.

15.

Principal	Interest Rate	Time (in years)	Interest	Final Balance
$300	4%	2		
$600	7%	2		
$500	8%	2		
$375	10%	4		
$1,000	3.5%	3		
$70,000	6.25%	30		

16.

Principal	Interest Rate	Time (in years)	Interest	Final Balance
$100	6%	5		
$800	4%	5		
$500	3%	10		
$800	6%	10		
$250	1.5%	2		
$300,000	4.25%	20		

Calculate the final balance after compounding the interest. Round to the nearest cent.

17.

Principal	Interest Rate	Time (in years)	Final Balance
$500	4%	2	
$6,200	3%	5	
$300	5%	8	
$20,000	4%	2	
$145	3.8%	3	
$810	2.9%	10	

18.

Principal	Interest Rate	Time (in years)	Final Balance
$300	6%	1	
$2,900	5%	4	
$800	3%	5	
$10,000	3%	4	
$250	4.1%	2	
$200	3.3%	5	

Mixed Practice

Complete each table. Round to the nearest whole percent.

19.

Original Value	New Value	Percent Decrease
$5	$4.50	

20.

Original Value	New Value	Percent Increase
$220	$300	

Complete each table. Round to the nearest cent.

21.

Original Price	Rate of Discount	Discount	Sale Price
$87.33	40%		

22.

Selling Price	Rate of Markup (based on selling price)	Markup	Cost
$1,824.00	20%		

23.

Selling Price	Rate of Sales Tax	Sales Tax
$200	7.25%	

24.

Sales	Rate of Commission	Commission
$537.14	10%	

25.

Principal	Interest Rate	Kind of Interest	Time (in years)	Interest	Final Balance
$3,000	5%	simple	5		

26.

Principal	Interest Rate	Kind of Interest	Time (in years)	Final Balance
$259.13	5.8%	compound	12	

Mixed Applications

Solve.

27. An upscale department-store chain reported that total sales this year were $2.3 billion—up from $1.8 billion last year. Find the percent increase in sales, to the nearest whole percent.

28. Last year, a local team won 20 games. This year, it won 15 games. What was the percent decrease of games won?

29. In 9 years, the number of elderly nursing home residents rose from 200,000 to 1.3 million. By what percent did the number of residents increase?

30. Due to a decrease in demand, a manufacturing plant decreased its production from 2,400 to 1,800 units per day. What was the percent decrease in the number of units produced per day?

31. The first commercial telephone exchange was set up in New Haven, Connecticut, in 1878. Between 1880 and 1890, the number of telephones in the United States increased from 50 thousand to 200 thousand, in round numbers. What percent increase was this? (***Source:*** U.S. Bureau of the Census)

32. A patient's medication was decreased from 250 milligrams to 200 milligrams per dose. What was the percent decrease in the dosage?

33. A customer paid 5% sales tax on a notebook computer that sold for $1,699. Calculate the amount of sales tax that she paid.

34. Last year, a town assessed the value of a residential property at $272,000. If the homeowner paid property tax of $3,264, what was the property tax rate?

35. A customer bought a cell phone for \$150. The total selling price of the phone, including sales tax, was \$159.75. What was the sales tax rate?

36. In a town, the sales tax rate is 7%. It costs \$5 to travel to a nearby town and back where the sales tax is only 5.8%. Is it worthwhile to make this trip to purchase an item that sells in both towns for \$800?

37. A pharmaceutical sales representative earns a 12% monthly commission on all her sales above \$5,000. Find her commission if her sales this month totaled \$27,500.

38. A sales assistant earns a flat salary of \$150 plus a 10% commission on sales of \$3,000. What were his total earnings?

39. On a restaurant table, a customer leaves \$1.35 as a tip. Assuming that the customer left a 15% tip, how much was the bill before the tip?

40. A salesperson receives a 5% commission on the first \$2,000 in sales and a 7% commission on sales above \$2,000. How much commission does she earn on sales of \$3,500?

41. An electronics store sells an MP3 player for \$59.95. If each MP3 player was marked up 40% based on the selling price, how much was the markup?

42. An antique store bought a table for \$90 and resold it for \$120. What is the markup percent, based on the selling price?

43. What is the markup percent, based on the selling price, on an item when the markup is \$8 and the selling price is \$15?

44. A store manager marked up the cost of merchandise by \$5. If the cost had been \$3, what percent of the selling price was the markup?

45. A store sells a television that lists for \$399 at a 35% discount rate. What is the sale price?

46. An appliance store has a sale on all its appliances. A washing machine that originally sold for \$800 is on sale for \$680. What is the discount rate?

47. A bank customer borrowed \$3,000 for 1 year at 5% simple interest to buy a computer. How much interest did the customer pay?

48. How much simple interest is earned on \$600 at an 8% annual interest rate for 2 years?

49. A couple deposited \$5,000 in a bank. How much interest will they have earned after 1 year if the interest rate is 5%?

50. A student borrowed \$2,000 from a friend, agreeing to pay her 4% simple interest. If he promised to repay her the entire amount at the end of 3 years, how much money must he pay her?

51. A home goods store had a "20% off" sale on all its merchandise. A customer bought a down comforter that originally cost \$180.

a. What was the sale price of the comforter?

b. Calculate the total amount the customer paid after 6% sales tax was added to the purchase.

52. During a sale, a shoe store marked down the price of a pair of sneakers that originally cost \$80 by 40%.

a. What was the sale price of the sneakers?

b. After two weeks, the store marked down the sale price by another 60%. What percent off the original price was the sale price after the second discount was applied?

53. An investor put \$3,000 in an account that pays 6% interest, compounded annually. Find the amount in the account after 2 years.

54. A bank pays 5.5% interest, compounded annually, on a 2-year certificate of deposit (CD) that initially costs \$500. What is the value of the CD at the end of the 2 years, rounded to the nearest cent?

55. A city had a population of 4,000. If the city's population increased by 10% per year, what was the population 4 years later?

56. An art dealer bought a painting for $10,000. If the value of the painting increased by 50% per year, what was its value 4 years later?

• *Check your answers on page A-10.*

MINDSTRETCHERS

Writing

1. Explain the difference between simple interest and compound interest.

Technology

2. Using a spreadsheet, construct a three-column table showing the original price, the 10% discount, and the selling price for items with an original price of any whole number of dollars between $1 and $100.

Mathematical Reasoning

3. If a quantity increases by a given percent and then decreases by the same percent, will the final value be the same as the original value? Explain.

KEY CONCEPTS AND SKILLS

CONCEPT SKILL

Concept/Skill	Description	Example
[6.1] **Percent**	A ratio or fraction with denominator 100. It is written with the % sign, which means divided by 100.	$7\% = \frac{7}{100}$ (7% ↑ Percent)
[6.1] **To change a percent to the equivalent fraction**	• Drop the % sign from the given percent and place the number over 100. • Simplify the resulting fraction, if possible.	$25\% = \frac{25}{100} = \frac{1}{4}$
[6.1] **To change a percent to the equivalent decimal**	• Drop the % sign from the given percent and divide the number by 100.	$23.5\% = .235$, or 0.235
[6.1] **To change a decimal to the equivalent percent**	• Multiply the number by 100 and insert a % sign.	$0.125 = 12.5\%$
[6.1] **To change a fraction to the equivalent percent**	• Multiply the fraction by 100 and insert a % sign.	$\frac{1}{5} = \frac{1}{5} \times 100\% = \frac{1}{\cancel{5}_1} \times \frac{\cancel{100}^{20}}{1}\%$ $= 20\%$
[6.2] **Base**	The number that we are taking the percent of. It always follows the word *of* in the statement of a percent problem.	50% of 8 is 4. (8 ↑ Base)
[6.2] **Amount**	The result of taking the percent of the base.	50% of 8 is 4. (4 ↑ Amount)
[6.2] **To solve a percent problem using the translation method**	• Translate as follows: What number, what percent → x is → = of → × or · % → decimal or fraction • Set up the equation. **The percent of the base is the amount.** • Solve.	What is 50% of 8? $x = 0.5 \cdot 8$ $x = 4$ 30% of what number is 6? $0.3 \cdot x = 6$ $\frac{\cancel{0.3}x}{\cancel{0.3}} = \frac{6}{0.3}$ $x = \frac{6}{0.3} = 20$ What percent of 8 is 2? $x \cdot 8 = 2$ $x = \frac{2}{8} = \frac{1}{4} = 25\%$

continued

CONCEPT SKILL

Concept/Skill	Description	Example
[6.2] **To solve a percent problem using the proportion method**	• Identify the amount, the base, and the percent, if known. • Set up and substitute into the proportion. $\frac{\text{Amount}}{\text{Base}} = \frac{\text{Percent}}{100}$ • Solve for the unknown quantity.	50% of 8 is what number? $\frac{x}{8} = \frac{50}{100}$ $100x = 400$ $x = 4$ 30% of what number is 6? $\frac{6}{x} = \frac{30}{100}$ $30x = 600$ $x = 20$ What percent of 8 is 2? $\frac{2}{8} = \frac{x}{100}$ $8x = 200$ $x = 25$ So the answer is 25%.
[6.3] **To find a percent increase or decrease**	• Compute the difference between the two given values. • Determine what percent this difference is of the *original value.*	Find the percent increase for a quantity that changes from 4 to 5. Difference: $5 - 4 = 1$ What percent of 4 is 1? $x \cdot 4 = 1$ $x = \frac{1}{4} = 0.25$, or 25%

Chapter 6 Review Exercises

To help you review this chapter, solve these problems.

[6.1] *Complete the following tables.*

1.

Fraction	Decimal	Percent
$\frac{1}{4}$		
	0.7	
		$\frac{3}{4}\%$
$\frac{5}{8}$		
		41%
$1\frac{1}{100}$		
		260%
	3.3	
	0.12	
		$66\frac{2}{3}\%$
$\frac{1}{6}$		

2.

Fraction	Decimal	Percent
$\frac{3}{8}$		
	0.49	
		0.1%
		150%
	0.875	
		$83\frac{1}{3}\%$
$2\frac{3}{4}$		
	1.2	
	0.75	
		10%
$\frac{1}{3}$		

[6.2] *Solve.*

3. What is 40% of 30?

4. What percent of 5 is 6?

5. 2 feet is what percent of 4 feet?

6. 30% of what number is 6?

7. What percent of 8 is 3.5?

8. Find 55% of 10.

9. $12 is 200% of what amount of money?

10. 2 is what percent of 10?

11. What is 1.2% of 25?

12. Find 115% of 400.

13. 35% of $200 is what?

14. $\frac{1}{2}\%$ of what number is 5?

15. 15 is what percent of 0.75?

16. 4.5 is what percent of 18?

17. Calculate $33\frac{1}{3}\%$ of $600.

18. What percent of $9 is $4?

19. Estimate 59% of $19.99.

20. 2.5% of how much money is $40?

21. What percent of $7.99 is $1.35, to the nearest whole percent?

22. 3.5 is $8\frac{1}{4}\%$ of what number, to the nearest hundredth?

[6.3] *Complete the following tables.*

23.

Original Value	New Value	Percent Decrease
24	16	

24.

Selling Price	Rate of Sales Tax	Sales Tax
$50	6%	

25.

Sales	Rate of Commission	Commission
$600	4%	

26.

Original Price	Rate of Discount	Discount	Sale Price
$200	15%		

27.

Selling Price	Rate of Markup (based on the selling price)	Markup	Cost
$51	50%		

28.

Principal	Interest Rate	Time (in years)	Simple Interest	Final Balance
$200	4%	2		

Mixed Applications

Solve.

29. On July 1, 2005, the sales tax rate in Chicago, already one of the highest among major U.S. cities, increased a quarter point to 9%. If the sales tax on a computer in Chicago amounted to $162, what was the selling price (before taxes) of the computer? (*Source:* The Tax Foundation)

30. Jonas Salk developed the polio vaccine in 1954. The number of reported polio cases in the United States dropped from 29,000 to 15,000 between 1955 and 1956. What was the percent drop, to the nearest whole percent? (*Source:* U.S. Bureau of the Census)

31. For their fees, one real estate agent charges 11% of a year's rent and another charges the first month's rent. Which agent charges more?

32. A particular community bank makes available loans with simple interest and with no prepayment penalty. How much interest is due on a five-year car loan of $24,000 based on a simple interest rate of 6%?

33. According to a city survey, 49% of respondents approve of how the mayor is handling his job and 31% disapprove. What percent neither approved nor disapproved?

34. Plastics make up about 11% and paper makes up $\frac{9}{25}$ of the solid municipal waste in the United States. Which makes up more of the solid municipal waste? (*Source:* Energy Information Administration)

35. According to a study, 25% of employees do not take all of their vacation time due to the demands of their jobs. Express this percent as a fraction. (*Source:* Families and Work Institute)

36. A typical markup rate in the hobby industry, based on selling price, is 40%. If a hobby shop sells a holiday train set that cost $120 for $220, was the markup rate on the train set above or below the typical markup rate? (*Source:* http://keytaps.com)

37. It takes a worker 50 minutes to commute to work. If he has been traveling for 20 minutes, what percent of his trip has been completed?

38. Approximately 1 of every 10 Americans is left-handed. What percent is this?

39. A couple financed a 30-year mortgage at a fixed interest rate of 6.29%. Express this rate as a decimal.

40. A clothing store places the following ad in a local newspaper:

At the store, what is the sale price of a suit that regularly sells for $230?

41. The following table deals with the oil reserves of two nations that are leading oil producers:

Country	Proven Oil Reserves (in billions of barrels)
Saudi Arabia	262
Canada	179

What percent of the size of Saudi Arabia's reserves is the size of Canada's reserves, rounded to the nearest percent? (*Source: Time Almanac 2006*)

42. The length of a person's thigh bone is usually about 27% of his or her height. Estimate someone's height whose thigh bone is 20 inches long. (*Source: American Journal of Physical Anthropology*)

43. The winner of a men's U.S. Open tennis match got $87\frac{1}{2}\%$ of his first serves in. If he had 72 first serves, how many went in?

44. In a scientific study that relates weight to health, people are considered overweight if their actual weight is at least 20% above their ideal weight. If you weigh 160 pounds and have an ideal weight of 130 pounds, are you considered overweight?

45. An airline oversold a flight to Los Angeles by nine seats, or 5% of the total number of seats available on the airplane. How many seats does the airplane have?

46. When an assistant became editor, the magazine's weekly circulation increased from 50,000 to 60,000. By what percent did the circulation increase?

47. The salary of an executive assistant had been $30,000 before she got a raise of $1,000. If the rate of inflation is 5%, has her salary kept pace with inflation?

48. The cat is the most popular pet in America. According to recent estimates, there are 53 million dogs. If there are 109% as many cats, how many cats are there?

49. At an auction, you bought a table for \$150. The auction house also charged a "buyer's premium"—an extra fee—of 10%. How much did you pay in all?

50. According to the news report, 80 tons of food met only 20% of the food needs in the refugee camp. How much additional food was needed?

51. A traveler needs 14,000 more frequent-flier miles to earn a free trip to Hawaii, which is 20% of the total number needed. How many frequent-flier miles in all does this award require?

52. A Sylvania compact fluorescent lightbulb (CFL) has a life of 8,000 hours. A Philips CFL has a life that is 25% longer. How long does a Philips CFL last? (*Source:* http://Sylvania.com and http://bulbs.com)

53. How much commission does a salesperson make on sales totaling \$5,000 at a 20% rate of commission?

54. At the end of the year, the receipts of a retail store amounted to \$200,000. Of these receipts, 85% went for expenses; the rest was profit. How much profit did the store make?

55. If a bank customer deposits \$7,000 in a bank account that pays a 6.5% rate of interest compounded annually, what will be the balance after 2 years?

56. Suppose that a country's economy expands by 2% per year. By what percent will it expand in 10 years, to the nearest whole percent?

57. Complete the following table, which describes a company's income for the four quarters of last year:

Quarter	Income	Percent of Total Income (rounded to the nearest whole percent)
1	\$375,129	
2	289,402	
3	318,225	
4	402,077	
Total		100%

58. The following graph shows the sources from which the federal government received income in a recent year:

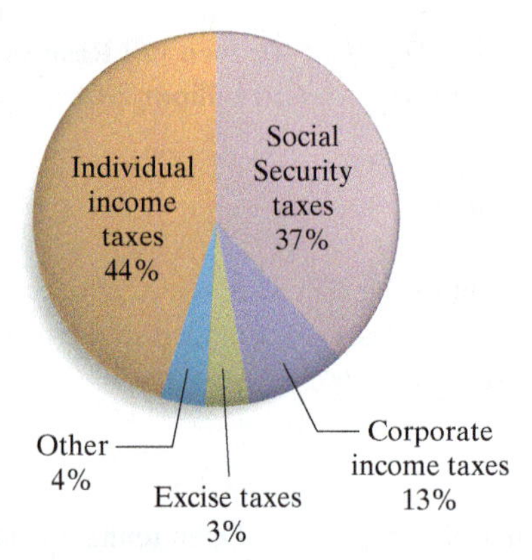

If the total amount of money taken in was \$2,154 billion, compute how much money was received from each source, to the nearest billion dollars. (*Source:* U.S. Office of Management and Budget)

• *Check your answers on page A-11.*

Chapter 6 POSTTEST

FOR EXTRA HELP

Test solutions are found on the enclosed CD.

To see if you have mastered the topics in this chapter, take this test.

Rewrite.

1. 4% as a fraction

2. $27\frac{1}{2}\%$ as a fraction

3. 174% as a decimal

4. 8% as a decimal

5. 0.009 as a percent

6. 10 as a percent

7. $\frac{5}{6}$ as a percent, rounded to the nearest whole percent

8. $2\frac{1}{5}$ as a percent

Solve.

9. What is 25% of 30 miles?

10. Find 120% of 40.

11. Estimate 32% of $20.77.

12. 8% of what number is 16?

13. What percent of 10 is 6?

14. What percent of 4 is 10?

15. To pay for tuition, a college student borrows $2,000 from a relative for 2 years at 5% simple interest. Find the amount of simple interest that is due.

16. In a parking lot that has 150 spaces, 4% are for handicap parking. How many handicap spaces are in the lot?

17. A customer paid $9.95 in sales tax on an iPod nano that cost $199. What was the sales tax rate?

18. Milk is approximately 50% cream. How much milk is needed to produce 2 pints of cream?

19. A department store sells a pair of shoes for $79 at a markup rate of 40% based on the selling price. How much money is the markup?

20. A college ended six straight years of tuition increases by raising its tuition from $3,000 to $3,100. Find the percent increase.

• *Check your answers on page A-12.*

Cumulative Review Exercises

To help you review, solve the following.

1. Divide: $1{,}962 \div 18$

2. Express $\frac{5}{6}$ as a decimal, rounded to the nearest hundredth.

3. Multiply: 0.2×3.5

4. Find the sum of $3\frac{4}{5}$ and $1\frac{9}{10}$.

5. Solve for x: $\frac{x}{3} = 2.5$

6. 20% of what amount is \$200?

Solve.

7. The government withdrew $\frac{1}{4}$ million of its 2 million troops. What fraction of the total is this?

8. In a recent survey of college students, 7 out of 10 said they used the Internet every day. At this rate, how many of the 9,160 students at a college would be expected to use the Internet every day?

9. Three FM stations are highlighted on the radio dial shown. These stations have frequencies 99.5 (WBAI), 96.3 (WQXR), and 104.3 (WAXQ). Label the three stations on the dial.

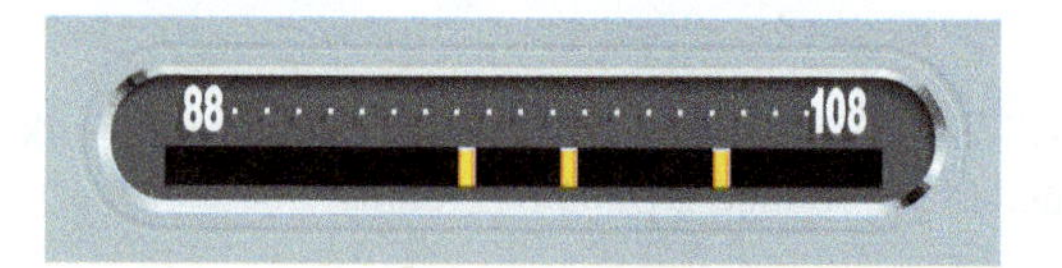

10. In a recent year, about 27% of the 885 thousand American doctors were female. How many female doctors were there, to the nearest thousand? (*Source:* American Medical Association)

• *Check your answers on page A-12.*

APPENDIX

Scientific Notation

Frequently, scientists deal with numbers that are either very large or very small. For instance, in astronomy, they study the distance to the nearest star; in biology, the length of a virus.

40,000,000,000,000 kilometers ← The distance between the Sun and Proxima Centauri

0.0000000000001 meters ← The length of a virus

Scientists commonly write such numbers not in standard notation, but rather in *scientific notation.* Also, scientific and graphing calculators generally show answers in scientific notation that are too long to fit in their display.

Scientific notation is based on powers of 10. ***A number is said to be in scientific notation if it is written as the product of a decimal between 1 and 10 (not including 10) and a power of 10.***

For instance, the number 7.35×10^5 is written in scientific notation, because the decimal factor 7.35 is between 1 and 10, and 10^5 is a power of 10. However, the number 81.45×2^6 is not in scientific notation, because the base of the exponent is not 10, and the decimal factor 81.45 is not less than 10.

Tip A number written in scientific notation has one nonzero digit in its decimal factor.

Let's consider how to change a number from scientific notation to standard notation and vice versa. First, we will look at *large numbers.*

EXAMPLE 1

Change the number 7.35×10^5 from scientific notation to standard notation.

Solution To express 7.35×10^5 in standard notation, we need to multiply 7.35 by 10^5 or 100,000.

$$7.35 \times 10^5 = 7.35 \times 100{,}000 = 735{,}000.$$

So 7.35×10^5 written in standard notation is 735,000.

PRACTICE 1

Express 2.539×10^7 in standard notation.

In Example 1, note that the power of 10 is *positive* and that the decimal point is moved five places *to the right*. So a shortcut for writing 7.35×10^5 in standard notation is to move the decimal point in 7.35 five places to the right.

$$7.35 \times 10^5 = 7\,3\,5\,0\,0\,0 = 735{,}000$$

To change a number from standard notation to scientific notation, the process is reversed.

EXAMPLE 2

Rewrite 37,000,000,000 in scientific notation.

Solution We know that for a number to be written in scientific notation, it must be the product of a decimal number between 1 and 10 and a power of 10. Recall that 37,000,000,000 and 37,000,000,000. are the same. We move the decimal point *to the left* so that there is one nonzero digit to its left. The number of places moved is the power of 10 by which we need to multiply.

$$37{,}000{,}000{,}000. = 3.7000000000 \times 10^{10} = 3.7 \times 10^{10}$$

Note that we dropped the extra zeros in 3.7000000000. So 37,000,000,000 expressed in scientific notation is 3.7×10^{10}.

PRACTICE 2

Write 8,000,000,000,000 in scientific notation.

Now, let's turn our attention to writing *small numbers* in scientific notation. The key is an understanding of *negative exponents*. Until now, we have only considered exponents that were either positive integers or 0. What meaning should we attach to a negative exponent? The following pattern, in which each number is $\frac{1}{10}$ of the previous number, suggests an answer.

$$10^3 = 1{,}000$$
$$10^2 = 100$$
$$10^1 = 10$$
$$10^0 = 1$$
$$10^{-1} = \frac{1}{10}$$
$$10^{-2} = \frac{1}{100}$$
$$10^{-3} = \frac{1}{1{,}000}$$

Notice that 10^{-1}, or $\frac{1}{10}$, is the reciprocal of 10^1 or 10. So in general, ***a number raised to a negative exponent is defined to be the reciprocal of that number raised to the corresponding positive exponent***.

When written in scientific notation, large numbers have positive powers of 10, whereas small numbers have negative powers of 10. For instance, 3×10^5 is large, but 3×10^{-5} is small.

Next, let's look at how we change *small* numbers from scientific notation to standard notation and vice versa.

EXAMPLE 3

Convert 3×10^{-5} to standard notation.

Solution Using the meaning of negative exponents, we get:

$$3 \times 10^{-5} = 3 \times \frac{1}{10^5}, \quad \text{or} \quad \frac{3}{10^5}$$

Since $10^5 = 100{,}000$, dividing 3 by 10^5 gives us:

$$\frac{3}{10^5} = \frac{3}{100{,}000} = 0.00003$$

So 3×10^{-5} written in standard notation is 0.00003.

PRACTICE 3

Change 4.3×10^{-9} to standard notation.

In Example 3, note that the power of 10 is *negative* and the decimal point, which is understood to be at the right end of a whole number, was moved five places *to the left*. So just as with 7.35×10^5, there is a shortcut for expressing 3×10^{-5} in standard notation. To do this, we move the decimal point five places *to the left*:

$$3 \times 10^{-5} = 3. \times 10^{-5} = .0\,0\,0\,0\,3 = .00003, \text{ or } 0.00003.$$

Tip When converting a number from scientific notation to standard notation, move the decimal point to the *left* if the power of 10 is *negative* and to the *right* if the power of 10 is *positive*.

EXAMPLE 4

Write 0.00000000000000002 in scientific notation.

Solution To write 0.00000000000000002 in scientific notation, we move the decimal point *to the right* until there is one nonzero digit to the left of the decimal point. The number of places moved, preceded by a *negative* sign, is the power of 10 that we need.

$$0.00000000000000002 = 0\,0\,0\,0\,0\,0\,0\,0\,0\,0\,0\,0\,0\,0\,0\,0\,0\,2. \times 10^{-17}$$
$$= 2 \times 10^{-17}$$

PRACTICE 4

Express 0.000000000071 in scientific notation.

Computation Involving Scientific Notation

Now we consider how to perform calculations on numbers written in scientific notation. We focus on the operations of multiplication and division.

Multiplying and dividing numbers written in scientific notation can best be understood in terms of two *laws of exponents*—the *product rule* and the *quotient rule*.

- The *product rule of exponents* states that when we multiply a base raised to a power by the same base raised to another power, we add the exponents and leave the base the same. For example,

$$10^3 \cdot 10^2 = 10^{3+2} = 10^5$$

Add the exponents.
Keep the base.

This result is reasonable, since $10^3 \times 10^2 = 1{,}000 \times 100 = 100{,}000 = 10^5$.

- The *quotient rule of exponents* states that when we divide a base raised to a power by the same base to another power, we subtract the second power from the first power, and leave the base the same. For instance,

Subtract the exponents.

$$10^5 \div 10^2 = 10^{5-2} = 10^3$$

Keep the base.

We would have expected this result even if we did not know the quotient rule,

since $\dfrac{10^5}{10^2} = \dfrac{100{,}000}{100} = \dfrac{1{,}000}{1} = 1{,}000 = 10^3$.

EXAMPLE 5

Calculate, writing the result in scientific notation:

a. $(4 \times 10^{-1})(2.1 \times 10^6)$

b. $(1.2 \times 10^5) \div (2 \times 10^{-4})$

Solution

a. $(4 \times 10^{-1})(2.1 \times 10^6) = (4 \times 2.1)(10^{-1} \times 10^6)$ Regroup the factors.

$= (8.4)(10^{-1} \times 10^6)$ Multiply the decimal factors.

$= 8.4 \times 10^{-1+6}$ Use the product rule of exponents.

$= 8.4 \times 10^5$ Simplify.

b. $(1.2 \times 10^5) \div (2 \times 10^{-4}) = \dfrac{1.2 \times 10^5}{2 \times 10^{-4}}$

$= \dfrac{1.2}{2} \times \dfrac{10^5}{10^{-4}}$ Write as the product of fractions.

$= 0.6 \times \dfrac{10^5}{10^{-4}}$ Divide the decimal factors.

$= 0.6 \times 10^{5-(-4)}$ Use the quotient rule of exponents.

$= 0.6 \times 10^9$ Simplify.

Note that 0.6×10^9 is not written in scientific notation, because 0.6 is not between 1 and 10, that is, it does not have one nonzero digit to the left of the decimal point. To write 0.6×10^9 in scientific notation, we convert 0.6 to scientific notation and simplify the product.

$0.6 \times 10^9 = 6 \times 10^{-1} \times 10^9$

$= 6 \times 10^{-1+9}$ Use the product rule of exponents.

$= 6 \times 10^8$

So the quotient, written in scientific notation, is 6×10^8.

PRACTICE 5

Calculate, expressing the answer in scientific notation:

a. $(7 \times 10^{-2})(3.52 \times 10^3)$

b. $(5.01 \times 10^3) \div (6 \times 10^{-9})$

Exercises

FOR EXTRA HELP

Express in scientific notation.

1. 400,000,000

2. 10,000,000

3. 0.0000035

4. 0.00017

5. 0.00000000031

6. 218,000,000,000

Express in standard notation.

7. 3.17×10^8

8. 9.1×10^5

9. 1×10^{-6}

10. 8.013×10^{-4}

11. 4.013×10^{-5}

12. 2.1×10^{-3}

Multiply, and write the result in scientific notation.

13. $(3 \times 10^2)(3 \times 10^5)$

14. $(5 \times 10^6)(1 \times 10^3)$

15. $(2.5 \times 10^{-2})(8.3 \times 10^{-3})$

16. $(2.1 \times 10^4)(8 \times 10^{-4})$

Divide, and write the result in scientific notation.

17. $(2.5 \times 10^8) \div (2 \times 10^{-2})$

18. $(3.0 \times 10^4) \div (1 \times 10^3)$

19. $(1.2 \times 10^5) \div (3 \times 10^3)$

20. $(4.88 \times 10^{-3}) \div (8 \times 10^2)$

Answers

Chapter 1

Pretest: Chapter 1, *p. 2*

1. Two hundred five thousand, seven **2.** 1,235,000 **3.** Hundred thousands **4.** 8,100 **5.** 8,226 **6.** 4,714 **7.** 185 **8.** 29,124 **9.** 260 **10.** 308 R6 **11.** 2^3 **12.** 36 **13.** 5 **14.** 43 **15.** 75 years old **16.** $675 **17.** 69 **18.** 156 sec **19.** $27 **20.** Room C, which measures 126 sq ft

Practices: Section 1.1, *pp. 4–8*

1, ***p. 4:*** **a.** Thousands **b.** Hundred thousands **c.** Ten millions **2,** ***p. 4:*** Eight billion, three hundred seventy-six thousand, fifty-two **3,** ***p. 4:*** $7,372,050 Seven million, three hundred seventy-two thousand, fifty dollars **4,** ***p. 5:*** $95,000,003 **5,** ***p. 5:*** $375,000 **6,** ***p. 6:*** **a.** 2 ten thousands + 7 thousands + 0 hundreds + 1 ten + 3 ones = 20,000 + 7,000 + 0 + 10 + 3 **b.** 1 million + 2 hundred thousands + 7 ten thousands + 9 tens + 3 ones = 1,000,000 + 200,000 + 70,000 + 90 + 3
7, ***p. 7:*** **a.** 52,000 **b.** 50,000 **8,** ***p. 8:*** 420,000,000
9, ***p. 8:*** **a.** Two hundred forty-eight thousand, seven hundred eighty-eight **b.** 400,000

Exercises 1.1, *pp. 9–13*

1. whole numbers **3.** odd **5.** standard form **7.** placeholder **9.** expanded form **11.** $\underline{4}$,867 **13.** 3$\underline{1}$6 **15.** 2$\underline{8}$,461,013 **17.** Hundred thousands **19.** Hundreds **21.** Billions **23.** Four hundred eighty-seven thousand, five hundred **25.** Two million, three hundred fifty thousand **27.** Nine hundred seventy-five million, one hundred thirty-five thousand **29.** Two billion, three hundred fifty-two **31.** One billion **33.** 10,120 **35.** 150,856 **37.** 6,000,055 **39.** 50,600,195 **41.** 400,072 **43.** 3 ones = 3 **45.** 8 hundreds + 5 tens + 8 ones = 800 + 50 + 8 **47.** 2 millions + 5 hundred thousands + 4 ones = 2,000,000 + 500,000 + 4 **49.** 670 **51.** 7,100 **53.** 30,000 **55.** 700,000 **57.** 30,000

59.

To the nearest	135,842	2,816,533
Hundred	135,800	2,816,500
Thousand	136,000	2,817,000
Ten thousand	140,000	2,820,000
Hundred thousand	100,000	2,800,000

61. 1 ten thousand + 2 thousands + 5 tens + 1 one = 10,000 + 2,000 + 50 + 1 **63.** $\underline{4}$0,059 **65.** 1,056,100; one million, fifty-six thousand, one hundred **67.** Nine hundred thousand **69.** Eight thousand, nine hundred fifty-nine **71.** Thirty-seven thousand, eight hundred forty-two **73.** 100,000,000,000 **75.** 3,288 **77.** 2,908,000 **79.** 150 ft **81.** 20,000 mi **83.** 500 g **85. a.** One million, three hundred ninety-nine thousand, five hundred forty-two **b.** 700,000

Practices: Section 1.2, *pp. 15–22*

1, ***p. 15:*** 385 **2,** ***p. 16:*** 10,436 **3,** ***p. 17:*** 16 mi **4,** ***p. 18:*** 651 **5,** ***p. 19:*** 4,747 **6,** ***p. 19:*** 765 plant species **7,** ***p. 21:*** **a.** 128,000 **b.** 233,000 **c.** Less **8,** ***p. 22:*** 9,477 **9,** ***p. 22:*** 2,791 **10,** ***p. 22:*** 87,000 mi

Calculator Practices, *p. 23*

11, ***p. 23:*** 49,532 **12,** ***p. 23:*** 31,899 **13,** ***p. 23:*** 2,499 ft

Exercises 1.2, *pp. 24–30*

1. right **3.** sum **5.** Associative Property of Addition **7.** subtrahend **9.** 177,778 **11.** 14,710 **13.** 14,002 **15.** 56,188 **17.** 6,978 **19.** 4,820 **21.** 413 **23.** 14,865 **25.** 15,509 m **27.** 82 hr **29.** $104,831 **31.** $12,724 **33.** 31,200 tons **35.** 13,296,657 **37.** 22,912,891

39.

+	400	200	1,200	300	Total
300	700	500	1,500	600	3,300
800	1,200	1,000	2,000	1,100	5,300
Total	1,900	1,500	3,500	1,700	8,600

41.

+	389	172	1,155	324	Total
255	644	427	1,410	579	3,060
799	1,188	971	1,954	1,123	5,236
Total	1,832	1,398	3,364	1,702	8,296

43. a; possible estimate: 12,800 **45.** a; possible estimate: $900,000 **47.** 217 **49.** 90 **51.** 362 **53.** 68,241 **55.** 2,285 **57.** 52,999 **59.** 2,943 **61.** 203,465 **63.** 368 **65.** 4,996 **67.** 982 **69.** 1,995 mi **71.** $669 **73.** $3,609 **75.** 273 books **77.** 209 m **79.** 2,001,000 **81.** 813,429 **83.** c; possible estimate: 40,000,000 **85.** a; possible estimate: $200,000 **87.** 7,065 **89.** 1,676 **91.** 5,186 **93.** 281,000,000 **95.** 2,600,000 sq mi **97.** **a.** Austria: 23; Canada: 24; Germany: 29; Russia: 22; United States: 25 **b.** Germany **99.** About 43 years old **101.** No, the elevator is not overloaded. The total weight of passengers is 963 lb. **103.** 180°F **105.** 12 mi **107.** 36 hr **109.** $2,951 **111.** **a.** No **b.** Yes **113.** **a.** 347,000,000 lb **b.** Possible estimate: 2,300,000,000 lb **c.** 2,260,000,000 lb **115.** (a) Addition; (b) yes, $1,563; (c) possible estimate: $1,600

Practices: Section 1.3, *pp. 34–37*

1, ***p. 34:*** 608 **2,** ***p. 34:*** 4,230 **3,** ***p. 35:*** 480,000 **4,** ***p. 35:*** 205,296 **5,** ***p. 36:*** 107 sq ft **6,** ***p. 36:*** 112,840 **7,** ***p. 37:*** No; possible estimate = 20,000

Calculator Practices, *p. 38*

8, ***p. 38:*** 1,026,015; **9,** ***p. 38:*** 345,546;

Exercises 1.3, *pp. 39–42*

1. product **3.** Identity Property of Multiplication **5.** addition **7.** 400 **9.** 142,000 **11.** 170,000 **13.** 7,000,000 **15.** 12,700 **17.** 418 **19.** 3,248,000 **21.** 65,268 **23.** 817 **25.** 34,032 **27.** 3,003 **29.** 3,612 **31.** 57,019 **33.** 243,456 **35.** 200,120 **37.** 149,916 **39.** 144,500 **41.** 123,830 **43.** 3,312 **45.** 2,106 **47.** 40,000 **49.** 23,085 **51.** 3,274,780 **53.** 54,998,850 **55.** c; possible estimate: 480,000 **57.** b; possible estimate: 80,000 **59.** 2,880 **61.** 230,520 **63.** 1,071,000 **65.** 300,000 **67.** 3,300 yr **69.** **a.** 3,000,000 **b.** 1,000,000 **71.** Yes **73.** 5,775 sq in. **75.** 1,750 mi **77.** $442 **79.** **a.** 294 mi **b.** 1,470 mi **81.** (a) Multiplication (b) Colorado; area ≈ 106,700 sq mi (c) possible estimate: 120,000 sq mi

Practices: Section 1.4, *pp. 47–51*

1, ***p. 47:*** 807 **2,** ***p. 47:*** 7,002 **3,** ***p. 48:*** 5,291 R1 **4,** ***p. 49:*** 79 R1 **5,** ***p. 49:*** 94 R10 **6,** ***p. 50:*** 607 R3 **7,** ***p. 50:*** 200 **8,** ***p. 51:*** 967 **9,** ***p. 51:*** 5 times

Calculator Practice, *p. 52:* 603

Exercises 1.4, *p. 53*

1. divisor **3.** multiplication **5.** 400 **7.** 560 **9.** 301 **11.** 3,003 **13.** 202 **15.** 500 **17.** 30 **19.** 14 **21.** 42 **23.** 400 **25.** 159 **27.** 5,353 **29.** 1,002 **31.** 6,944 **33.** 1,001 **35.** 3,050 **37.** 651 R2 **39.** 11 R7 **41.** 116 R83 **43.** 700 R2 **45.** 723 R19 **47.** 428 R8 **49.** 721 **51.** 155 **53.** a; possible estimate: 7,000 **55.** a; possible estimate: 400 **57.** 907 R1 **59.** 2,000 **61.** 2,400 **63.** 370 **65.** $135 **67.** 2 times **69.** 300 people per square mile **71.** 6 calories **73.** **a.** 304 tiles **b.** 26 boxes **c.** $468 **75.** (a) Division, (b) more than 4 times, since the quotient of 1,306,313,800 and 295,734,100 is 4 with a remainder, (c) possible estimate of the quotient: 4 with a remainder, so that China's population is more than 4 times that of the United States.

Practices: Section 1.5, *pp. 57–62*

1, ***p. 57:*** $5^5 \cdot 2^2$ **2,** ***p. 58:*** **a.** 1 **b.** 1,331 **3,** ***p. 58:*** 784 **4,** ***p. 58:*** 10^9 **5,** ***p. 59:*** 28 **6,** ***p. 60:*** 146 **7,** ***p. 60:*** 4 **8,** ***p. 60:*** 23 **9,** ***p. 61:*** 60 ft **10,** ***p. 61:*** $40 **11,** ***p. 62:*** **a.** 46 fatalities **b.** 2004 and 2005

Calculator Practices, *p. 63*

12, ***p. 63:*** 140,625; **13,** ***p. 63:*** 131

Exercises 1.5, *pp. 64–68*

1. base **3.** adding

5.

n	0	2	4	6	8	10	12
n^2	0	4	16	36	64	100	144

7.

n	0	2	4	6	8
n^3	0	8	64	216	512

9. 10^2 **11.** 10^4 **13.** 10^6 **15.** $2^2 \cdot 3^2$ **17.** $4^3 \cdot 5^1$ **19.** 900 **21.** 1,568 **23.** 18 **25.** 2 **27.** 35 **29.** 343 **31.** 250 **33.** 36 **35.** 8 **37.** 92 **39.** 60 **41.** 28 **43.** 6 **45.** 99 **47.** 99 **49.** 4 **51.** 39 **53.** 1 **55.** 93 **57.** 18 **59.** 419 **61.** 137,088 **63.** $\boxed{4} \cdot 3 + \boxed{6} \cdot 5 + \boxed{8} \cdot 7 = 98$ **65.** $(\boxed{8})(3 + \boxed{4}) - 2 \cdot \boxed{6} = 44$ **67.** $\boxed{8} + 10 \times \boxed{4} - \boxed{6} \div 2 = 45$ **69.** $(5 + 2) \cdot 4^2 = 112$ **71.** $(5 + 2 \cdot 4)^2 = 169$ **73.** $(8 - 4) \div 2^2 = 1$ **75.** 242 sq cm **77.** 3,120 sq in.

79.

Input	Output
0	
1	
2	

81. 25 **83.** 40 **85.** 4 **87.** 2,412 mi **89.** 8 **91.** 10^8 **93.** 289 **95.** 48 **97.** 8 **99.** 625 sq ft **101.** $5^2 + 12^2 = 13^2$; $25 + 144 = 169$ **103.** 10^6 **105.** **a.** $21,500 **b.** $1,050 **107.** **a.** 69 **b.** At home; the average score for home games was higher than the average score for away games. **109.** **a.** Broadcast television:

791 hr; cable and satellite television: 976 hr **b.** Cable and satellite television; by 185 hr. **111.** (a) Addition, division, and subtraction. (b) The daily average circulation of newspaper B was greater by 11,553. (c) possible estimate: 10,000

Practices: Section 1.6, *pp. 71–73*

1, ***p. 71:*** 10,670 employees **2,** ***p. 72:*** 2 yr **3,** ***p. 72:*** 1,551 students **4,** ***p. 73:*** 180 lb

Exercises 1.6, *pp. 74–75*

1. $2,150 **3.** 27 mi **5.** 75 times **7.** 5,882 mi **9.** 528,179 immigrants **11.** 300¢, or $3 **13.** $17,000 **15.** $6,036 **17.** $1,458 **19.** 8 extra pens **21.** 1952 was closer by 31 votes. **23.** (a) Subtraction and division, (b) $983, (c) possible estimate: $1,000

Review Exercises: Chapter 1, *pp. 79–83*

1. Ones **2.** Ten thousands **3.** Hundred millions **4.** Ten billions **5.** Four hundred ninety-seven **6.** Two thousand, fifty **7.** Three million, seven **8.** Eighty-five billion **9.** 251 **10.** 9,002 **11.** 14,000,025 **12.** 3,000,003,000 **13.** 2 millions + 5 hundred thousands = 2,000,000 + 500,000 **14.** 4 ten thousands + 2 thousands + 7 hundreds + 7 ones = 40,000 + 2,000 + 700 + 7 **15.** 600 **16.** 1,000 **17.** 380,000 **18.** 70,000 **19.** 9,486 **20.** 65,692 **21.** 173,543 **22.** 150,895 **23.** 1,957,825 **24.** $223,067 **25.** 445 **26.** 10,016 **27.** 11,109 **28.** 5,510 **29.** 11,042,223 **30.** $2,062,852 **31.** 11,006 **32.** 2,989 **33.** 432 **34.** 1,200 **35.** 149,073 **36.** 12,000,000 **37.** 477,472 **38.** 1,019,000 **39.** 1,397,508 **40.** 188,221,590 **41.** 39 **42.** 307 R3 **43.** 37 R10 **44.** 680 R8 **45.** 25,625 **46.** 957 **47.** 343 **48.** 1 **49.** 72 **50.** 300,000 **51.** 5 **52.** 169 **53.** 5 **54.** 19 **55.** 12 **56.** 18 **57.** 10,833,312 **58.** 2,694 **59.** $7^2 \cdot 5^2$ **60.** $2^2 \cdot 5^3$ **61.** 39 **62.** 7 **63.** 6 **64.** 5 **65.** Two million, four hundred thousand **66.** 150,000,000 **67.** $307 per week **68.** 1758 **69.** 300,000 sq mi **70.** 22,000,000 iPods **71.** 9 **72.** 272 legs **73.** 509 m **74.** 23 flats **75.** $604,015,000,000

76.

Net sales	$430,000
− Cost of merchandise sold	− 175,000
Gross margin	$255,000
− Operating expenses	− 135,000
Net profit	$120,000

77. 6,675 sq m **78.** Possible answer: 20 **79.** 1968 to 1972 (15,379,754 votes) **80.** 4,341 points **81. a.** 1,832 km **b.** 1,800 km **82. a.** 12,677,500 **b.** The average would increase by 239,475 **83.** 29 sq mi **84.** 162 cm

Posttest: Chapter 1, *p. 84*

1. 225,067 **2.** 1,768,405 **3.** One million, two hundred five thousand, seven **4.** 200,000 **5.** 1,894 **6.** 607 **7.** 147 **8.** 297,496 **9.** 509 **10.** 622 R19 **11.** 625 **12.** $4^3 \cdot 5^2$ **13.** 84 **14.** 2 **15.** 5,700,000 sq mi **16.** 46,848,000 acres **17.** $469 **18.** Below; $1,000,000 is smaller than the average, which was $1,380,468. **19.** $1,380 **20.** 12 mg

Chapter 2

Pretest: Chapter 2, *p. 86*

1. 1, 2, 4, 5, 10, 20 **2.** $2 \times 2 \times 2 \times 3 \times 3$, or $2^3 \times 3^2$ **3.** $\frac{2}{5}$ **4.** $\frac{61}{3}$ **5.** $1\frac{1}{30}$ **6.** $\frac{3}{4}$ **7.** 20 **8.** $\frac{1}{8}$ **9.** $1\frac{1}{5}$ **10.** $12\frac{5}{6}$ **11.** $2\frac{1}{4}$ **12.** $4\frac{5}{8}$ **13.** $3\frac{1}{2}$ **14.** 60 **15.** $\frac{2}{3}$ **16.** $3\frac{2}{3}$ **17.** $\frac{1}{8}$ **18.** 6 students **19.** $20\frac{7}{8}$ ft **20.** 66 g

Practices: Section 2.1, *pp. 87–93*

1, ***p. 87:*** 1, 7 **2,** ***p. 88:*** 1, 3, 5, 15, 25, 75 **3,** ***p. 89:*** 1, 2, 3, 5, 6, 9, 10, 15, 18, 30, 45, 90 **4,** ***p. 89:*** Yes; 24 is a multiple of 3. **5,** ***p. 90:*** **a.** Prime **b.** Composite **c.** Prime **d.** Composite **e.** Prime **6,** ***p. 91:*** $2^3 \times 7$ **7,** ***p. 91:*** 3×5^2 **8,** ***p. 92:*** 18 **9,** ***p. 93:*** 66 **10,** ***p. 93:*** 12 **11,** ***p. 93:*** 6 yr

Exercises 2.1, *pp. 94–95*

1. factors **3.** prime **5.** prime factorization **7.** 1, 3, 7, 21 **9.** 1, 17 **11.** 1, 2, 3, 4, 6, 12 **13.** 1, 31 **15.** 1, 2, 3, 4, 6, 9, 12, 18, 36 **17.** 1, 29 **19.** 1, 2, 4, 5, 10, 20, 25, 50, 100 **21.** 1, 2, 4, 7, 14, 28 **23.** Prime **25.** Composite (2, 4, 8) **27.** Composite (7) **29.** Prime **31.** Composite (3, 9, 27) **33.** 2^3 **35.** 7^2 **37.** $2^3 \times 3$ **39.** 2×5^2 **41.** 7×11 **43.** 3×17 **45.** 5^2 **47.** 2^5 **49.** 3×7 **51.** $2^3 \times 13$ **53.** 11^2 **55.** 2×71 **57.** $2^2 \times 5^2$ **59.** 5^3 **61.** $3^3 \times 5$ **63.** 15 **65.** 40 **67.** 90 **69.** 110 **71.** 72 **73.** 360 **75.** 300 **77.** 84 **79.** 105 **81.** 60 **83.** 3×5^2 **85.** 1, 2, 3, 4, 6, 8, 9, 12, 18, 24, 36, and 72 **87. a.** No, because 1995 is not a multiple of 10 **b.** Yes, because 1990 is a multiple of 10 **89.** No **91.** 99 students **93.** 30 days

Practices: Section 2.2, *pp. 98–106*

1, ***p. 98:*** $\frac{5}{8}$ **2,** ***p. 98:*** $\frac{7}{30}$ **3,** ***p. 98:*** $\frac{41}{101}$

4, ***p. 99:*** =

5, ***p. 99:*** **a.** $\frac{16}{3}$ **b.** $\frac{102}{5}$ **6,** ***p. 100:*** **a.** 2 **b.** $5\frac{5}{9}$ **c.** $2\frac{2}{3}$ **7,** ***p. 102:*** Possible answer: $\frac{4}{10}, \frac{6}{15}, \frac{8}{20}$ **8,** ***p. 102:*** $\frac{45}{72}$ **9,** ***p. 103:*** $\frac{2}{3}$ **10,** ***p. 103:*** $\frac{7}{3}$ **11,** ***p. 104:*** $\frac{5}{16}$ **12,** ***p. 105:*** $\frac{11}{16}$ **13,** ***p. 106:*** $\frac{8}{15}, \frac{23}{30}, \frac{9}{10}$ **14,** ***p. 106:*** Country stations

Exercises 2.2, *pp. 107–112*

1. proper fraction **3.** equivalent **5.** like fractions **7.** $\frac{1}{3}$ **9.** $\frac{3}{6}$ **11.** $1\frac{1}{4}$ **13.** $3\frac{2}{4}$ **15.**

17. **19.**

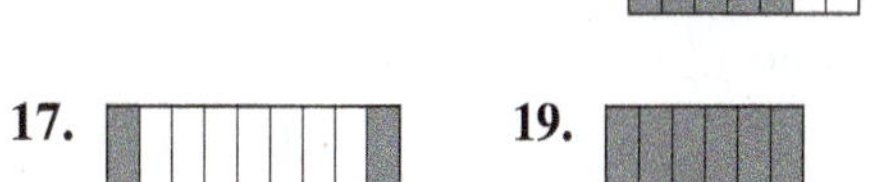

21. **23.**

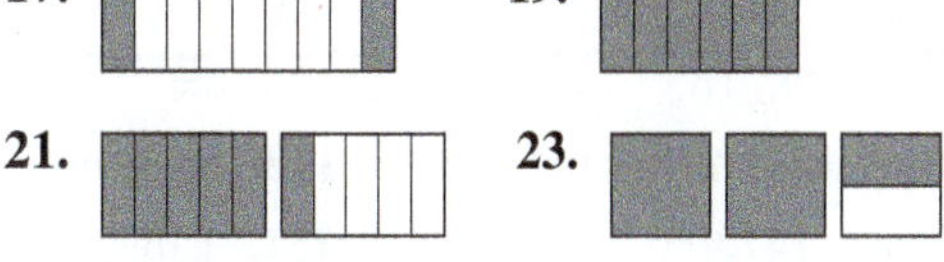

25.

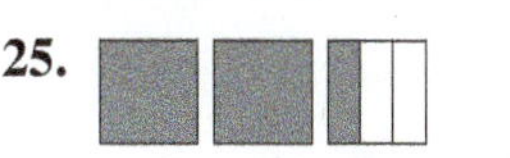

27. Proper **29.** Improper **31.** Mixed **33.** Improper **35.** Proper **37.** Mixed **39.** $\frac{13}{5}$ **41.** $\frac{55}{9}$ **43.** $\frac{57}{5}$ **45.** $\frac{5}{1}$ **47.** $\frac{59}{8}$ **49.** $\frac{88}{9}$ **51.** $\frac{27}{2}$

53. $\frac{98}{5}$ **55.** $\frac{14}{1}$ **57.** $\frac{54}{11}$ **59.** $\frac{115}{14}$ **61.** $\frac{202}{25}$ **63.** $1\frac{1}{3}$
65. $1\frac{1}{9}$ **67.** 3 **69.** 1 **71.** $19\frac{4}{5}$ **73.** $9\frac{1}{9}$ **75.** 1
77. $8\frac{2}{9}$ **79.** $13\frac{1}{2}$ **81.** $11\frac{1}{9}$ **83.** 27 **85.** 8
87. Possible answers: $\frac{2}{16}, \frac{3}{24}$ **89.** Possible answers: $\frac{4}{22}, \frac{6}{33}$
91. Possible answers: $\frac{6}{8}, \frac{9}{12}$ **93.** Possible answers: $\frac{2}{18}, \frac{3}{27}$
95. 9 **97.** 15 **99.** 40 **101.** 36 **103.** 40 **105.** 54
107. 36 **109.** 42 **111.** 6 **113.** 49 **115.** 32
117. 30 **119.** $\frac{2}{3}$ **121.** 1 **123.** $\frac{1}{3}$ **125.** $\frac{9}{20}$ **127.** $\frac{1}{4}$
129. $\frac{1}{8}$ **131.** $\frac{5}{4}$, or $1\frac{1}{4}$ **133.** $\frac{33}{16}$, or $2\frac{1}{16}$ **135.** $\frac{9}{16}$
137. $\frac{7}{24}$ **139.** 3 **141.** $\frac{1}{7}$ **143.** $3\frac{2}{3}$ **145.** 3 **147.** $<$
149. $>$ **151.** $=$ **153.** $<$ **155.** $\frac{1}{4}, \frac{1}{3}, \frac{1}{2}$ **157.** $\frac{7}{12}, \frac{2}{3}, \frac{5}{6}$
159. $\frac{3}{5}, \frac{2}{3}, \frac{8}{9}$ **161.** $\frac{5}{6}$ **163.** Possible answers: $\frac{4}{18}, \frac{6}{27}$
165. $\frac{12}{15}$ **167.** $2\frac{1}{5}$ hr per day **169. a.** $\frac{33}{758}$ **b.** $\frac{725}{758}$
171. $\frac{50}{103}$ **173.** No; $\frac{1}{4} = \frac{25}{100}$, which is greater than $\frac{23}{100}$
175. a. Petroleum products **b.** Natural gas **177.** $190\frac{1}{3}$ lb

Practices: Section 2.3, *pp. 113–127*

1, *p. 113:* $\frac{2}{3}$ **2, *p. 114:*** $1\frac{7}{40}$ **3, *p. 114:*** $\frac{2}{5}$
4, *p. 114:* a. $\frac{3}{5}$ g **b.** $\frac{2}{5}$ g **5, *p. 116:*** $1\frac{2}{3}$ **6, *p. 116:*** $\frac{3}{10}$
7, *p. 116:* $\frac{71}{72}$ **8, *p. 117:*** 2 mi **9, *p. 118:*** $34\frac{2}{5}$
10, *p. 118:* $7\frac{1}{2}$ **11, *p. 118:*** 4 lengths **12, *p. 119:*** $7\frac{5}{8}$
13, *p. 119:* $11\frac{5}{24}$ **14, *p. 120:*** $4\frac{2}{5}$ **15, *p. 121:*** $2\frac{1}{8}$ in.
16, *p. 121:* $4\frac{7}{12}$ **17, *p. 122:*** $6\frac{1}{4}$ mi **18, *p. 123:*** $1\frac{2}{7}$
19, *p. 124:* $5\frac{1}{6}$ **20, *p. 124:*** $13\frac{1}{2}$ **21, *p. 125:*** No, there will be only $3\frac{5}{8}$ yd left. **22, *p. 126:*** $10\frac{11}{20}$ **23, *p. 126:*** $1\frac{3}{8}$
24, *p. 127:* $6\frac{3}{4}$

Exercises 2.3, *pp. 128–132*

1. numerators **3.** borrow **5.** $1\frac{1}{4}$ **7.** $1\frac{1}{2}$ **9.** $\frac{4}{5}$ **11.** $\frac{3}{5}$
13. $1\frac{1}{6}$ **15.** $\frac{7}{8}$ **17.** $\frac{77}{100}$ **19.** $\frac{37}{40}$ **21.** $1\frac{5}{18}$ **23.** $1\frac{17}{100}$
25. $\frac{3}{4}$ **27.** $\frac{53}{80}$ **29.** $1\frac{7}{72}$ **31.** $1\frac{13}{40}$ **33.** $3\frac{1}{3}$ **35.** $15\frac{2}{5}$
37. $14\frac{1}{5}$ **39.** 15 **41.** $10\frac{5}{12}$ **43.** $3\frac{11}{15}$ **45.** $13\frac{13}{15}$
47. $6\frac{19}{24}$ **49.** $20\frac{1}{4}$ **51.** $10\frac{3}{100}$ **53.** $11\frac{3}{8}$ **55.** $36\frac{3}{50}$
57. $91\frac{7}{12}$ **59.** $6\frac{1}{2}$ **61.** $10\frac{33}{40}$ **63.** $11\frac{3}{8}$ **65.** $\frac{1}{5}$ **67.** $\frac{2}{5}$
69. $\frac{4}{25}$ **71.** $\frac{1}{2}$ **73.** 2 **75.** $\frac{1}{12}$ **77.** $\frac{5}{18}$ **79.** $\frac{1}{20}$ **81.** $\frac{1}{14}$
83. $\frac{5}{72}$ **85.** $\frac{1}{4}$ **87.** $4\frac{2}{5}$ **89.** $1\frac{3}{4}$ **91.** 20 **93.** $4\frac{1}{10}$
95. $3\frac{1}{3}$ **97.** $3\frac{3}{10}$ **99.** $6\frac{1}{3}$ **101.** $5\frac{1}{2}$ **103.** $4\frac{1}{2}$
105. $3\frac{1}{4}$ **107.** $11\frac{4}{5}$ **109.** $6\frac{2}{3}$ **111.** $7\frac{5}{6}$ **113.** $3\frac{13}{24}$
115. $15\frac{7}{18}$ **117.** $2\frac{29}{30}$ **119.** $\frac{1}{4}$ **121.** $5\frac{5}{12}$ **123.** $13\frac{39}{40}$
125. $\frac{3}{8}$ **127.** $1\frac{11}{40}$ **129.** $16\frac{23}{30}$ **131.** $5\frac{1}{5}$ **133.** $18\frac{11}{20}$
135. $4\frac{1}{8}$ **137.** $8\frac{1}{3}$ **139.** $2\frac{14}{15}$ **141.** $\frac{1}{8}$ in. **143. a.** $1\frac{1}{2}$ mi **b.** $\frac{1}{4}$ mi **145.** 5 hr **147.** $5\frac{5}{6}$ ft **149.** $\frac{1}{10}$ **151.** 1 lb

Practices: Section 2.4, *pp. 134–143*

1, *p. 134:* $\frac{15}{28}$ **2, *p. 134:*** $\frac{81}{100}$ **3, *p. 134:*** 20 **4, *p. 134:*** $\frac{7}{22}$
5, *p. 135:* $\frac{2}{9}$ **6, *p. 135:*** $5\frac{1}{4}$ hr **7, *p. 135:*** \$20,769
8, *p. 136:* $7\frac{7}{8}$ **9, *p. 136:*** 28 **10, *p. 137:*** $25\frac{1}{2}$ sq in.
11, *p. 138:* $18\frac{1}{4}$ **12, *p. 139:*** 6 **13, *p. 140:*** 8
14, *p. 140:* $2\frac{2}{3}$ yr **15, *p. 141:*** $1\frac{3}{5}$ **16, *p. 141:*** $\frac{7}{16}$
17, *p. 141:* 6 lb **18, *p. 142:*** 6 **19, *p. 143:*** $4\frac{1}{2}$

Exercises 2.4, *pp. 144–148*

1. multiply **3.** reciprocal **5.** $\frac{2}{15}$ **7.** $\frac{5}{12}$ **9.** $\frac{9}{16}$ **11.** $\frac{8}{25}$
13. $\frac{35}{32} = 1\frac{3}{32}$ **15.** $\frac{45}{16} = 2\frac{13}{16}$ **17.** $\frac{2}{9}$ **19.** $\frac{7}{12}$ **21.** $\frac{3}{40}$
23. $\frac{31}{30} = 1\frac{1}{30}$ **25.** $\frac{40}{3} = 13\frac{1}{3}$ **27.** $\frac{40}{3} = 13\frac{1}{3}$ **29.** 16
31. 4 **33.** 4 **35.** $\frac{35}{4} = 8\frac{3}{4}$ **37.** $1\frac{5}{16}$ **39.** $2\frac{1}{8}$ **41.** $\frac{25}{27}$
43. $2\frac{2}{3}$ **45.** 1 **47.** $\frac{7}{8}$ **49.** $1\frac{13}{35}$ **51.** $4\frac{41}{100}$ **53.** $7\frac{4}{5}$
55. 375 **57.** 8 **59.** 3 **61.** $41\frac{2}{3}$ **63.** $113\frac{1}{3}$ **65.** $1\frac{1}{6}$
67. $\frac{7}{12}$ **69.** $\frac{77}{100}$ **71.** $3\frac{3}{8}$ **73.** $\frac{9}{10}$ **75.** $\frac{32}{35}$ **77.** $3\frac{1}{2}$
79. $4\frac{4}{9}$ **81.** $1\frac{1}{2}$ **83.** $2\frac{1}{3}$ **85.** $1\frac{1}{5}$ **87.** $\frac{1}{4}$ **89.** $\frac{2}{21}$
91. $\frac{1}{9}$ **93.** 40 **95.** $16\frac{1}{3}$ **97.** $13\frac{1}{3}$ **99.** 7 **101.** $6\frac{11}{18}$
103. $1\frac{2}{3}$ **105.** $9\frac{22}{27}$ **107.** $100\frac{1}{2}$ **109.** $\frac{7}{90}$ **111.** $\frac{5}{26}$
113. $3\frac{1}{5}$ **115.** $\frac{21}{200}$ **117.** $\frac{35}{44}$ **119.** $1\frac{47}{115}$ **121.** $\frac{14}{27}$
123. $2\frac{1}{4}$ **125.** $1\frac{7}{18}$ **127.** $4\frac{13}{15}$ **129.** $\frac{87}{160}$ **131.** $3\frac{19}{27}$
133. $4\frac{1}{5}$ **135.** $3\frac{1}{8}$ **137.** $11\frac{1}{6}$ **139.** $20\frac{13}{16}$ **141.** $2\frac{5}{22}$
143. $\frac{21}{40}$ **145.** 8 **147.** $\frac{7}{12}$ **149.** \$12,000 **151.** $6\frac{1}{4}$
153. \$$191\frac{1}{4}$ **155.** $\frac{27}{64}$ **157.** 7 times **159. a.** The scented candle **b.** The unscented candle

Review Exercises: Chapter 2, *pp. 152–157*

1. 1, 2, 3, 5, 6, 10, 15, 25, 30, 50, 75, 150 **2.** 1, 2, 3, 4, 5, 6, 9, 10, 12, 15, 18, 20, 30, 36, 45, 60, 90, 180
3. 1, 3, 19, 57 **4.** 1, 2, 5, 7, 10, 14, 35, 70 **5.** Prime
6. Composite **7.** Composite **8.** Prime **9.** $2^2 \times 3^2$
10. 3×5^2 **11.** $3^2 \times 11$ **12.** 2×3^3 **13.** 42 **14.** 10
15. 72 **16.** 60 **17.** $\frac{2}{4}$ **18.** $\frac{6}{12}$ **19.** $1\frac{1}{6}$ **20.** $2\frac{3}{5}$
21. Mixed **22.** Proper **23.** Improper **24.** Improper
25. $\frac{23}{3}$ **26.** $\frac{9}{5}$ **27.** $\frac{91}{10}$ **28.** $\frac{59}{7}$ **29.** $6\frac{1}{2}$ **30.** $4\frac{2}{3}$
31. $2\frac{3}{4}$ **32.** 1 **33.** 84 **34.** 4 **35.** 5 **36.** 27
37. $\frac{1}{2}$ **38.** $\frac{5}{7}$ **39.** $\frac{2}{3}$ **40.** $\frac{3}{4}$ **41.** $5\frac{1}{2}$ **42.** $8\frac{2}{3}$ **43.** $6\frac{2}{7}$
44. $8\frac{5}{7}$ **45.** $>$ **46.** $>$ **47.** $<$ **48.** $>$ **49.** $>$
50. $>$ **51.** $>$ **52.** $>$ **53.** $\frac{2}{7}, \frac{3}{8}, \frac{1}{2}$ **54.** $\frac{2}{15}, \frac{1}{5}, \frac{1}{3}$
55. $\frac{3}{4}, \frac{4}{5}, \frac{9}{10}$ **56.** $\frac{13}{18}, \frac{7}{9}, \frac{7}{8}$ **57.** $\frac{6}{5} = 1\frac{1}{5}$ **58.** $\frac{3}{4}$ **59.** $\frac{15}{8} = 1\frac{7}{8}$
60. $\frac{3}{5}$ **61.** $\frac{11}{15}$ **62.** $1\frac{17}{24}$ **63.** $1\frac{4}{5}$ **64.** $1\frac{37}{40}$ **65.** $5\frac{7}{8}$
66. $9\frac{1}{2}$ **67.** $10\frac{3}{5}$ **68.** 8 **69.** $12\frac{1}{3}$ **70.** $4\frac{3}{10}$ **71.** $5\frac{7}{10}$
72. $17\frac{13}{24}$ **73.** $23\frac{5}{12}$ **74.** $46\frac{3}{8}$ **75.** $20\frac{3}{4}$ **76.** $56\frac{1}{24}$ **77.** $\frac{1}{4}$
78. $\frac{2}{3}$ **79.** 1 **80.** 0 **81.** $\frac{1}{4}$ **82.** $\frac{3}{8}$ **83.** $\frac{7}{20}$ **84.** $\frac{7}{30}$
85. $7\frac{1}{2}$ **86.** $2\frac{3}{10}$ **87.** $3\frac{3}{4}$ **88.** $18\frac{1}{2}$ **89.** $6\frac{1}{2}$ **90.** $1\frac{7}{10}$
91. $2\frac{2}{3}$ **92.** $\frac{1}{5}$ **93.** $1\frac{4}{5}$ **94.** $\frac{3}{4}$ **95.** $2\frac{1}{2}$ **96.** $3\frac{1}{3}$ **97.** $\frac{3}{10}$
98. $2\frac{7}{8}$ **99.** $\frac{7}{12}$ **100.** $3\frac{8}{9}$ **101.** $\frac{2}{3}$ **102.** $9\frac{9}{20}$ **103.** $\frac{3}{16}$
104. $\frac{7}{16}$ **105.** $\frac{5}{8}$ **106.** $\frac{1}{6}$ **107.** $5\frac{1}{3}$ **108.** $\frac{7}{10}$ **109.** $\frac{1}{125}$
110. $\frac{8}{27}$ **111.** $\frac{1}{4}$ **112.** $\frac{7}{120}$ **113.** $\frac{24}{25}$ **114.** $1\frac{5}{9}$ **115.** $2\frac{2}{3}$
116. $\frac{2}{3}$ **117.** 6 **118.** $18\frac{5}{12}$ **119.** $8\frac{7}{16}$ **120.** $21\frac{1}{4}$

121. $\frac{9}{20}$ **122.** $1\frac{9}{16}$ **123.** $37\frac{1}{27}$ **124.** $3\frac{3}{8}$ **125.** $3\frac{1}{8}$
126. $1\frac{41}{90}$ **127.** $2\frac{1}{10}$ **128.** $7\frac{1}{5}$ **129.** $\frac{3}{2}$ **130.** $\frac{2}{3}$ **131.** $\frac{1}{8}$
132. 4 **133.** $\frac{7}{40}$ **134.** $\frac{5}{81}$ **135.** $\frac{2}{15}$ **136.** $\frac{1}{200}$
137. $\frac{3}{4}$ **138.** $1\frac{1}{3}$ **139.** 30 **140.** $8\frac{3}{4}$ **141.** $1\frac{1}{6}$ **142.** $1\frac{4}{5}$
143. 2 **144.** 4 **145.** $1\frac{3}{4}$ **146.** $\frac{4}{7}$ **147.** $1\frac{7}{12}$
148. $\frac{12}{19}$ **149.** $5\frac{1}{2}$ **150.** $2\frac{11}{20}$ **151.** 2 **152.** 3 **153.** $9\frac{3}{4}$
154. $1\frac{3}{10}$ **155.** $5\frac{1}{3}$ **156.** $7\frac{5}{9}$ **157.** No **158.** 50¢
159. $\frac{1}{4}$ **160.** $\frac{2}{9}$ **161.** The Filmworks camera **162.** $\frac{39}{40}$
163. The patient got back more than $\frac{1}{3}$, because $\frac{275}{700} = \frac{11}{28} = \frac{33}{84}$ which is greater than $\frac{1}{3} = \frac{28}{84}$. **164.** Yes it should, because $\frac{23}{32}$ is greater than $\frac{2}{3}$. $\frac{23}{32} = \frac{69}{96}$, whereas $\frac{2}{3} = \frac{64}{96}$ **165. a.** $\frac{12}{23}$ **b.** $\frac{3}{4}$
166. a. Lisa Gregory **b.** Monica Yates **167.** $\frac{3}{4}$ **168.** $\frac{11}{12}$ oz
169. $\frac{1}{4}$ carat **170.** $\frac{3}{5}$ **171.** 12 women **172.** 2,685 undergraduate students **173.** $1,050 **174.** 7 lb **175.** $18
176. 2 times **177.** 19 fish **178.** $11\frac{3}{4}$ mi **179.** $62\frac{5}{6}$ ft
180. $\frac{5}{6}$ hr **181.** 1,500 fps **182.** 500 lb/sq in.
183. $281\frac{1}{4}$ lb **184.** $46,000 **185.** 8 orbits
186. $29\frac{5}{9}$ sq yd
187.

Employee	Saturday	Sunday	Total
L. Chavis	$7\frac{1}{2}$	$4\frac{1}{4}$	$11\frac{3}{4}$
R. Young	$5\frac{3}{4}$	$6\frac{1}{2}$	$12\frac{1}{4}$
Total	$13\frac{1}{4}$	$10\frac{3}{4}$	24

188.

Worker	Hours per Day	Days Worked	Total Hours	Wage per Hour	Gross Pay
Maya	5	3	15	$7	$105
Noel	$7\frac{1}{4}$	4	29	$10	$290
Alisa	$4\frac{1}{2}$	$5\frac{1}{2}$	$24\frac{3}{4}$	$9	$222\frac{3}{4}$

189. $10\frac{10}{11}$ lb **190.** $22\frac{1}{2}$ cups

Posttest: Chapter 2, *p. 158*

1. 1, 3, 7, 9, 21, 63 **2.** 2×3^3 **3.** $\frac{4}{9}$ **4.** $\frac{12}{1}$
5. $10\frac{1}{4}$ **6.** $\frac{7}{8}$ **7.** $\frac{5}{10}$ **8.** 24 **9.** $1\frac{13}{24}$ **10.** $8\frac{7}{40}$
11. $4\frac{2}{7}$ **12.** $5\frac{23}{30}$ **13.** $\frac{1}{81}$ **14.** 12 **15.** $\frac{7}{9}$ **16.** $7\frac{5}{6}$
17. $\frac{5}{6}$ **18.** $19\frac{1}{5}$ mi **19.** $90\frac{2}{3}$ sq ft **20.** $16\frac{1}{4}$ in.

Cumulative Review: Chapter 2, *p. 159*

1. Five million, three hundred fifteen **2.** 581,400
3. 908 **4.** $\frac{3}{4}$ **5.** $6\frac{2}{5}$ **6.** 7 **7.** $\frac{3}{8}$ **8.** 1 million times
9. 549 copies **10.** $\frac{1}{3}$

Chapter 3

Pretest: Chapter 3, *p. 162*

1. Hundredths **2.** Four and twelve thousandths **3.** 3.1
4. 0.0029 **5.** 21.52 **6.** 7.3738 **7.** 11.69 **8.** 9.81
9. 8,300 **10.** 18.423 **11.** 0.0144 **12.** 7.1 **13.** 0.00605
14. 32.7 **15.** 0.875 **16.** 2.83 **17.** One with a pH value of 2.95 **18.** $39.788 billion **19.** 3 times
20. $3.74

Practices: Section 3.1, *pp. 164–170*

1, ***p. 164:*** **a.** The tenths place **b.** The ten-thousandths place **c.** The thousandths place **2,** ***p. 165:*** $\frac{7}{8}$ **3,** ***p. 165:*** $2\frac{3}{100}$
4, ***p. 165:*** **a.** $5\frac{3}{5}$ **b.** $5\frac{3}{5}$ **5,** ***p. 166:*** **a.** $7\frac{3}{1,000}$ **b.** $4\frac{1}{10}$
6, ***p. 166:*** **a.** Sixty-one hundredths **b.** Four and nine hundred twenty-three thousandths **c.** Seven and five hundredths
7, ***p. 166:*** **a.** 0.043 **b.** 10.26 **8,** ***p. 167:*** 3.14
9, ***p. 167:*** 0.8297 **10,** ***p. 167:*** 3.51, 3.5, 3.496
11, ***p. 168:*** The one with the rating of 8.1, because $9 > 8.2 > 8.1$ **12,** ***p. 169:*** **a.** 748.1 **b.** 748.08 **c.** 748.077 **d.** 748 **e.** 700 **13,** ***p. 170:*** 7.30
14, ***p. 170:*** 11.4 m

Exercises 3.1, *pp. 171–175*

1. right **3.** hundredths **5.** greater **7.** $2.\underline{7}8$
9. $2.001\underline{7}5$ **11.** $3\underline{5}8.02$ **13.** $0.7\underline{7}2$ **15.** Tenths
17. Hundredths **19.** Thousandths **21.** Ones **23.** $\frac{3}{5}$
25. $\frac{39}{100}$ **27.** $1\frac{1}{2}$ **29.** 8 **31.** $5\frac{3}{250}$ **33.** Fifty-three hundredths **35.** Three hundred five thousandths
37. Six tenths **39.** Five and seventy-two hundredths
41. Twenty-four and two thousandths **43.** 0.8
45. 1.041 **47.** 60.01 **49.** 4.107 **51.** 3.2 m **53.** $>$
55. $<$ **57.** $>$ **59.** $=$ **61.** $<$ **63.** 7, 7.07, 7.1
65. 4.9, 5.001, 5.2 **67.** 9.1 mi, 9.38 mi, 9.6 mi **69.** 17.4
71. 3.591 **73.** 37.1 **75.** 0.40 **77.** 7.06 **79.** 9 mi
81.

To the Nearest	8.0714	0.9916
Tenth	8.1	1.0
Hundredth	8.07	0.99
Ten	10	0

83. $0.\underline{0}24$ **85.** 870.06 **87.** 2.04 m, 2.14 m, 2.4 m
89. Twenty-three and nine hundred thirty-four thousandths
91. Eighteen and seven tenths; eighteen and eight tenths
93. Three hundred one and three tenths, Fifty-five and nine tenths, Two hundred sixty-eight and two tenths, Forty-six and six tenths, Forty-three and six tenths **95.** One hundred-thousandth; eight hundred-thousandths **97.** 1.2 acres
99. 74.59 mph **101.** 14.7 lb **103.** $0.005
105. 352.1 kWh **107.** Less **109.** Last winter
111. Yes **113.** No **115.** $57.03 **117.** 0.001 **119.** 1.8

Practices: Section 3.2, *pp. 177–182*

1, ***p. 177:*** 10.387 **2,** ***p. 177:*** 39.3 **3,** ***p. 178:*** 102.1°F
4, ***p. 178:*** 46.2125 **5,** ***p. 179:*** $485.43
6, ***p. 179:*** 13.5 mi **7,** ***p. 179:*** 22.13 mi **8,** ***p. 180:*** 0.863
9, ***p. 180:*** 0.079 **10,** ***p. 180:*** 0.5744
11, ***p. 181:*** Possible estimate: $480
12, ***p. 181:*** Possible estimate: $2 million

Calculator Practices, *p. 182*

13, *p. 182*: 79.23; **14, *p. 182*:** 0.00002

Exercises 3.2, *pp. 183–186*

1. decimal points **3.** sum **5.** 9.33 **7.** 0.9 **9.** 8.13 **11.** 21.45 **13.** 7.67 **15.** $77.21 **17.** 1.08993 **19.** 24.16 **21.** 44.422 **23.** 20.32 mm **25.** 16.682 kg **27.** 23.30595 **29.** 0.7 **31.** 16.8 **33.** 18.41 **35.** 75.63 **37.** 22.324 **39.** 0.17 **41.** 0.1142 **43.** 6.2 **45.** 15.37 **47.** 5.9 **49.** 6.21 **51.** 1.85 lb **53.** 4.9°F **55.** 39.752 **57.** 27.9 mg **59.** 3.205 **61.** 21.19896 **63.** c; possible estimate: 0.083 **65.** b; possible estimate: 0.06 **67.** 7.771 **69.** 7.75 lb **71.** 11.6013 **73.** $1.03 **75.** 56.8 centuries **77.** $1.7 million **79.** 6.84 in. **81.** Yes; 2.8 + 2.9 + 2.6 + 1.6 = 9.9

83. a.

Gymnast	VT	UB	BB	FX	AAS
Madeline Whiteman	9.2	9.275	8.6	8.05	35.125
Jordyn Stengel	9	9	8.65	8.45	35.1

b. Madeline Whiteman **85.** (a) Addition, subtraction; (b) total 13.2 mg iron; no, she needs 4.8 mg more; (c) possible estimates: 1 + 2 + 0 + 2 + 1 + 1 + 1 + 1 + 1 + 2 + 1 + 0 = 13; 18 − 13 = 5

Practices: Section 3.3, *pp. 188–192*

1, *p. 188*: 9.835 **2, *p. 189*:** 1.4 **3, *p. 189*:** 0.01 **4, *p. 189*:** 0.024 **5, *p. 189*:** 9.91 **6, *p. 190*:** 325 **7, *p. 190*:** 327,000 **8, *p. 190*:** **a.** 18.015 **b.** 18 **9, *p. 191*:** 0.0003404; possible estimate: 0.004 × 0.09 = 0.00036 **10, *p. 191*:** 3.6463 **11, *p. 192*:** Possible answer: 1,200 mi

Calculator Practices, *p. 192*

12, *p. 192*: 815.6 **13, *p. 192*:** 9.261

Exercises 3.3, *pp. 193–195*

1. multiplication **3.** two **5.** square **7.** 2.99212 **9.** 204.360 **11.** 2,492.0 **13.** 0.0000969 **15.** 2,870.00 **17.** $0.73525 **19.** 0.54 **21.** 0.4 **23.** 0.02 **25.** 0.0028 **27.** 0.765 **29.** 2.016 **31.** 7.602 **33.** 0.5 **35.** 5.852 **37.** 151.14 **39.** 3.7377 **41.** 1.7955 **43.** 8,312.7 **45.** 23 **47.** 0.09 **49.** 1.05 **51.** 0.000000001 **53.** 42.5 ft **55.** 1.4 mi **57.** 3.29025 **59.** 272,593.75 **61.** 70 **63.** 25.75 **65.** 1.09 **67.** 2.86

69.

Input	Output
1	3.8 × **1** − 0.2 = 3.6
2	3.8 × **2** − 0.2 = 7.4
3	3.8 × **3** − 0.2 = 11.2
4	3.8 × **4** − 0.2 = 15

71. a; possible estimate: 50 **73.** b; possible estimate: 0.014 **75.** 8.75 **77.** 0.068 **79.** 4.48 **81.** 2,900 fps **83.** 57,900,000 km **85.** 19.6 sq ft **87.** 1.25 mg **89.** 1,308 calories

91. a.

Purchase	Quantity	Unit Price	Price
Belt	1	$11.99	$11.99
Shirt	3	$16.95	$50.85
Total Price			$62.84

b. $17.16 **93.** (a) Multiplication and addition; (b) 88.81 in.; (c) possible estimate: 90 in.

Practices: Section 3.4, *pp. 198–205*

1, *p. 198*: 0.375 **2, *p. 198*:** 7.625 **3, *p. 199*:** 83.3 **4, *p. 199*:** 0.8 **5, *p. 201*:** 18.04 **6, *p. 201*:** 2,050 **7, *p. 202*:** 73.4 **8, *p. 202*:** 0.0341 **9, *p. 203*:** 0.00086 **10, *p. 203*:** 1.5 **11, *p. 204*:** 21.1; possible estimate: 20 **12, *p. 204*:** 295.31 **13, *p. 205*:** 8 times as great

Calculator Practices, *pp. 205–206*

14, *p. 205*: 0.2 **15, *p. 206*:** 4.29

Exercises 3.4, *pp. 207–210*

1. decimal **3.** right **5.** quotient **7.** 0.5 **9.** 0.25 **11.** 3.7 **13.** 1.625 **15.** 2.875 **17.** 21.03 **19.** 4.25 **21.** 4.2 **23.** 1.375 **25.** 8.5 **27.** 0.67 **29.** 0.78 **31.** 3.11 **33.** 5.06 **35.** 3.286 **37.** 0.273 **39.** 6.571 **41.** 70.077 **43.** 58.82 **45.** 0.0663 **47.** 2.8875 **49.** 0.286 **51.** 4.3 **53.** 0.0015 **55.** 1.73 **57.** 2.875 **59.** 4 **61.** 70.4 **63.** 94 **65.** 12.5 **67.** 0.3 **69.** 0.2 **71.** 0.952 **73.** 0.00082 **75.** 383.88 **77.** 0.01 **79.** 9.23 **81.** 9,666.67 **83.** 1,952.38 **85.** 325.18 **87.** 67.41 **89.** 41.61 **91.** 0.17136 **93.** 0.13 **95.** 52.2 **97.** 4.05

99.

Input	Output
1	**1** ÷ 5 − 0.2 = 0
2	**2** ÷ 5 − 0.2 = 0.2
3	**3** ÷ 5 − 0.2 = 0.4
4	**4** ÷ 5 − 0.2 = 0.6

101. c; possible estimate: 50 **103.** b; possible estimate: 0.2 **105.** 0.8 **107.** 1.17 **109.** 0.45 **111.** 0.0037 in. per yr **113. a.** 0.6 **b.** 0.55 **c.** The women's team has a better record. The team won $\frac{3}{5}$, or 0.6, of the games played, and the men's team won $\frac{11}{20}$, or 0.55, of the games played.

115. a.

SUVs	Distance Driven (in miles)	Gasoline Used (in gallons)	Miles per Gallon
A	17.4	1.2	14.5
B	8.4	0.6	14
C	23.4	1.2	19.5

b. SUV C **117.** 2,000 shares **119.** 13 times **121.** 0.4 lb **123.** (a) Division; (b) .366; (c) .4

Review Exercises: Chapter 3, *pp. 214–216*

1. Hundredths **2.** Tenths **3.** Tenths **4.** Ten-thousandths **5.** $\frac{7}{20}$ **6.** $8\frac{1}{5}$ **7.** $4\frac{7}{1{,}000}$ **8.** 10 **9.** Seventy-two hundredths **10.** Five and six tenths **11.** Three and nine ten-thousandths **12.** Five hundred ten and thirty-six thousandths **13.** 0.007 **14.** 2.1 **15.** 0.09 **16.** 7.041 **17.** $<$ **18.** $>$ **19.** $>$ **20.** $>$ **21.** 1.002, 0.8, 0.72 **22.** 0.004, 0.003, 0.00057 **23.** 7.3 **24.** 0.039 **25.** 4.39 **26.** $899 **27.** 12.11 **28.** 52.75 **29.** $24.13 **30.** 12 m **31.** 28.78 **32.** 87.752 **33.** 1.834 **34.** 48.901 **35.** 98.2033 **36.** $90,948.80 **37.** 2.912 **38.** 1,008 **39.** 0.00001 **40.** 13.69 **41.** 2,710 **42.** 0.034 **43.** 5.75 **44.** 13.5 **45.** 1,569.36846 **46.** 441.760662 **47.** 0.625 **48.** 90.2 **49.** 4.0625 **50.** 0.045 **51.** 0.17 **52.** 0.29 **53.** 8.33 **54.** 11.22 **55.** 0.65 **56.** 1.6 **57.** 0.175 **58.** 0.277 **59.** 5.2 **60.** 3.2 **61.** 23.7 **62.** 16,358.3 **63.** 1.9 **64.** 360.7 **65.** 3.0 **66.** 0.3 **67.** 1.18 **68.** 117 **69.** 34.375 **70.** 1.4 **71.** Four ten-millionths **72.** $57.86 **73.** 54.49 sec **74.** 1.647 in. **75.** No, it would have traveled 0.585 mi, which is less than 0.75 mi. **76.** 4.35 times **77.** $0.06 **78.** Possible estimate: 15 in. **79.** 7.19 g **80.** 3.5°C **81.** 36,162.45
82.

Quarter	Google	Yahoo!
1st	1.257	1.174
2nd	1.384	1.253
3rd	1.578	1.330
4th	1.919	1.501
Total:	6.138	5.258

$0.88 billion, or $880,000,000

Posttest: Chapter 3, *p. 217*

1. 6 **2.** Five and one hundred two thousandths **3.** 320.15 **4.** 0.00028 **5.** $3\frac{1}{25}$ **6.** 0.004 **7.** 4.354 **8.** $5.66 **9.** 20.9 **10.** 5.72 **11.** 0.001 **12.** 3.36 **13.** 0.0029 **14.** 32.7 **15.** 0.375 **16.** 4.17 **17.** 0.01 lb **18.** 2.6 ft **19.** Belmont Stakes **20.** $2,807.21

Cumulative Review: Chapter 3, *p. 218*

1. 1,000,000 **2.** 32 **3.** $1\frac{1}{2}$ **4.** 27,403 **5.** $2\frac{2}{3}$ **6.** $\frac{17}{30}$ **7.** 610 **8.** 325 **9.** 26,000 mi **10.** $193.86

Chapter 4

Pretest: Chapter 4, *p. 220*

1. Possible answer: four less than t **2.** Possible answer: quotient of y and three **3.** $m + 8$ **4.** $2n$ **5.** 4 **6.** $1\frac{1}{2}$ **7.** $x + 3 = 5$ **8.** $4y = 12$ **9.** $x = 6$ **10.** $t = 10$ **11.** $n = 13$ **12.** $a = 12$ **13.** $m = 6.1$ **14.** $n = 30$ **15.** $m = \frac{1}{2}$, or 0.5 **16.** $n = 15$ **17.** $63 = x + 36$; $x = 27$ moons **18.** $6.75 = x - 2.75$; $x = \$9.50$ **19.** $\frac{2}{5}x = 39{,}900$; $x = 99{,}750$ sq mi **20.** $40 = 10x$; $x = 4$ mg

Practices: Section 4.1, *pp. 222–225*

1, ***p. 222:*** Answers may vary **a.** One-half of p **b.** x less than 5 **c.** y divided by 4 **d.** 3 more than n **e.** $\frac{3}{5}$ of b
2, ***p. 223:*** **a.** $x + 9$ **b.** $10y$ **c.** $n - 7$ **d.** $p \div 5$ **e.** $\frac{2}{5}v$
3, ***p. 223:*** **a.** $q + 12$ **b.** $\frac{9}{a}$ **c.** $\frac{2}{7}c$ **4,** ***p. 223:*** $\frac{h}{4}$ hr
5, ***p. 223:*** $s - 3$ **6,** ***p. 224:*** **a.** 25 **b.** 0.38 **c.** 4.8 **d.** 26.6 **7,** ***p. 224:*** $\frac{1}{5}p$; $69,800 **8,** ***p. 225:*** The total amount is $(15.45 + t)$ dollars; $18.45 for $t = \$3$

Exercises 4.1, *pp. 226–229*

1. variable **3.** algebraic **5.** 9 more than t; t plus 9 **7.** c minus 12; 12 subtracted from c **9.** c divided by 3; the quotient of c and 3; **11.** 10 times s; the product of 10 and s **13.** y minus 10; 10 less than y **15.** 7 times a; the product of 7 and a **17.** x divided by 6; the quotient of x and 6 **19.** x minus $\frac{1}{2}$; $\frac{1}{2}$ less than x **21.** $\frac{1}{4}$ times w; $\frac{1}{4}$ of w **23.** 2 minus x; the difference between 2 and x **25.** 1 increased by x; x added to 1 **27.** 3 times p; the product of 3 and p **29.** n decreased by 1.1; n minus 1.1 **31.** y divided by 0.9; the quotient of y and 0.9 **33.** $x + 10$ **35.** $n - 1$ **37.** $y + 5$ **39.** $t \div 6$ **41.** $10y$ **43.** $w - 5$ **45.** $n + \frac{4}{5}$ **47.** $z \div 3$ **49.** $\frac{2}{7}x$ **51.** $k - 6$ **53.** $n + 12$ **55.** $n - 5.1$ **57.** 26 **59.** 2.5 **61.** 15 **63.** $1\frac{1}{6}$ **65.** 1.1 **67.** $\frac{1}{5}$

69.

x	$x + 8$
1	9
2	10
3	11
4	12

71.

n	$n - 0.2$
1	0.8
2	1.8
3	2.8
4	3.8

73.

x	$\frac{3}{4}x$
4	3
8	6
12	9
16	12

75.

z	$\frac{z}{2}$
2	1
4	2
6	3
8	4

77. $x - 7$ **79.** Possible answers: n over 2; n divided by 2 **81.** $3.5t$ **83.** Possible answers: 6 more than x; the sum of x and 6 **85.** $(m - 25)$ mg **87.** $30° + 90° + d°$, or $120° + d°$ **89.** 220 mi **91.** **a.** $1.5w$ dollars **b.** $13.50

Practices: Section 4.2, *pp. 231–235*

1, ***p. 231:*** **a.** $n - 5.1 = 9$ **b.** $y + 2 = 12$ **c.** $n - 4 = 11$ **d.** $n + 5 = 7\frac{3}{4}$ **2,** ***p. 231:*** $p - 6 = 49.95$, where p is the regular price. **3,** ***p. 232:*** $x = 9$; **4,** ***p. 233:*** $t = 2.7$; **5,** ***p. 233:*** $m = 5\frac{1}{4}$; **6,** ***p. 234:*** **a.** $11 = m - 4$; $m = 15$

b. $12 + n = 21; n = 9$ **7,** ***p. 234:*** $x + 3.99 = 27.18;$ $x = \$23.19$ **8,** ***p. 235:*** $x - 262{,}000 = 308{,}000;$ $x = 570{,}000$ sq mi

Exercises 4.2, *pp. 236–239*

1. equation **3.** subtract **5.** $z - 9 = 25$ **7.** $7 + x = 25$ **9.** $t - 3.1 = 4$ **11.** $\frac{3}{2} + y = \frac{9}{2}$ **13.** $n - 3\frac{1}{2} = 7$ **15. a.** Yes **b.** No **c.** Yes **d.** No **17.** Subtract 4. **19.** Add 11. **21.** Add 7. **23.** Subtract 2. **25.** $a = 31$ **27.** $y = 2$ **29.** $x = 12$ **31.** $n = 4$ **33.** $m = 2$ **35.** $y = 90$ **37.** $z = 2.9$ **39.** $n = 8.9$ **41.** $y = 0.9$ **43.** $x = 8\frac{2}{3}$ **45.** $m = 5\frac{1}{3}$ **47.** $x = 3\frac{3}{4}$ **49.** $c = 47\frac{1}{5}$ **51.** $x = 13$ **53.** $y = 6\frac{1}{4}$ **55.** $a = 3\frac{5}{12}$ **57.** $x = 8.2$ **59.** $y = 19.91$ **61.** $x = 4.557$ **63.** $y = 10.251$ **65.** $n + 3 = 11; n = 8$ **67.** $y - 6 = 7; y = 13$ **69.** $n + 10 = 19; n = 9$ **71.** $x + 3.6 = 9; x = 5.4$ **73.** $n - 4\frac{1}{3} = 2\frac{2}{3}; n = 7$ **75.** Equation c **77.** Equation a **79.** $a = 14.5$ **81.** Equation b **83.** Yes **85.** Subtract 2. **87.** Equation a **89.** $621{,}000 = x - 13{,}000; x = \$634{,}000$ **91.** $40° + x = 90°; 50°$ **93.** $x + 12 = 96$; \$84 **95.** $45 = x - 20$; 65 mph **97. a.** $x + 794{,}000{,}000 = 1{,}324{,}089{,}000$ **b.** \$530,089,000 **c.** \$500,000,000

Practices: Section 4.3, *pp. 241–246*

1, ***p. 241:*** **a.** $2x = 14$ **b.** $\frac{a}{6} = 1.5$ **c.** $\frac{n}{0.3} = 1$ **d.** $10 = \frac{1}{2}n$ **2,** ***p. 241:*** $15 = 3w$ **3,** ***p. 243:*** $x = 5$ **4,** ***p. 243:*** $a = 6$ **5,** ***p. 243:*** $x = 4$ **6,** ***p. 244:*** $a = 2.88$ **7,** ***p. 244:*** $x = 16$ **8,** ***p. 245:*** **a.** $12 = \frac{z}{6}, z = 72; 12 \stackrel{?}{=} \frac{72}{6}, 12 \stackrel{\checkmark}{=} 12$ **b.** $16 = 2x, 8 = x$, or $x = 8; 16 \stackrel{?}{=} 2(8), 16 \stackrel{\checkmark}{=} 16$ **9,** ***p. 245:*** $1.6 = 5x;$ $x = 0.32$ km **10,** ***p. 246:*** $\frac{3}{8}x = 525$; \$1,400

Exercises 4.3, *pp. 247–251*

1. divide **3.** substituting **5.** equation **7.** $\frac{3}{4}y = 12$ **9.** $\frac{x}{7} = \frac{7}{2}$ **11.** $\frac{1}{3}x = 2$ **13.** $\frac{n}{3} = \frac{1}{3}$ **15.** $9a = 27$ **17. a.** Yes **b.** No **c.** No **d.** No **19.** Divide by 3. **21.** Multiply by 2. **23.** Divide by $\frac{3}{4}$ or multiply by $\frac{4}{3}$ **25.** Divide by 1.5. **27.** $x = 6$ **29.** $x = 18$ **31.** $n = 4$ **33.** $x = 91$ **35.** $y = 4$ **37.** $b = 20$ **39.** $m = 157.5$ **41.** $t = 0.4$ **43.** $x = \frac{3}{2}$, or $1\frac{1}{2}$ **45.** $x = 36$ **47.** $t = 3$ **49.** $y = \frac{2}{5}$ **51.** $n = 700$ **53.** $x = 12.5$ **55.** $x = \frac{1}{2}$ **57.** $m = 6$ **59.** $x = 6.8$ **61.** $x = 4.9$ **63.** $8n = 56;$ $n = 7$ **65.** $\frac{3}{4}y = 18; y = 24$ **67.** $\frac{x}{5} = 11; x = 55$ **69.** $2x = 36; x = 18$ **71.** $\frac{1}{2}a = 4; a = 8$ **73.** $\frac{n}{5} = 1\frac{3}{5};$ $n = 8$ **75.** $\frac{n}{2.5} = 10; n = 25$ **77.** Equation d **79.** Equation a **81.** $x = 5.5$ **83.** Equation d **85.** $\frac{y}{3} = 6$ **87.** $2x = 5$ **89.** Yes **91.** $4s = 60;$ $s = 15$ units **93.** $56 = \frac{1}{2}x$; 112 mi **95.** $12x = 119.88$; \$9.99 **97. a.** $\frac{2}{5}x = 60$; 150 ml **b.** 90 ml **99. a.** $79.6 = \frac{x}{3{,}537{,}441}$ **b.** 281,580,304 people **c.** 280,000,000 people

Review Exercises: Chapter 4, *pp. 253–255*

1. x plus 1 **2.** Four more than y **3.** w minus 1 **4.** Three less than s **5.** c divided by 7 **6.** The quotient of a and 10 **7.** Two times x **8.** The product of 6 and y **9.** y divided by 0.1 **10.** The quotient of n and 1.6 **11.** One-third of x **12.** One-tenth of w **13.** $m + 9$ **14.** $b + \frac{1}{2}$ **15.** $y - 1.4$ **16.** $z - 3$ **17.** $\frac{3}{x}$ **18.** $n \div 2.5$ **19.** $3n$ **20.** $12n$ **21.** 12 **22.** 19 **23.** 0 **24.** 6 **25.** 0.3 **26.** 6.5 **27.** $1\frac{1}{2}$ **28.** $\frac{5}{12}$ **29.** 0.4 **30.** $4\frac{1}{2}$ **31.** 1.6 **32.** 9 **33.** $x = 9$ **34.** $y = 9$ **35.** $n = 26$ **36.** $b = 20$ **37.** $a = 3.5$ **38.** $c = 7.5$ **39.** $x = 11$ **40.** $y = 2$ **41.** $w = 1\frac{1}{2}$ **42.** $s = \frac{1}{3}$ **43.** $c = 6\frac{3}{4}$ **44.** $p = 11\frac{2}{3}$ **45.** $m = 5$ **46.** $n = 0$ **47.** $c = 78$ **48.** $y = 90$ **49.** $n = 11$ **50.** $x = 25$ **51.** $x = 31.0485$ **52.** $m = 26.6225$ **53.** $n - 19 = 35$ **54.** $a - 37 = 234$ **55.** $9 + n = 5\frac{1}{2}$ **56.** $s + 26 = 30\frac{1}{3}$ **57.** $2y = 16$ **58.** $25t = 175$ **59.** $34 = \frac{n}{19}$ **60.** $17 = \frac{z}{13}$ **61.** $\frac{1}{3}n = 27$ **62.** $\frac{2}{5}n = 4$ **63. a.** No **b.** Yes **c.** Yes **d.** No **64. a.** Yes **b.** No **c.** No **d.** Yes **65.** $x = 5$ **66.** $t = 2$ **67.** $a = 105$ **68.** $n = 54$ **69.** $y = 9$ **70.** $r = 10$ **71.** $w = 90$ **72.** $x = 100$ **73.** $y = 20$ **74.** $a = 120$ **75.** $n = 32$ **76.** $b = 32$ **77.** $m = 3.15$ **78.** $z = 0.57$ **79.** $x = \frac{2}{5}$, or 0.4 **80.** $t = \frac{1}{2}$, or 0.5 **81.** $m = 1.2$ **82.** $b = 9.8$ **83.** $x = 12.5$ **84.** $x = 1.4847$ **85.** $2h$ degrees; 6 degrees **86.** $\frac{d}{20}$ dollars per hr; \$9.55 per hr **87.** $89p$ cents; \$2.67 **88.** $(3{,}000 + d)$ dollars; \$3,225 **89.** $x + 238 = 517$; \$279 **90.** $\frac{1}{4}x = 500{,}000;$ $x = 2{,}000{,}000$ **91.** $177 = 2.5x$; 71 L **92.** $225 = x + 50; x = 175$ **93.** $\frac{x}{6} = 30$; 180 lb **94.** $1.8x = 6{,}696; x = 3{,}720$ km **95.** $98.6 + x = 101;$ $x = 2.4$°F **96.** $x - 256 = 8{,}957$; 9,213 applications

Posttest: Chapter 4, *p. 256*

1. Possible answer: x plus $\frac{1}{2}$ **2.** Possible answer: the quotient of a and 3 **3.** $n - 10$ **4.** $\frac{8}{p}$ **5.** 0 **6.** $\frac{1}{4}$ **7.** $x - 6 = 4\frac{1}{2}$ **8.** $\frac{y}{8} = 3.2$ **9.** $x = 0$ **10.** $y = 12$ **11.** $n = 27$ **12.** $a = 738$ **13.** $m = 7.8$ **14.** $n = 50$ **15.** $x = \frac{11}{20}$ **16.** $n = 760$ **17.** $1\frac{3}{4} + x = 2\frac{1}{4}; x = \frac{1}{2}$ lb **18.** $\frac{1}{3}x = 30{,}000; x = 90{,}000$ elephants **19.** $6 = \frac{2}{3}x;$ 9 billion people **20.** $x - 19.8 = 7.6$; 27.4 degrees Celsius

Cumulative Review: Chapter 4, *p. 257*

1. $5\frac{3}{8}$ **2.** 0.0075 **3.** Yes **4.** 23,316 **5.** 3.14 **6.** $n = 7.8$ **7.** $x = 32$ **8.** 7,200 cartoons **9.** 55,000 beehives **10.** He got back $\frac{2}{7}$ of his money, which is less than $\frac{1}{3}$.

Chapter 5

Pretest: Chapter 5, *p. 260*

1. $\frac{3}{4}$ **2.** $\frac{2}{5}$ **3.** $\frac{5}{3}$ **4.** $\frac{19}{51}$ **5.** $\frac{16 \text{ gal}}{5 \text{ min}}$ **6.** $\frac{5 \text{ mg}}{3 \text{ hr}}$ **7.** $\frac{2 \text{ dental assistants}}{1 \text{ dentist}}$ **8.** $\frac{1 \text{ calculator}}{1 \text{ student}}$ **9.** $\frac{\$230}{\text{box}}$ **10.** $\frac{\$0.50}{\text{bottle}}$ **11.** True **12.** False **13.** $x = 9$ **14.** $x = 31\frac{1}{2}$ **15.** $x = 16$ **16.** $x = 160$ **17.** $\frac{4}{5}$ **18.** 200 lb/min **19.** \$264,000 **20.** 87 mi

Practices: Section 5.1, *pp. 261–265*

1, ***p. 261:*** $\frac{2}{3}$ **2,** ***p. 262:*** $\frac{9}{5}$ **3,** ***p. 262:*** $\frac{1}{3}$
4, ***p. 263:*** **a.** $\frac{5 \text{ ml}}{2 \text{ min}}$ **b.** $\frac{3 \text{ lb}}{2 \text{ wk}}$ **5,** ***p. 263:*** **a.** 48 ft/sec
b. 0.375 hit per time at bat **6,** ***p. 264:*** $\frac{500 \text{ beats}}{1 \text{ min}}$
7, ***p. 264:*** **a.** \$174/flight **b.** \$2.75/hr **c.** \$0.99/download
8, ***p. 265:*** The 39-oz can

Exercises 5.1, *pp. 266–271*

1. quotient **3.** simplest form **5.** denominator
7. $\frac{2}{3}$ **9.** $\frac{2}{3}$ **11.** $\frac{11}{7}$ **13.** $\frac{3}{2}$ **15.** $\frac{1}{4}$ **17.** $\frac{4}{3}$ **19.** $\frac{1}{1}$ **21.** $\frac{5}{3}$
23. $\frac{7}{24}$ **25.** $\frac{20}{1}$ **27.** $\frac{8}{7}$ **29.** $\frac{4}{5}$ **31.** $\frac{5 \text{ calls}}{2 \text{ days}}$ **33.** $\frac{36 \text{ cal}}{5 \text{ min}}$
35. $\frac{1 \text{ million hits}}{3 \text{ mo}}$ **37.** $\frac{17 \text{ baskets}}{30 \text{ attempts}}$ **39.** $\frac{37 \text{ points}}{2 \text{ games}}$ **41.** $\frac{100 \text{ sq ft}}{\$329}$
43. $\frac{16 \text{ males}}{3 \text{ females}}$ **45.** $\frac{8 \text{ Democrats}}{7 \text{ Republicans}}$ **47.** $\frac{1 \text{ lb}}{8 \text{ servings}}$ **49.** $\frac{307 \text{ flights}}{3 \text{ days}}$
51. $\frac{1 \text{ lb}}{200 \text{ sq ft}}$ **53.** 225 revolutions/min **55.** 8 gal/day
57. 0.3 tank/acre **59.** 1.6 yd/dress **61.** 2 hr/day
63. 0.25 km/min **65.** 70 fat calories/tbsp **67.** \$0.45/bar
69. \$2.95/roll **71.** \$66.67/plant **73.** \$99/night

75.

Number of Units	Total Price	Unit Price
125	\$6.69	\$0.05
500	\$15.49	\$0.03

500 envelopes

77.

Number of Units	Total Price	Unit Price
180	\$12.99	\$0.072
250	\$17.49	\$0.070

250 tablets

79.

Number of Units	Total Price	Unit Price
25	\$14.99	\$0.60
50	\$26.55	\$0.53
100	\$54.99	\$0.55

50 discs

81. \$0.16/oz **83.** 2 tutors/15 students **85.** $\frac{5}{1}$ **87.** $\frac{2}{3}$
89. 170 cal/oz **91.** 25 times/min **93.** $\frac{1}{2}$ **95.** 8.4 people/sq km **97.** Lower **99.** **a.** $\frac{62}{67}$ **b.** $\frac{7}{8}$ **101.** (a) Division; (b) 0.5131495; (c) possible estimate: 0.5

Practices: Section 5.2, *pp. 272–276*

1, ***p. 272:*** Yes **2,** ***p. 272:*** Not a true proportion
3, ***p. 273:*** No **4,** ***p. 274:*** $x = 8$ **5,** ***p. 274:*** $x = 12$
6, ***p. 274:*** 64,000 flowers **7,** ***p. 275:*** 810 mi
8, ***p. 276:*** 160 ft

Exercises 5.2, *pp. 277–280*

1. proportion **3.** as **5.** True **7.** False **9.** True
11. False **13.** True **15.** True **17.** $x = 20$ **19.** $x = 38$
21. $x = 4$ **23.** $x = 13$ **25.** $x = 8$ **27.** $x = 4$
29. $x = 20$ **31.** $x = 15$ **33.** $x = 21$ **35.** $x = 13\frac{1}{3}$
37. $x = 100$ **39.** $x = 1.8$ **41.** $x = 21$ **43.** $x = 280$
45. $x = 300$ **47.** $x = 20$ **49.** $x = 10$
51. $x = \frac{1}{5}$, or 0.2 **53.** $x = 0.005$ **55.** $x = \frac{2}{5}$
57. $x = 1\frac{3}{5}$ **59.** False **61.** Not the same **63.** $1\frac{7}{8}$ gal
65. 54.5 g **67.** 100 hydrogen atoms **69.** $41\frac{2}{3}$ in.
71. 0.25 ft **73.** \$600 **75.** 12,000 fish **77.** 280 times
79. 90 mg and 50 mg **81.** **a.** 92 g **b.** 4 g **83.** (a) Multiplication, division; (b) 835,312.5 gal; (c) possible estimate: 880,000 gal

Review Exercises: Chapter 5, *pp. 283–284*

1. $\frac{2}{3}$ **2.** $\frac{1}{2}$ **3.** $\frac{3}{4}$ **4.** $\frac{25}{8}$ **5.** $\frac{8}{5}$ **6.** $\frac{3}{4}$ **7.** $\frac{44 \text{ ft}}{5 \text{ sec}}$
8. $\frac{9 \text{ applicants}}{2 \text{ positions}}$ **9.** 0.0025 lb/sq ft **10.** 500,000,000 calls/day
11. 8 yd/down **12.** 400 sq ft/gal **13.** 10,500,000 vehicles/yr **14.** 76,000 commuters/day **15.** \$118.75/night **16.** \$3.89/rental **17.** \$1,250/station
18. \$93.64/share

19.

Number of Units	Total Price	Unit Price
47	\$11.95	\$0.25
92	\$29.90	\$0.33

47 issues

20.

Number of Units	Total Price	Unit Price
300	\$59.99	\$0.20
525	\$74.99	\$0.14

525 checks

21.

Number of Units	Total Price	Unit Price
90	\$7.19	\$0.08
180	\$7.43	\$0.04
360	\$17.91	\$0.05

180 capsules

22.

Number of Units (Fluid Ounces)	Total Price	Unit Price
4	\$1.89	\$0.47
14	\$3.59	\$0.26
20	\$4.69	\$0.23

20 fl ounce-bottle

23. True **24.** False **25.** False **26.** True
27. $x = 6$ **28.** $x = 3$ **29.** $x = 32$ **30.** $x = 30$
31. $x = 2$ **32.** $x = 8$ **33.** $x = \frac{7}{10}$ **34.** $x = \frac{2}{5}$
35. $x = 67\frac{1}{2}$ **36.** $x = 45$ **37.** $x = \frac{3}{4}$, or 0.75
38. $x = \frac{3}{20}$, or 0.15 **39.** $x = 28$ **40.** $x = 0.14$
41. $\frac{1}{15}$ **42.** $\frac{23}{45}$ **43.** \$90/day **44.** 0.125 in./mo **45.** $\frac{2}{3}$
46. 50,000 books **47.** No **48.** $2\frac{1}{7}$ hr **49.** 55 cc
50. 1.25 in. **51.** 0.68 g/cc **52.** 1,175.4 people/sq mi

Posttest: Chapter 5, *p. 285*

1. $\frac{2}{3}$ **2.** $\frac{5}{14}$ **3.** $\frac{55}{31}$ **4.** $\frac{12}{1}$ **5.** $\frac{13 \text{ revolutions}}{12 \text{ sec}}$ **6.** $\frac{1 \text{ cm}}{25 \text{ km}}$ **7.** 68 mph **8.** 8 m/sec **9.** $136/day **10.** $0.80/greeting card **11.** False **12.** True **13.** $x = 25$ **14.** $x = 6$ **15.** $x = 28$ **16.** $x = 1$ **17.** 5 million e-mail addresses **18.** $\frac{3}{19}$ **19.** 25 ft **20.** 48 beats/min

Cumulative Review: Chapter 5, *p. 286*

1. $\frac{2}{5}$ **2.** 8,200 **3.** $x = 2.5$ **4.** $\frac{1}{4}$ **5.** $4 per yd **6.** Possible answer: 3 **7.** $x = \frac{3}{4}$ **8.** 2,106 sq ft **9.** 7.5 in. **10.** $180

Chapter 6

Pretest: Chapter 6, *p. 288*

1. $\frac{1}{20}$ **2.** $\frac{3}{8}$ **3.** 2.5 **4.** 0.03 **5.** 0.7% **6.** 800% **7.** 67% **8.** 110% **9.** $37\frac{1}{2}$ ft **10.** 55 **11.** Possible estimate: $48 **12.** 250 **13.** 40% **14.** 250% **15.** $14 **16.** 8% **17.** $\frac{6}{25}$ **18.** 25% **19.** $61.11 **20.** $10,000

Practices: Section 6.1, *pp. 290–296*

1, ***p. 290:*** $\frac{21}{100}$ **2,** ***p. 291:*** $\frac{9}{4}$, or $2\frac{1}{4}$ **3,** ***p. 291:*** $\frac{1}{8}$ **4,** ***p. 291:*** $\frac{43}{50}$ **5,** ***p. 292:*** 0.31 **6,** ***p. 292:*** 0.05 **7,** ***p. 293:*** 0.482 **8,** ***p. 293:*** 0.6225 **9,** ***p. 293:*** 1.12 **10,** ***p. 294:*** 2.5% **11,** ***p. 294:*** 9% **12,** ***p. 294:*** 70% **13,** ***p. 294:*** 300% **14,** ***p. 294:*** 71% **15,** ***p. 295:*** Nitrogen; 78% > 0.93%, or 0.78 > 0.0093. **16.** ***p. 295:*** 16% **17.** ***p. 296:*** True. $\frac{2}{3} \approx 67\% > 60\%$ **18.** ***p. 296:*** 27%

Exercises 6.1, *pp. 297–301*

1. percent **3.** left **5.** $\frac{2}{25}$ **7.** $2\frac{1}{2}$ **9.** $\frac{33}{100}$ **11.** $\frac{9}{50}$ **13.** $\frac{7}{50}$ **15.** $\frac{13}{20}$ **17.** $\frac{3}{400}$ **19.** $\frac{3}{1{,}000}$ **21.** $\frac{3}{40}$ **23.** $\frac{1}{7}$ **25.** 0.06 **27.** 0.72 **29.** 0.001 **31.** 1.02 **33.** 0.425 **35.** 5 **37.** 1.069 **39.** 0.035 **41.** 0.009 **43.** 0.0075 **45.** 31% **47.** 17% **49.** 30% **51.** 4% **53.** 12.5% **55.** 129% **57.** 290% **59.** 287% **61.** 101.6% **63.** 900% **65.** 30% **67.** 10% **69.** 16% **71.** 90% **73.** 6% **75.** $55\frac{5}{9}\%$ **77.** $11\frac{1}{9}\%$ **79.** 600% **81.** 150% **83.** $216\frac{2}{3}\%$ **85.** < **87.** < **89.** 44% **91.** 225%

93.

Fraction	Decimal	Percent
$\frac{1}{3}$	0.333 ...	$33\frac{1}{3}\%$
$\frac{2}{3}$	0.666 ...	$66\frac{2}{3}\%$
$\frac{1}{4}$	0.25	25%
$\frac{3}{4}$	0.75	75%
$\frac{1}{5}$	0.2	20%
$\frac{2}{5}$	0.4	40%
$\frac{3}{5}$	0.6	60%

95. $1\frac{1}{25}$ **97.** $316\frac{2}{3}\%$ **99.** 0.275 **101.** 310% **103.** 254% **105.** 0.79 **107.** $\frac{9}{10}$ **109.** $\frac{1}{10}$ **111.** 900% **113.** $1\frac{7}{20}$ **115.** 0.845 **117.** Among men; $\frac{1}{3} = 33\frac{1}{3}\% < 40\%$ **119. a.** 0.4% **b.** 99.6% **121. a.** Division, multiplication; **b.** 55.27...%; **c.** possible estimate: 50%

Practices: Section 6.2, *pp. 303–309*

1, ***p. 303:*** **a.** $x = 0.7 \cdot 80$ **b.** $0.5 \cdot x = 10$ **c.** $x \cdot 40 = 20$ **2,** ***p. 303:*** 8 **3,** ***p. 303:*** 12 **4,** ***p. 304:*** Possible estimate: 200 **5,** ***p. 304:*** 51 workers **6,** ***p. 305:*** 50 **7,** ***p. 305:*** 7.2 **8,** ***p. 305:*** 2,500,000 sq ft **9,** ***p. 306:*** $83\frac{1}{3}\%$ **10,** ***p. 306:*** $112\frac{1}{2}\%$ **11,** ***p. 307:*** 30% **12,** ***p. 307:*** 270 **13,** ***p. 308:*** 1,080 **14,** ***p. 308:*** $33\frac{1}{3}\%$ **15,** ***p. 308:*** $98 **16,** ***p. 309:*** $340,000 **17,** ***p. 309:*** 105% **18,** ***p. 309:*** $5.97

Exercises 6.2, *pp. 310–313*

1. base **3.** percent **5.** 6 **7.** 23 **9.** 2.87 **11.** $140 **13.** 0.62 **15.** 0.1 **17.** 4 **19.** $18.32 **21.** 32 **23.** $120 **25.** 2.5 **27.** $250 **29.** 45 **31.** 1.75 **33.** 4,600 **35.** $49,230.77 **37.** 50% **39.** 75% **41.** $83\frac{1}{3}\%$ **43.** 25% **45.** 150% **47.** $112\frac{1}{2}\%$ **49.** 62.5% **51.** 31% **53.** 60 **55.** $66\frac{2}{3}\%$ **57.** 175 mi **59.** 5% **61.** $500 **63.** 10 **65.** $66\frac{2}{3}\%$ **67.** 10.2 gal **69.** $600 **71.** 25% **73.** 54 tables **75.** 30% **77.** 40 questions **79.** 6.8 million **81. a.** Approximately 4 million **b.** 75% **83.** $30,000,000 **85.** 5,100 employees **87.** $9,000

Practices: Section 6.3, *pp. 314–320*

1, ***p. 314:*** 300% **2,** ***p. 315:*** 1929 **3,** ***p. 316:*** $555 **4,** ***p. 316:*** **a.** $1,125 **b.** $2,625 **5,** ***p. 317:*** $69.60 **6,** ***p. 317:*** 50% **7,** ***p. 318:*** $1,450 **8,** ***p. 318:*** $1,792 **9,** ***p. 320:*** $2,524.95

Exercises 6.3, *pp. 321–326*

1. original **3.** discount

5.

Original Value	New Value	Percent Increase or Decrease
$10	$12	20% increase
$10	$8	20% decrease
$6	$18	200% increase
$35	$70	100% increase
$14	$21	50% increase
$10	$1	90% decrease
$8	$6.50	$18\frac{3}{4}\%$ decrease
$6	$5.25	$12\frac{1}{2}\%$ decrease

7.

Selling Price	Rate of Sales Tax	Sales Tax
$30.00	5%	$1.50
$24.88	3%	$0.75
$51.00	$7\frac{1}{2}\%$	$3.83
$196.23	4.5%	$8.83

9.

Sales	Rate of Commission	Commission
$700	10%	$70.00
$450	2%	$9.00
$870	$4\frac{1}{2}$%	$39.15
$922	7.5%	$69.15

11.

Original Price	Rate of Discount	Discount	Sale Price
$700.00	25%	$175.00	$525.00
$18.00	10%	$1.80	$16.20
$43.50	20%	$8.70	$34.80
$16.99	5%	$0.85	$16.14

13.

Selling Price	Rate of Markup	Markup	Cost
$10.00	50%	$5.00	$5.00
$23.00	70%	$16.10	$6.90
$18.40	10%	$1.84	$16.56
$13.55	60%	$8.13	$5.42

15.

Principal	Interest Rate	Time (in years)	Interest	Final Balance
$300	4%	2	$24.00	$324.00
$600	7%	2	$84.00	$684.00
$500	8%	2	$80.00	$580.00
$375	10%	4	$150.00	$525.00
$1,000	3.5%	3	$105.00	$1,105.00
$70,000	6.25%	30	$131,250.00	$201,250.00

17.

Principal	Interest Rate	Time (in years)	Final Balance
$500	4%	2	$540.80
$6,200	3%	5	$7,187.50
$300	5%	8	$443.24
$20,000	4%	2	$21,632.00
$145	3.8%	3	$162.17
$810	2.9%	10	$1,078.05

19.

Original Value	New Value	Percent Decrease
$5	$4.50	10%

21.

Original Price	Rate of Discount	Discount	Sale Price
$87.33	40%	$34.93	$52.40

23.

Selling Price	Rate of Sales Tax	Sales Tax
$200	7.25%	$14.50

25.

Principal	Interest Rate	Kind of Interest	Time (in years)	Interest	Final Balance
$3,000	5%	simple	5	$750.00	$3,750.00

27. 28% **29.** 550% **31.** 300% **33.** $84.95 **35.** 6.5% **37.** $2,700 **39.** $9 **41.** $23.98 **43.** $53\frac{1}{3}$% **45.** $259.35 **47.** $150 **49.** $250 **51. a.** $144 **b.** $152.64 **53.** $3,370.80 **55.** 5,856

Review Exercises: Chapter 6, *pp. 329–332*

1.

Fraction	Decimal	Percent
$\frac{1}{4}$	0.25	25%
$\frac{7}{10}$	0.7	70%
$\frac{3}{400}$	0.0075	$\frac{3}{4}$%
$\frac{5}{8}$	0.625	62.5%
$\frac{41}{100}$	0.41	41%
$1\frac{1}{100}$	1.01	101%
$2\frac{3}{5}$	2.6	260%
$3\frac{3}{10}$	3.3	330%
$\frac{3}{25}$	0.12	12%
$\frac{2}{3}$	0.66 ...	$66\frac{2}{3}$%
$\frac{1}{6}$	0.166 ...	$16\frac{2}{3}$%

2.

Fraction	Decimal	Percent
$\frac{3}{8}$	0.375	37.5%
$\frac{49}{100}$	0.49	49%
$\frac{1}{1,000}$	0.001	0.1%
$1\frac{1}{2}$	1.5	150%
$\frac{7}{8}$	0.875	87.5%
$\frac{5}{6}$	0.833 ...	$83\frac{1}{3}$%
$2\frac{3}{4}$	2.75	275%
$1\frac{1}{5}$	1.2	120%
$\frac{3}{4}$	0.75	75%
$\frac{1}{10}$	0.1	10%
$\frac{1}{3}$	0.33 ...	$33\frac{1}{3}$%

3. 12 **4.** 120% **5.** 50% **6.** 20 **7.** 43.75% **8.** 5.5 **9.** $6 **10.** 20% **11.** 0.3 **12.** 460 **13.** $70 **14.** 1,000 **15.** 2000% **16.** 25% **17.** $200 **18.** $44\frac{4}{9}$% **19.** $12 **20.** $1,600 **21.** 17% **22.** 42.42

23.

Original Value	New Value	Percent Decrease
24	16	$33\frac{1}{3}$%

24.

Selling Price	Rate of Sales Tax	Sales Tax
$50	6%	$3.00

25.

Sales	Rate of Commission	Commission
$600	4%	$24

26.

Original Price	Rate of Discount	Discount	Sale Price
$200	15%	$30	$170

27.

Selling Price	Rate of Markup (based on the selling price)	Markup	Cost
$51	50%	$25.50	$25.50

28.

Principal	Interest Rate	Time (in years)	Simple Interest	Final Balance
$200	4%	2	$16	$216

29. $1,800 **30.** 48% **31.** The agent that charges 11% **32.** $7,200 **33.** 20% **34.** Paper **35.** $\frac{1}{4}$ **36.** Above the typical markup **37.** 40% **38.** 10% **39.** 0.0629 **40.** $207 **41.** 68% **42.** Possible estimate: 80 in. **43.** 63 first serves **44.** Yes **45.** 180 **46.** 20% **47.** No **48.** 57,770,000 cats **49.** $165 **50.** 320 tons **51.** 70,000 mi **52.** 10,000 hr **53.** $1,000 **54.** $30,000 **55.** $7,939.58 **56.** 22%

57.

Quarter	Income	Percent of Total Income (rounded to the nearest whole precent)
1	$375,129	27%
2	289,402	21%
3	318,225	23%
4	402,077	29%
Total	$1,384,833	100%

58. Individual income taxes: $948 billion; Social Security taxes: $797 billion; corporate income taxes: $280 billion; excise taxes: $65 billion; other: $86 billion

Posttest: Chapter 6, *p. 333*

1. $\frac{1}{25}$ **2.** $\frac{11}{40}$ **3.** 1.74 **4.** 0.08 **5.** 0.9% **6.** 1,000% **7.** 83% **8.** 220% **9.** 7.5 mi **10.** 48 **11.** Possible estimate: $7 **12.** 200 **13.** 60% **14.** 250% **15.** $200 **16.** 6 spaces **17.** 5% **18.** 4 pt **19.** $31.60 **20.** $3\frac{1}{3}$%

Cumulative Review: Chapter 6, *p. 334*

1. 109 **2.** 0.83 **3.** 0.7 **4.** $5\frac{7}{10}$ **5.** 7.5 **6.** $1,000 **7.** $\frac{1}{8}$ **8.** 6,412 students

9.

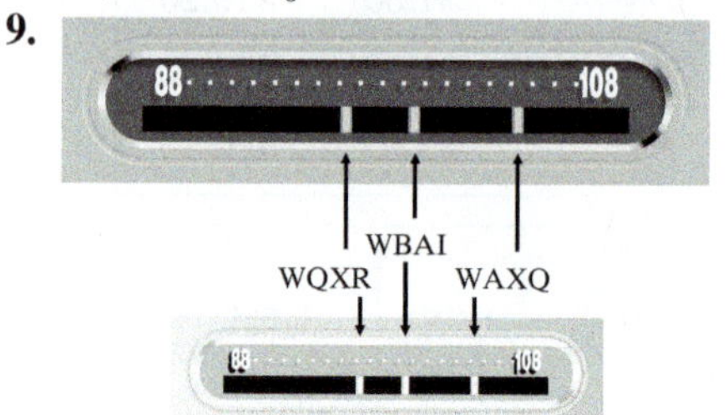

10. 239 thousand

Appendix

Practices: Appendix, *pp. 335–338*

1, ***p. 335:*** 25,390,000 **2,** ***p. 336:*** 8.0×10^{12}, or 8×10^{12} **3,** ***p. 337:*** 0.0000000043 **4,** ***p. 337:*** 7.1×10^{-11} **5,** ***p. 338:*** **a.** 2.464×10^{2} **b.** 8.35×10^{11}

Exercises p. *339*

1. 4×10^{8} **3.** 3.5×10^{-6} **5.** 3.1×10^{-10} **7.** 317,000,000 **9.** 0.000001 **11.** 0.00004013 **13.** 9×10^{7} **15.** 2.075×10^{-4} **17.** 1.25×10^{10} **19.** 4×10^{1}

Glossary

The numbers in brackets following each glossary term represent the section that term is discussed in.

addends [1.2] In an addition problem, the numbers being added are called addends.

algebraic expression [4.1] An algebraic expression is an expression that combines variables, constants, and arithmetic operations.

amount (percent) [6.2] The amount is the result of taking the percent of the base.

associative property of addition [1.2] The associative property of addition states that when adding three numbers, regrouping the addends gives the same sum.

associative property of multiplication [1.3] The associative property of multiplication states that when multiplying three numbers, regrouping the factors gives the same product.

average (or mean) [1.5] An average of a set of numbers is the sum of those numbers, divided by however many numbers are on the list.

base (exponent) [1.5] The base is the number that is a repeated factor when written with an exponent.

base (percent) [6.2] The base is the number that we take the percent of. It always follows the word "of" in the statement of a percent problem.

circumference [5.1] The distance around a circle is called the circumference.

commission [6.3] Salespeople may work on commission instead of receiving a fixed salary. This means that the amount of money that they earn is a specified percent of the total sales for which they are responsible.

commutative property of addition [1.2] The commutative property of addition states that changing the order in which two numbers are added does not affect the sum.

commutative property of multiplication [1.3] The commutative property of multiplication states that changing the order in which two numbers are multiplied does not affect the product.

composite number [2.1] A composite number is a whole number that has more than two factors.

constant [4.1] A constant is a known number.

decimal [3.1] A decimal is a number written with three parts: a whole number, the decimal point, and a fraction whose denominator is a power of 10.

decimal places [3.1] The decimal places are the places to the right of the decimal point.

denominator [2.2] The number below the fraction line in a fraction is called the denominator. It stands for the number of parts into which the whole is divided.

diameter [5.1] A line segment that passes through the center of a circle and has both endpoints on the circle is called the diameter of the circle.

difference [1.2] The result of a subtraction problem is called the difference.

digits [1.1] Digits are the numbers 0, 1, 2, 3, 4, 5, 6, 7, 8, and 9.

discount [6.3] When buying or selling merchandise, the term "discount" refers to a reduction on the merchandise's original price.

distributive property [1.3] The distributive property states that multiplying a factor by the sum of two numbers gives us the same result as multiplying the factor by each of the two numbers and then adding.

dividend [1.4] In a division problem, the number into which another number is being divided is called the dividend.

divisor [1.4] In a division problem, the number that is being used to divide another number is called the divisor.

equation [4.2] An equation is a mathematical statement that two expressions are equal.

equivalent fractions [2.2] Equivalent fractions are fractions that represent the same quantity.

evaluate [4.1] To evaluate an algebraic expression, substitute the given value for each variable and carry out the computation.

exponent (or power) [1.5] An exponent (or power) is a number that indicates how many times another number is used as a factor.

exponential form [1.5] Exponential form is a shorthand way of representing a repeated multiplication of the same factor.

factors [1.3] In a multiplication problem, the numbers being multiplied are called the factors.

fraction [2.2] A fraction is any number that can be written in the form $\frac{a}{b}$, where a and b are whole numbers and b is nonzero.

fraction line [2.2] The fraction line separates the numerator from the denominator, and stands for "out of" or "divided by."

identity property of addition [1.2] The identity property of addition states that the sum of a number and zero is the original number.

identity property of multiplication [1.3] The identity property of multiplication states that the product of any number and 1 is that number.

improper fraction [2.2] An improper fraction is a fraction greater than or equal to 1, that is, a fraction whose numerator is larger than or equal to its denominator.

least common denominator (LCD) [2.2] The least common denominator (LCD) for two or more fractions is the least common multiple of their denominators.

least common multiple (LCM) [2.1] The least common multiple (LCM) of two or more whole numbers is the smallest nonzero whole number that is a multiple of each number.

like fractions [2.2] Like fractions are fractions with the same denominator.

like quantities [5.1] Like quantities are quantities that have the same unit.

magic square [1.2] A magic square is a square array of numbers in which the sum of every row, column, and diagonal is the same number.

markup [6.3] The markup on an item is the difference between the selling price and the cost.

minuend [1.2] In a subtraction problem, the number that is being subtracted from is called the minuend.

mixed number [2.2] A mixed number is a number greater than 1 with a whole number part and a fractional part.

multiplication property of 0 [1.3] The multiplication property of 0 states that the product of any number and 0 is 0.

numerator [2.2] The number above the fraction line in a fraction is called the numerator. It tells us how many parts of the whole the fraction contains.

percent (or rate) [6.1] A percent is a ratio or fraction with denominator 100. A number written with the % sign means "divided by 100."

percent decrease [6.3] In a percent problem, if the quantity is decreasing, it is called a percent decrease.

percent increase [6.3] In a percent problem, if the quantity is increasing, it is called a percent increase.

perfect square [1.5] A perfect square is a number that is the square of any whole number.

period [1.1] A period is a group of three digits, which are separated by commas, when writing a large whole number in standard form.

place value [1.1] Each of the digits in a whole number in standard form has place value.

prime factorization [2.1] Prime factorization is the process of writing a whole number as a product of its prime factors.

prime number [2.1] A prime number is a whole number that has exactly two different factors, itself and 1.

principal [6.3] The principal is the amount of money borrowed.

product [1.3] The result of a multiplication problem is called the product.

proper fraction [2.2] A proper fraction is a fraction less than 1, that is, a fraction whose numerator is smaller than its denominator.

proportion [5.2] A proportion is a statement that two ratios are equal.

quotient [1.4] The result of a division problem is called the quotient.

rate [5.1] A rate is a ratio of unlike quantities.

ratio [5.1] A ratio is a comparison of two quantities expressed as a quotient.

reciprocal [2.4] The reciprocal of the fraction $\frac{a}{b}$ is $\frac{b}{a}$.

reduced to lowest terms (or simplest form) [2.2] A fraction is said to be reduced to lowest terms when the only common factor of its numerator and its denominator is 1.

rounding [1.1] Rounding is the process of approximating an exact answer by a number that ends in a given number of zeros.

simplest form (or reduced to lowest terms) [2.2] A fraction is said to be in simplest form when the only common factor of its numerator and its denominator is 1.

subtrahend [1.2] In a subtraction problem, the number that is being subtracted is called the subtrahend.

sum [1.2] The result of an addition problem is called the sum.

unit fraction [2.3] A fraction with 1 as the numerator is called a unit fraction.

unit price [5.1] The unit price is the price of one item, or one unit.

unit rate [5.1] A unit rate is a rate in which the number in the denominator is 1.

unlike fractions [2.2] Unlike fractions are fractions with different denominators.

unlike quantities [5.1] Unlike quantities are quantities that have different units.

variable [4.1] A variable is a letter that represents an unknown number.

Index

P

Q

R

S

T

U

V

W

U.S. Customary Units

Length
12 in. = 1 ft
3 ft = 1 yd
5,280 ft = 1 mi

Weight
16 oz = 1 lb
2,000 lb = 1 ton

Capacity
16 fl oz = 1 pt
2 pt = 1 qt
4 qt = 1 gal

Time
60 sec = 1 min
60 min = 1 hr
24 hr = 1 day
7 days = 1 wk
52 wk = 1 yr
12 mo = 1 yr
365 days = 1 yr

Metric Units

Length
1,000 mm = 1 m
100 cm = 1 m
1,000 m = 1 km

Weight
1,000 mg = 1 g
1,000 g = 1 kg

Capacity
1,000 ml = 1 L
1,000 L = 1 kl

Key U.S./Metric Conversions

Length
1 in. ≈ 2.5 cm
1 ft ≈ 30 cm
39 in. ≈ 1 m
3.3 ft ≈ 1 m
1 mi ≈ 1,600 m
1 mi ≈ 1.6 km

Weight
1 oz ≈ 28 g
1 lb ≈ 450 g
2.2 lb ≈ 1 kg
1 ton ≈ 910 kg

Capacity
1 pt ≈ 470 ml
1.1 qt ≈ 1 L
1 gal ≈ 3.8 L